D0759015

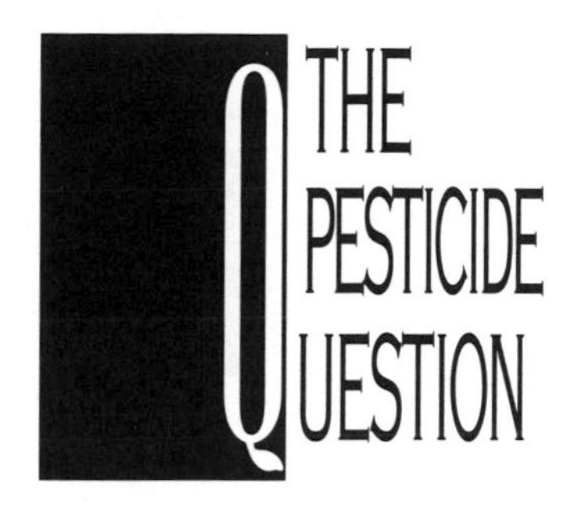
THE
PESTICIDE
QUESTION

ENVIRONMENT, ECONOMICS, AND ETHICS

DAVID PIMENTEL
HUGH LEHMAN
EDITORS

CHAPMAN & HALL
New York and London

First published in 1993 by
Chapman and Hall
an imprint of
Routledge, Chapman & Hall, Inc.
29 West 35th Street
New York, NY 10001-2291

Published in Great Britain by
Chapman and Hall
2-6 Boundary Row
London SE1 8HN

Printed in the United States of America on acid free paper.

Library of Congress Cataloging-in-Publication Data

The Pesticide question : environment, economics, and ethics / editors,
David Pimentel and Hugh Lehman.
p. cm.
Includes bibliographical references and index.
ISBN 0-412-03581-2
1. Pesticides—Environmental aspects. 2. Pesticides—Health aspects. 3. Pesticides—Government policy—United States.
I. Pimentel, David, 1925- . II. Lehman, Hugh.
QH545.P4P4793 1992
363.17'92—dc20 92-13910
CIP

British Library Cataloguing in Publication Data also available.

Contributors

Acquay, H.
Department of Natural Resources
Cornell University
Ithaca, NY

Biltonen, M.
Department of Horticulture
Cornell University
Ithaca, NY

Buttel, F. H.
Biology and Society Program
Cornell University
Ithaca, NY

Culliney, T. W.
State of Hawaii
Department of Agriculture
Division of Plant Industry
Plant Pest Control Branch
Honolulu, HI

Dahlberg, K. A.
Department of Political Science
Western Michigan University
Kalamazoo, MI

D'Amore, M.
New York State College of Agriculture
and Life Sciences
Cornell University
Ithaca, NY

Edwards, C. A.
Department of Entomology
Ohio State University
Columbus, OH

Giordano, S.
New York State College of Agriculture
and Life Sciences
Cornell University
Ithaca, NY

Graap, E.
New York State College of Agriculture
and Life Sciences
Cornell University
Ithaca, NY

Hathaway, J. S.
Natural Resources Defense Council
Washington, DC

Holochuck, N.
Department of History
Evergreen State College
Olympia, WA

Horowitz, A.
New York State College of Agriculture
and Life Sciences
Cornell University
Ithaca, NY

Keeton, W. S.
New York State College of Agriculture and Life Sciences
Cornell University
Ithaca, NY

Kirby, C.
Section of Genetics
Cornell University
Ithaca, NY

Kleinman, P.
Department of Natural Resources
Cornell University
Ithaca, NY

Kraus, T.
New York State College of Agriculture and Life Sciences
Cornell University
Ithaca, NY

Lakitan, B.
Department of Vegetable Science
Cornell University
Ithaca, NY

Lehman, H.
Department of Philosophy
University of Guelph
Guelph, Ontario

Lipner, V.
College of Arts and Sciences
Cornell University
Ithaca, NY

McLaughlin, L.
Department of Entomology
Cornell University
Ithaca, NY

Metcalf, R. L.
Department of Entomology
University of Illinois
Urbana, IL

Nelson, J.
New York State College of Agriculture and Life Sciences
Cornell University
Ithaca, NY

Osteen, C.
Economic Research Service
U.S. Department of Agriculture
Washington, DC

Perkins, J. H.
Department of History
Evergreen State College
Olympia, WA

Pettersson, O.
Swedish University of Agricultural Sciences
Uppsala, Sweden

Pimentel, D.
Department of Entomology and Section of Ecology and Systematics
New York State College of Agriculture and Life Sciences
Cornell University
Ithaca, NY

Pimentel, M.
Division of Nutritional Sciences
College of Human Ecology
Cornell University
Ithaca, NY

Reding, N. L.
Monsanto
St. Louis, MI

Rice, P.
International Agriculture
Cornell University
Ithaca, NY

Roach, W. J.
College of Arts and Sciences
Cornell University
Ithaca, NY

Roberts, W.
Plant Industry Branch
Ontario Ministry of Agriculture and Food
Guelph, Ontario

Sachs, C. E.
Department of Rural Sociology
Pennsylvania State University
State College, PA

Schuman, S.
Agricultural Medicine
College of Medicine
University of South Carolina
Charleston, SC

Selig, G.
New York State College of Agriculture and Life Sciences
Cornell University
Ithaca, NY

Shroff, A.
New York State College of Agriculture and Life Sciences
Cornell University
Ithaca, NY

Silva, M.
Department of Microbiology
Cornell University
Ithaca, NY

Surgeoner, G. A.
Department of Environmental Biology
University of Guelph
Guelph, Ontario

Vancini, F.
New York State College of Agriculture and Life Sciences
Cornell University
Ithaca, NY

Zepp, A.
Department of Natural Resources
Cornell University
Ithaca, NY

Contents

II Methods and Effects of Reducing Pesticide Use

III Government Policy and Pesticide Use

IV History, Public Attitudes, and Ethics in Regard to Pesticide Use

Preface

Pesticides have contributed to impressive agricultural productivity, but at the same time their use has caused serious health and environmental problems. Increasing concerns about health and the environment have led many in the general public to wonder whether the benefits of pesticides, e.g., the "perfect red apple," are worth the costs of environmental pollution, human illness, bird kills, and other natural biota destruction. A growing number of scientists and government officials have been viewed as primarily concerned with promoting commercial interests rather than protecting public welfare. Indeed, much of the public has lost faith in science and the government.

Serious ethical investigations of the social policies related to pesticide use must consider risks as well as benefits of pesticide use. To regain public trust, difficult questions regarding our moral obligations concerning pesticide use need to be investigated and appropriate actions need to be taken to protect public welfare and the environment.

To lessen our reliance on pesticides, proven nonchemical pest management technologies as well as technologies that reduce pesticide use need to be implemented. Using these technologies plus relying on a multidisciplinary approach for assessing the risks and benefits of pest control will help make agriculture now and in the future environmentally sound. In addition, this will improve the productivity and profitability of agriculture for farmers.

In this book, we present the results of more than a year's study by scientists and scholars who addressed the ethical, economic, environmental, and health issues related to pesticides and, more broadly, related to pest management. These investigators present their recommendations and methods for implementing environmentally sound pest management practices in agriculture.

Support for this investigation was provided in part by the EVS Program, National Science Foundation; the Agricultural Research Service, U.S. Department of Agriculture; the Office of Regulatory Affairs, Department of Health & Human Services, U.S. Food and Drug Administration; and the College of Agriculture and Life Sciences, Cornell University and College of Arts and Sciences, University of Guelph. We want to thank Dr. Rachelle D. Hollander (National Science Foundation); Dr. W. H. Tallent (ARS, U.S. Department of Agriculture); Dr. John R. Wessel (ORA, Food and Drug Administration); Dr. Allen L. Jennings (U.S. Environmental Protection Agency); Dr. William Gahr (U.S. Government Accounting Office); and Dr. John McCarthy (National Agricultural Chemicals Association) for their interest and encouragement for this study. Also, we want to thank Ms. Nancy Sorrells for helping with administrative and technical aspects of the study and the book. At the same time, we greatly appreciate Dr. Gregory Payne of Chapman & Hall for his interest in our project and assisting us with the details of publishing this book.

David Pimentel

Hugh Lehman

Introduction

1

New Directions for Pesticide Use

Hugh Lehman

In the final chapter of her book *Silent Spring,* Rachel Carson referred to Robert Frost's poem "The Road Not Taken."[1] She suggested there that human societies, through their extensive and intensive use of chemical pesticides, were following a road that, while attractive at the start, could lead to disaster. She recommended taking the other road. Frost suggested that he would have liked to travel on both roads but, being only one person, he could not do so. Frost's traveler took "the one less traveled by." Someone might say that, having explored one of the roads some distance, he could have returned to the other road and explored that also. Frost doubted that he ever would.

In a way, Frost's thought that he would never take the other road is correct. We cannot return the world to the condition it was in in the early 1940s and follow the road which we could have followed had we abstained from synthetic pesticide use. The world has changed too much and many of the changes are not within our power to reverse. The road that we might then have followed no longer exists and cannot be recreated. Yet we can change our course and there is reason to say that we have indeed begun to do so. From the 1940s to the 1980s the amounts of synthetic pesticides used increased rapidly. In the 1980s the amount of pesticide used stabilized.[2] There is strong evidence now that the amount of synthetic pesticides used in agriculture will begin to be significantly reduced.[3]

---

[1]*Silent Spring* by Rachel Carson (Cambridge, The Riverside Press, 1962), 277f. Frost's poem is contained in *The Poems of Robert Frost* (New York, Modern Library, 1946).

[2]Trends in pesticide use are discussed in a number of chapters in this volume. See chapters 2, 12, and 16. Also of interest is "Controlling Toxic Chemicals" by S. Postel, in *State of the World 1988*, ed. Lester Brown, et al. (New York, W.W. Norton and Company, 1988), p. 118f.

[3]See chapters 8 and 9 in this volume.

The other road, as conceived by Rachel Carson, involved use of biological and other nonchemical controls rather than chemical controls of pest insects, diseases, and weeds. However, she claimed that she was not advocating total elimination of chemical methods of control.[4] Thus, if we pursue the metaphor of a road, we should not think of the other road as one on which chemical methods of pest control are prohibited. Indeed, the new road we appear now to be following includes the judicious use of synthetic pesticides. Reductions in pesticide use envisaged in the next decades are to be obtained through a number of approaches including applications of chemical pesticides which are more precisely targeted, applications timed so as to be more effective, use of newer pesticides which are effective at smaller dosages, development of varieties of crops that are genetically resistant to pests, and other methods of nonchemical control including crop diversification, careful timing of planting, rotations, education of farm workers, etc.[5]

The intended destination of the new road is no longer the achievement of total elimination of any pest.[6] The objective now is "integrated pest management."[7] The basic idea of integrated pest management better reflects an awareness of our limitations. We do not (yet) have the capacity to gain total control over pests. Hopefully, we can learn to mitigate the damage that pests do to our crops through "management" of the crops, of the environment, of the pests, and of ourselves. Regulating our activities with due regard for the limitations which derive from our capacities and circumstances is, in some views, a moral virtue. The idea that we can exercise complete control over any aspect of the natural world has been described as arrogant.[8]

The suggestion that following the new road is the path of moral virtue should not be interpreted as implying that everyone who advocates following the old road is inherently evil and selfish, is secretly working to advance the interests of chemical corporations or, at least, is a naive dupe of such corporations—that ecologists and environmentalists are good while agriculturalists are evil. There is no doubt that many agricultural scientists

---

[4]R. Carson, *Silent Spring*, p. 12. Also see "Many Roads and Other Worlds" by G.J. Marco et al., in *Silent Spring Revisited*, ed. G.J. Marco et al. (Washington, D.C., American Chemical Society, 1987).

[5]Some discussion of these alternative methods is found in chapter 10 in this volume. See also chapter 9.

[6]Carson suggested the objective of entomologists on the old road was "to create a chemically sterile, insect-free world," *Silent Spring*, p. 12.

[7]Chapter 9 in this volume.

[8]See "A Summary of Silent Spring" by Gino Marco in *Silent Spring Revisited*, ed. G.J. Marco, et al. (Washington, American Chemical Society, 1987), p. xviii.

who advocated intensive and extensive pesticide use were motivated to eliminate hunger and malnutrition and, in general, to improve the quality of life of people all over the world. While we now recognize that we ought not destroy our life-support system by overuse of pesticides and other chemicals, we must not forget that we must continue to produce food and other agricultural products in sufficient quantities so that all people can afford to obtain such necessities of life. If environmentalists are wiser about the limitations of pesticide use, it is in part because they have learned from the experience of agricultural scientists during the years since World War II. To assume that advocates of following the new road are the sole possessors of virtue and wisdom reflects a naivete and hubris as pervasive as found among advocates of extensive pesticide use. In any case, the disposition to criticize the intentions of agriculturalists or environmentalists is largely irrelevant to the problem of determining what road we ought to take. In trying to establish the claim that we ought to follow the one road rather than another, we ought to avoid ad hominem arguments.

Discerning the direction of the new road, and, where we do perceive the direction, staying on the road, will not be easy tasks. Elimination or eradication of pests is a clear, if unobtainable, objective. Management of pests is not so clear. How much pest damage should we tolerate? Pimentel et al. argue that we ought to reduce "cosmetic standards" in foods—that is, tolerate increased presence of insect parts and insect damage.[9] A higher tolerance on our part, in these respects, would lead to reductions in the amounts of pesticide required for food production and, hopefully, to less damage to our environment—the life-support system on which our existence and the quality of our lives depend. But what criteria should determine new cosmetic standards? Some guidelines are suggested in chapter 3. Where the modifications of the fruit or vegetable due to the pest do not reduce nutritional quality or palatability and do not either produce or allow the production of substances toxic to humans, then such modifications should be permitted.

An objective of the old road was to reduce the costs of food and fiber to the consumer to the lowest possible point. This again is a relatively clear objective. On the new road, our objectives are more complex. The consumer of food and fiber is also a resident of the environment. As such, he or she may be adversely affected by the introduction of pesticides into air, water, and food. Food containing residues, even in minute amounts, may cause illness. Even if such food does not cause serious illness it often provokes concern as to whether it is causing serious illness and consequently

---

[9]See chapter 4 in this volume.

provokes distress. An excellent review of the complexities involved in evaluating health questions is found in chapter 6.[10]

Whereas cheap and abundant food is a great benefit to people, illness and concern about possible illness are harmful.[11] Yet, total elimination of synthetic pesticide residues would lead, at least in the near future, to higher food prices and smaller harvests. It is reasonable to suppose that this effect would produce illness, death, and other distress (due to malnutrition, etc.). It is not as if we can replace the clear objective (cheapest possible food) with another equally clear objective. Somehow, we have to strike a balance amongst competing objectives. We want abundant, inexpensive food but we want it not to cause illness. Further, as noted by Pimentel, pesticide use gives rise to many other losses. Reduction in pesticide usage, even given somewhat increased costs of food due to reductions in food production, could be more than offset by decreases in other losses due to pesticide damage.[12] How do we define this more complex objective? What losses to fisheries or birds, etc., should we tolerate in the course of our agricultural production?

Our agricultural production systems have changed rapidly over the past half century. These changes are described in a number of chapters in this book, including, "Pesticides: Historical Changes Demand Ethical Choices."[13] Awareness of the many facets of change in our agricultural production systems may help, through providing a historical perspective, in determining the destination of the new road more wisely. In chapter 16 changes in agricultural practices involving the use of pesticides are described as one of a number of technological changes affecting agriculture. The rationale for taking a new course in regard to the use of pesticides will probably have implications for other aspects of agricultural technology also. After all, all aspects of a production system are integrated with respect to the goals of the system. Change to one aspect of the system, namely, pesticide use, will require changes in other aspects, e.g., development and use of machinery, fertilizer, crop varieties, etc. Such perspective may help us to see what changes we have to make in our social practices in order to stay on the new road.

---

[10]The difficulties involved in determining undesirable health effects due to pesticide use are also discussed in chapter 5 of this volume. Also of interest in this regard are "Human Health Effects of Pesticides" by J.E. Davies and R. Doon in *Silent Spring Revisited*, (see footnote 8), and chapter 3 of this volume.

[11]Concern about harmful health effects in foods is discussed in chapter 15 of this volume.

[12]For extensive discussion of harmful effects of pesticides on the environment see chapter 2. Also see chapter 3 and chapter 16.

[13]Some history of the use of pesticides in agriculture in the U.S. is also contained in chapters 11 and 12.

However, historical perspective should be supplemented by sociological analyses of the causes for taking the new road as well as of social and economic consequences that may result therefrom. Such consequences are discussed in "Socioeconomic Impacts and Social Implications of Reducing Pesticide and Agricultural Chemical Use in the United States," by F. Buttel.[14] Buttel relates the taking of the new road in regard to pesticide use to the broader environmental movement which has grown throughout the industrialized world. The environmental movement reflects, at least in part, widespread dissatisfaction with many changes in modern industrialized, commercialized life which are widely perceived as losses. Among these are perceptions of loss of genetic diversity, loss of "purity" of water and food, loss of a healthful workplace for farm workers, and loss of autonomy in some respects by producers.[15] These efforts to regain what we have (allegedly) lost in these regards have given rise to the "Green" movement. Over time, this movement, as Buttel notes, has led to some greening even of colleges of agriculture.[16] Incorporating values advocated by the Greens without losing value overall will, however, require understanding why farmers use pesticides. Buttel's chapter 7 contributes to our understanding of such motivations.

In the past, as a result of our desire for low-cost agriculture, we evolved many government policies which furthered this objective. If we are to modify our destination, we may well have to modify government policies that foster pesticide use. These policies are carefully analyzed in "Government Policies That Encourage Pesticide Use in the United States," by K. Dahlberg.[17] Further, we have to review government practices in our efforts to stay on course. One such review is provided in "Alar: The EPA's Mismanagement of an Agricultural Chemical," by J. Hathaway.[18]

The decision to take a new road should be attended by careful ethical analysis. An effort to indicate how such an analysis might proceed is given in "Values, Ethics, and the Use of Synthetic Pesticides in Agriculture," by H. Lehman.[19] Rachel Carson appealed to ethical premises. Explicit among her premises is the proposition that "the methods (of controlling pests) must be such that they do not destroy us along with the insects."[20] Implicit in her charges are other ethical premises. Among these are claims

---

[14]Chapter 7.

[15]See chapter 3.

[16]The Greening of colleges of agriculture is one aspect of the Greening of America. See *The Greening of America* by Charles A. Reich (New York, Random House, 1970).

[17]Chapter 11.

[18]Chapter 13.

[19]Chapter 14. Also see chapter 17.

[20]*Silent Spring*, p. 9.

that large-scale introduction of pesticides into our environment should not occur without "advance investigation of their effects on soil, water, wildlife, and man himself,"[21] and that people ought not to be subjected to poisonous substances without their consent or knowledge.[22] Further, she challenged the assumption, which she attributed to many people of our era, that people have a moral right to make a profit at virtually any cost and that it is acceptable for those with economic or political power to mislead members of the public in regard to potential harm from use of agricultural chemicals.[23]

Carson's ethical assumptions are not all that unusual. Indeed, it would appear that she is largely in agreement with the principles expressed in the "Code of Ethics for Registered Professional Entomologists."[24] Among the principles therein expressed are the claims that

> 1.1 The Professional Entomologist's knowledge and skills will be used for the betterment of human welfare.
>
> 2.1 The Professional Entomologist will have proper regard for the safety, health, and welfare of the public in the performance of professional duties.
>
> 2.2 The Professional Entomologist will be honest and impartial, and will preface any one-sided statements, criticisms, or arguments by clearly indicating on whose behalf they are made.[25]

If we destroy ourselves along with the insects, weeds, fungi, etc., we have hardly bettered human welfare. It would appear that having a proper regard for safety, etc., would require advanced investigation of effects of pesticides on water, etc. However, although the entomologist's code commands honesty, I see nothing in it that suggests that people are entitled to be informed about the pesticides to which they are exposed. It appears that Carson's principles reflect a greater influence of Kantian ethical thought than do those of the entomologists.[26]

---

[21]*Silent Spring*, p. 13.

[22]*Silent Spring*, p. 12.

[23]*Silent Spring*, p. 13.

[24]"Code of Ethics for Registered Professional Entomologists of the American Registry of Professional Entomologists."

[25]We shall not here undertake a critical analysis of the entomologist's code of ethics. Questions could be raised concerning the clarity of expressions such as "proper regard" and "betterment of human welfare." Further, assuming that the principles containing these terms could be clarified, one could ask whether such principles can be established by reference to basic moral principles (be they consequentialist, Kantian, or other principles). Further, assuming that the entomologist's principles can be sufficiently clarified, one may ask whether the practices of entomologists working for chemical companies or the government are in accord with requirements stipulated by the principles.

[26]For a brief explanation of Kantian ethical principles see part III, section B, of chapter 14.

It is generally accepted that Carson's work has changed our social practices in regard to the use of pesticides.[27] Concerns about human health and the quality of our environment have induced politicians in several countries to adopt more stringent controls on the use of pesticides[28] as well as to adopt policies intended to lead to reductions in pesticide use. Are such changes warranted on ethical grounds? If we look at the new road from an ethical perspective, such as is indicated in "Values, Ethics, and the Use of Synthetic Pesticides in Agriculture," it appears that taking the new road is ethically required.[29] Given that the reduction in losses attendant on decreased pesticide usage more than offsets any reduction in agricultural production, the new road is required on utilitarian or consequentialist grounds. Further, given that the losses attendant on pesticide usage are often borne by individuals other than those who benefit from the use of pesticide, the new road is ethically required in light of the principle that we ought not treat others merely as means to our own ends.

---

[27]See "Many Roads and Other Worlds," *Silent Spring Revisited*.

[28]See "The Not So Silent Spring" by J. Moore, in *Silent Spring Revisited*.

[29]Chapter 14.

Part I

Social and Environmental Effects of Pesticides

2

The Impact of Pesticides on the Environment

Clive A. Edwards

Introduction

The use of chemicals to control pests which harm crops, annoy humans, and transmit diseases of both animals and humans is not a new practice. Homer described how Odysseus fumigated the hall, house, and the court with burning sulfur to control pests. As long ago as A.D. 70, Pliny the Elder recommended the use of arsenic to kill insects, and the Chinese used arsenic sulfide for the same purpose as early as the 16th century. By the early 20th century, inorganic chemicals such as lead arsenate and copper acetoarsenite were in common use to control insect pests.

However, until 50 years ago, most arthropod pests, diseases, and weeds were still controlled mainly by cultural methods. The era of synthetic chemical pesticides truly began about 1940 when the organochlorine and organophosphorus insecticides were discovered and synthetic-hormone-based herbicides were developed. These chemicals and others that were developed subsequently seemed to be so successful in controlling pests that there was extremely rapid adoption of their use and the buildup of a large multibillion-dollar agrochemical industry. There are currently more than 1,600 pesticides available (Hayes and Lawes, 1991) and their world-wide use is still increasing (Edwards, 1985); about 4.4 million tons of pesticides are used annually with a value of $20 billion (E.P.A., 1989). The United States accounts for 27% of this market, exporting about 450 million pounds and importing about 150 million pounds.

In the early years of the expansion of pesticide use the effectiveness of these chemicals on a wide range of arthropods, pathogens, and weeds was so spectacular that they were applied widely and often relatively indiscriminately in developed countries. There was little anxiety as to possible human, ecological, or environmental hazards until the late 1950s and early

1960s, when attention was attracted to the issue by the publication of Rachel Carson's *Silent Spring* (1962), followed shortly by *Pesticides and the Living Landscape* (Rudd, 1964). Although these publications tended to overdramatize the potential hazards of pesticides to humans and the environment they effectively focused attention on relevant issues. These included the acute and chronic toxicity of pesticides to humans, domestic animals, and wildlife; their phytotoxicity to plants; the development of new pest species; the development by pests of resistance to these chemicals; and the persistence of pesticides in soils and water and their potential for global transport and environmental contamination.

In response to the recognition of such potential environmental hazards from pesticides, most developed countries and relevant international agencies set up complex registration systems, monitoring organizations, and requirements that had to be met before a new pesticide could be released for general use. Data were requested on toxicity to mammals and other organisms, degradation pathways, and fate. Monitoring programs were instituted to determine residues of pesticides in soil, water, and food, as well as in flora and fauna (Carey et al., 1979). However, many pesticides are still used without adequate registration requirements or suitable precautions in many developing countries, so potential environmental problems are much greater in these countries. And in the United States there are still environmental problems due to pesticides.

The pesticides in current use vary greatly in structure, toxicity, persistence, and environmental impact. They include the following:

Organochlorine Insecticides

These insecticides, which are very persistent in soil and are very toxic to many arthropods, were used widely in the 25 years after the Second World War. They include compounds as DDT, benzene hexachloride, chlordane, heptachlor, toxaphene, methoxychlor, aldrin, dieldrin, endrin, and endosulfan, which are relatively non-soluble, have a low volatility, and are lipophilic. They do not have very high acute mammalian toxicities but their persistence and their tendency to be bioconcentrated into living tissues and move through food chains has meant that, with the exception of lindane, their use has been largely phased out, other than in certain developing countries. However, many soils and rivers are still contaminated with DDT, endrin, and dieldrin (White et al., 1983a; White and Krynitsky, 1986), the most persistent of these compounds, and there are still reports of their residues in wildlife (Riseborough, 1986), so these chemicals still present an environmental hazard. Unfortunately, most monitoring for these chemicals was phased out after their use was banned or restricted.

Organophosphate Insecticides

Some of the organophosphate insecticides were first developed as nerve gases during the Second World War. They include parathion, diazinon, trichlorfon, phorate, carbophenothion, disulfoton, dimethoate, fenthion, thionazin, menazon, dyfonate, and chlorfenvinphos. Although these chemicals are much less persistent than the organochlorines, many of them have much higher mammalian toxicities and potential to kill birds and other wildlife. Some of them are systemic and taken up into plants. They have sometimes caused severe local environmental problems in contamination of water and local kills of wildlife, but most of their environmental effects have not been drastic, although they do sometimes contaminate human food.

Pyrethroid Insecticides

These are insecticides of very low mammalian toxicity and persistence. Since they are very toxic to insects they can be used at low dosages. Their main environmental impact occurs because they are broad-spectrum toxicants and are very toxic to fish and aquatic organisms. Since they affect a broad range of insects they may affect beneficial species, lessen natural control, and increase the need for chemical control measures.

Carbamates

These chemicals include not only insecticides, such as carbaryl, but also acaricides, fungicides, such as benomyl, and nematicides. They tend to be rather more persistent than the organophosphates in soil and they differ considerably in their mammalian toxicity. However, most of them tend to be broad-spectrum toxicants affecting quite different groups of organisms, so they have the potential for considerable environmental impact, particularly in soils.

Nematicides

Some soil nematicides, such as dichlopropene, methyl isocyanate, chloropicrin, and methyl bromide, act as soil fumigants. Others, such as aldicarb, dazomet, and metham sodium, are effective mainly through contact. All of them are of very high mammalian toxicity and have a broad spectrum of toxicity, killing organisms from an extremely wide range of taxa belonging to both the plant and animal kingdoms. Although very transient in soil, they can cause drastic localized ecological effects that may persist for several seasons.

Molluscicides

Two molluscicides are in common use against terrestrial molluscs, metaldehyde and methiocarb. Although the former is of high mammalian toxicity, both are used mainly as baits and hence cause few environmental problems other than occasional deaths of wild mammals that eat these baits. Several molluscides used to control aquatic snails, *N*-trityl morpholine, copper sulfate, niclosamine, and sodium pentachlorophenate, are toxic to fish.

Systemic Herbicides

These include the "hormone"-type herbicides such as 2,4,5-T; 2,4-D; MCPA; and CMPP. They are not persistent in soil, are quite selective in their action on various species of plants, and are of low mammalian toxicity, so they cause few serious direct environmental problems. However, most are relatively soluble and reach waterways and groundwater relatively easily. Most of these herbicides are not very toxic to fish although some herbicides that are used to control aquatic weeds are moderately toxic to fish (Edwards, 1977).

Contact Herbicides

These chemicals tend to kill weeds when they contact the upper surface of the foliage. They include chemicals such as the dinitrophenols, cyanophenols, and pentachlorophenol; paraquat is also sometimes classed as a contact herbicide. They are usually nonpersistent; the triazines are of low mammalian toxicity but can persist in soil for several years. They are slightly toxic to soil organisms and moderately toxic to aquatic organisms. They cause few direct environmental problems other than affecting the growth of crops in certain years when their residues have not broken down from the previous year's treatments. However, their indirect effects, such as leaving bare soil which is susceptible to erosion, can have considerable environmental impact.

Fungicides

Many different types of fungicides are used, of widely differing chemical structures. Most have relatively low mammalian toxicities, and with the exception of the carbamates, a relatively narrow spectrum of toxicity to soil and aquatic organisms. Their most serious environmental impact is on soil microorganisms.

As a result of the appreciation that pesticides could have serious ecological side effects, that could create new pests by killing their natural enemies, the concept of minimizing pesticide use and complementing it with cultural and biological control techniques was developed. This was termed integrated pest control (IPC) or management (IPM) and the idea was first suggested by Stern et al. (1959) and modified by other workers (Smith and Reynolds, 1965; Norton and Holling, 1975).

During the last 20 years two new concepts have been developed. The first of these is *ecotoxicology*, or *environmental ecotoxicology*; in this field holistic studies are made of the effects of toxic substances (including pesticides) in both natural and manmade environments, the environmental risks are assessed, and measures to prevent or minimize environmental damage are made (Truhart, 1975; Duffus, 1980; Butler, 1978). There has also been great progress in the area of *agroecology*, which aims to understand the ecological processes that drive agricultural ecosystems. Such knowledge is the key to being able to minimize the use of synthetic pesticides to manage pests (Carroll et al., 1990).

There have been a number of books and major reviews of the environmental impacts of pesticides. These include *Organic Pesticides in the Environment* (Rosen and Kraybill, 1966); *Pesticides in the Environment and their Effects on Wildlife* (Moore, 1966); *Pesticides and Pollution* (Mellanby, 1967); *Organochlorine Pesticides in the Environment* (Stickel, 1968); *Since Silent Spring* (Graham, 1970); *Persistent Pesticides in the Environment* (Edwards, 1973a); *Environmental Pollution by Pesticides* (Edwards, 1973b); *Pesticide Residues in the Environment in India* (Edwards et al., 1980); *Pollution and the Use of Chemicals in Agriculture* (Irvine and Knights, 1974); *Ecology of Pesticides* (Brown, 1978); *Use and Significance of Pesticides in the Environment* (McEwen and Stevenson, 1979); *Chemicals in the Environment: Distribution, Transport, Fate and Analysis* (Neely, 1980); *Ecotoxicology: The Study of Pollutants in Ecosystems* (Moriarty, 1983).

Acute Effects of Pesticides on Living Organisms

Since pests, whether they are arthropods, nematodes, fungi, bacteria, or weeds, are all living organisms, all of the pesticides that are designed to control them are of necessity biocides. Indeed, some of the organophosphate insecticides that are effective insect control agents were developed originally during the Second World War as human nerve-gas agents. However, pesticides differ greatly in the mammalian toxicity; toxic doses range from amounts as low as 1 mg/kg in the diet of a vertebrate animal to practically harmless. Some, such as the soil fumigants dichlorpropene and metham sodium, are extremely broad biocides that are toxic to most living organisms. Others, including many herbicides, are highly specific in their action on plants and have little toxicity to other organisms. Some pesticides, particularly the organochlorine insecticides, are extremely stable compounds and persist in the environment for many years. Others, such as the fumigant nematicides break down in a few hours or days.

There is increasing pressure on pesticide manufacturers from national and international pesticide registration authorities to provide comprehensive data on the acute toxicity of their chemicals for humans, rats or mice, fish, aquatic crustacea, plants, and other selected organisms. However, such data can only indicate the possible toxicity of a chemical to related organisms which may actually differ greatly in their susceptibility to particular chemicals. No data at all are available on the toxicity of most chemicals to the countless species of unrelated organisms. Some of these may include endangered species or may play important roles in dynamic biological processes or food chains.

There has been some progress in recent years in developing predictive models of the likely toxicity of a particular chemical to different organisms based on data on the behavior and toxicity of related chemicals, the structure of the chemical, its water solubility and volatility, its lipid/water partition coefficient, and other properties (Moriarty, 1983).

Living organisms differ greatly in their susceptibility to pesticides and we are gradually accumulating a data bank on which chemicals present the greatest acute toxic hazard to the various groups of organisms. The characteristics of some of these organisms and their possible susceptibility to pesticides will be summarized briefly.

Microorganisms

The numbers of microorganisms in all of the physical compartments of the environment are extremely large and they have immense diversity in form, structure, physiology, food sources, and life cycles. This diversity makes it almost impossible to assess or predict the effects of pesticides upon them. Moreover, the situation is even more complex because microorganisms can utilize most pesticides as food sources upon which to grow; indeed, microorganisms are the main agents of degradation of pesticides.

We still know relatively little of the complex ecology of microorganisms in soil and water. Clearly, they can utilize many substances as food sources and are involved in complex food chains. Most of the evidence is that if an ecological niche is made unsuitable for particular microorganisms by environmental or chemical factors, some other microorganism that can withstand these factors will fill the niche. Moreover, unless a pesticide is very persistent, any effect it may have on particular microorganisms is relatively transient, so populations usually recover in 2–8 weeks after exposure.

Since there are such enormous numbers of microorganisms, it is impossible to test the acute toxicity of pesticides to them individually, and it is possible to generalize only in the broadest terms as to the acute toxicity of pesticides to particular soil- and water-inhabiting microorganisms, based on tests on groups of organisms.

Most of those workers who have reviewed the effects of pesticides on soil microorganisms (Parr, 1974; Brown, 1978; Edwards, 1989; Domsch, 1963, 1983) have reported that soil fumigants and fungicides have much more drastic effects on soil microorganisms than insecticides or herbicides. These chemicals are applied to control certain specific soil microorganisms, they are usually used at high doses, and they are usually broad-spectrum biocides. There is little evidence that other pesticides have serious effects.

There are relatively few data on the toxicity of pesticides to microorganisms in aquatic environments (Parr, 1974). Much of the microbial activity is in the bottom sediments, and this is where pesticides in aquatic systems became concentrated, through runoff and erosion from agricultural land.

There is a considerable amount of literature on the effects of pesticides on aquatic algae that are a major part of the phytoplankton in aquatic systems. Herbicides such as simazine and terbutryn can have drastic effects on these organisms (Gurney and Robinson, 1989).

Invertebrates

The invertebrates that inhabit soil or water or live above ground are extremely diverse, belonging to a wide range of taxa. There are extremely large numbers of species, with many species still to be described, and their populations are enormous. We still know relatively little of the biology and ecology of many of the invertebrate species that inhabit soil and water. Thompson and Edwards (1974) reviewed the effects of pesticides on soil and aquatic invertebrates but there have been few comprehensive reviews of the effects of pesticides on particular groups of invertebrates, an exception being a review of the effects of pesticides on earthworms (Edwards and Bohlen, 1991). Because of the diversity of the invertebrate fauna it is extremely difficult to make any generalizations on the acute toxicity of pesticides in individual species.

Soil-Inhabiting Invertebrates

There has been a review of the effects of pesticides on soil- inhabiting invertebrates (Edwards and Thompson, 1973). This reported that there are relatively few data on the acute toxicity of pesticides to individual species of soil-inhabiting invertebrates; most studies have involved studying the effects of pesticides on mixed populations of invertebrates in soil in the laboratory or field. Nevertheless, it is possible to make some empirical assessments of the susceptibility of different groups of soil-inhabiting invertebrates to different groups of chemicals.

Nematodes

Nematodes, which are extremely numerous in most soils and include parasites of plants and animals as well as free-living saprophagous species, are not susceptible to most pesticides. Only those pesticides that are designed to kill nematodes, i.e., fumigant and contact nematicides, seem to be acutely toxic to nematodes and other pesticides have little direct effects on them.

Mites (Acarina)

Populations of mites are large and occur in most soils. The different taxa differ greatly in susceptibility to pesticides but all tend to be sensitive to acaricides and some insecticides as well as soil-fumigant nematicides and fungicides, but most fungicides and herbicides have no direct effects on them. The more active predatory species of mites tend to be more susceptible to pesticides than the sluggish saprophagous species.

Springtails (Collembola)

These arthropods, which are closely related to insects, are extremely numerous in most soils. They are susceptible to many insecticides but not to other pesticides. Their susceptibility to different insecticides has not been well documented and is extremely difficult to predict. Few herbicides and fungicides are directly toxic to them. There seems to be a positive correlation between the degree of activity of springtails and their susceptibility to insecticides.

Pauropods (Pauropoda)

These small animals, which are common in many soils but occur in smaller numbers than mites or springtails, seem to be very sensitive to many insecticides but not to other pesticides. Little is known about their ecological importance.

Symphylids (Symphyla)

These arthropods, some of which are pests and others saprophagous and all of which are common in many soils, tend to not be very susceptible to pesticides, not even to insecticides; moreover, they penetrate deep into the soil, where their exposure to these chemicals is minimized.

Millipedes (Diplopoda)

These common soil-inhabiting arthropods, which live mainly on decaying organic matter and seedlings, are intermediate in their susceptibility to insecticides between pauropods and symphylids. Since they live on or near the soil surface, they are exposed to many chemicals that occur as surface residues.

Centipedes (Chilopoda)

These predatory invertebrates, which are often predators of pests, are common in most soils and tend to be susceptible to many insecticides and nematicides but not to other pesticides. Since they are very active their exposure is considerable as they move through contaminated soil.

Earthworms (Lumbricidae)

These invertebrates are important in breaking down organic matter and in the maintenance of soil structure and fertility. Because of their importance in soils, and because of their selection as key-indicator organisms for soil contamination, there is a great deal more information available on the acute toxicity of pesticides to them than to any other group of soil-inhabiting invertebrates (Edwards and Bohlen, 1991). The chemicals that are acutely toxic to earthworms include endrin, heptachlor, chlordane, parathion, phorate, aldicarb, carbaryl, bendiocarb, methomyl, chloropicrin, dichloropene, benomyl, thiabendazole, thiophanate methyl, and chloroacetamide. This is a relatively small number of pesticides out of the more than 200 that have been tested for acute toxicity. Carbamates seem to be particularly toxic to earthworms.

Molluscs (Mollusca)

Very few pesticides are toxic to slugs or snails, probably because of their protective coating of mucus. Copper sulfate, metaldehyde, and methiocarb are all toxic to molluscs.

Insects (Insecta)

Soil-inhabiting insects and larvae can be pests, predators of pests, or important in breaking down organic matter in soil. They are susceptible to many insecticides but the variability in susceptibility between species and chemicals is much too great for any general trends to emerge. Few insects are affected by fungicides or herbicides. There is some tendency for the more active predatory species to be more susceptible to insecticides than the more nonmotile species.

Pesticides, particularly insecticides, also affect aerial insects, particularly bees. Bees are extremely important, not only in providing honey but also in pollination. Data on acute toxicity of pesticides to bees have been required by most pesticide registration authorities, but this does not avoid considerable mortality in the field.

Aquatic Invertebrates

In general, aquatic invertebrates are much more susceptible to pesticides than soil-inhabiting invertebrates, particularly if the pesticide is water-soluble. Lethal doses of a pesticide are readily picked up as water passes over the respiratory surfaces of the invertebrates, and it is difficult for aquatic invertebrates to escape such exposure.

Although it is much easier to assess the acute toxicity of pesticides to individual species of aquatic invertebrates than to species of soil organisms, in laboratory tests, not many aquatic species have been tested extensively in this way. The invertebrate species that have been tested most commonly have been *Daphnia pulex, D. magna, Simocephalus*, mosquito larvae (*Aedes*), *Chironomus* larvae, stonefly nymphs (*Pteronarcys californica* and *Acroneuria pacifica*) (Jensen and Gaufin, 1964), mayfly nymphs (*Hexagenia*), caddis fly larvae (*Hydropsyche*) (Carlson, 1966), copepods (*Cyclops*), ostracods, and the amphipod *Gammarus*.

It is extremely difficult to differentiate between the different taxa of aquatic invertebrates in terms of their susceptibility to different pesticides. However, there is little doubt that insecticides have much more effect on most aquatic invertebrates than fungicides or herbicides, although there have been reports of some herbicides killing them. There have been several reviews of the effects of pesticides on aquatic organisms (Muirhead-Thomson, 1971; Edwards, 1977; Thompson and Edwards, 1974; Parr, 1974; Brown, 1978). To briefly review the aquatic invertebrate fauna in terms of susceptibility to pesticides:

Crustacea

Crustacea differ greatly in size and numbers. They include small, swimming crustaceans such as *Cyclops* and *Daphnia*; intermediate-sized organisms such as shrimps and prawns, and larger invertebrates including crabs and lobsters. They all seem to be relatively susceptible to many insecticides but herbicides and fungicides are not very toxic to them.

Molluscs and Annelids

These are bottom-living organisms such as oysters, clams, and other shellfish or small worms that live in the bottom mud or sediment in salt- and freshwater systems. Most of them tend to have much lower sensitivities to pesticides than the arthropods, although they take up and bioconcentrate some persistent pesticides into their tissues, and these may eventually accumulate to a toxic level.

Insects

A wide range of insect larvae inhabit water, particularly freshwater. Some, such as mosquito larvae, are free-living in water, but the majority live on or in the bottom sediment. These include chironomid, mayfly, dragonfly, stonefly, and caddis fly larvae. These insects are very susceptible to many insecticides, particularly the more persistent ones which tend to concentrate in the bottom sediment and remain there for considerable periods, but herbicides and fungicides are not usually toxic to aquatic insects.

In conclusion, the organochlorine insecticides are moderately toxic to not only insect larvae but also to many other aquatic invertebrates. Organophosphate and carbamate insecticides tend to be less toxic than organochlorines to insect larvae but very toxic to some species (Poirier and Surgeoner, 1988). Carbamates are probably the least toxic. The most toxic group of insecticides to aquatic invertebrates is the pyrethroids, which have a broad spectrum of activity and affect most species. For instance, Anderson (1989) reported that pyrethroids were very toxic to mosquitoes, blackfly, and chironomid larvae, and Day (1989) reported the same for the zooplankton.

Fish

All aquatic organisms tend to be much more susceptible to pesticides than terrestrial ones. There are many reasons for this but the most important is that the contamination can spread rapidly through an aquatic system, and there is no escape for fish and other organisms. In most developed countries, reports of fish kills by pesticides are very common, particularly in summer (Muirhead-Thomson, 1971). There have been no good estimates of overall losses of fish due to pesticides but there is little doubt that such losses must be enormous world-wide.

There is a considerable data bank on the acute toxicity of pesticides to fish since, in the developed countries, a major requirement before a pesticide can be registered is to provide data on its acute toxicity to fish. However, these data tend to be confined to assays on a few species of fish that are easy to breed and culture, and may not always be relevant to field populations.

Pesticides applied to agricultural land can fall out from aerial sprays on to water or eventually reach aquatic systems such as rivers or lakes through drainage or by runoff and soil erosion. Another source of contamination of water is the disposal of pesticides and their containers in aquatic systems and industrial effluents from pesticide factories. Fish are particularly susceptible to poisonous chemicals since they are exposed to such chemicals in solution, in the water in which they live, or in suspension absorbed on to sediments, as the water passes over the fishes' gills.

Birds

Many birds are susceptible to many pesticides and we have a great deal of information on the acute toxicity of different pesticides to them since not only are many pesticides tested for their effects on indicator bird species during the registration process, but also there are monitoring schemes for recording numbers of birds killed by pesticides in many countries (Riseborough, 1986; Hardy, 1990). Most toxic pesticides for birds are encountered through feeding on contaminated food, such as seed dressed with pesticides or plants treated with pesticides or animals fed upon that have died from pesticides. Avian toxicity is sometimes a reason for registration of a pesticide to be refused, particularly if the use pattern, e.g., as a seed dressing, would expose large numbers of birds to the pesticide.

Mammals

All pesticides are tested for acute toxicity to representative mammals during their development and registration phases. The species that are normally tested for acute toxic responses to pesticides are mice or rats, and there are comprehensive lists of the toxicity of virtually all pesticides in these animals. Any pesticide with a high mammalian toxicity is more difficult to register for general use unless it is very effective and provides good benefits. However, it is difficult to use such specific data to predict harm to other mammals with quite different habits and susceptibilities in the field. Data on the toxicity of pesticides to mammals in the field is relatively scarce.

Humans

It is impossible to obtain direct acute toxicity data for human beings. Hence, data from animal toxicity testing is used to predict the potential acute toxicity of pesticides to human beings. This is far from satisfactory as a toxicity index since different groups of mammals have different susceptibilities to pesticides. However, since pesticides have important effects on humans such data are used with added safety factors. It has been estimated that there are between 850,000 and 1.5 million pesticide poisonings of humans annually worldwide, from which between 3,000 and 20,000 people die. There have been many serious accidents in which many people have died from pesticides (Hayes and Lawes, 1991).

Indirect Effects of Pesticides on Living Organisms

The indirect effects of pesticides on living organisms are probably more important than their direct toxicity (McEwen and Stephenson, 1979). Moreover, we know very much less about the indirect effects of pesticides on living organisms because it is very much more difficult to assess such effects reliably. The indirect effects of pesticides on organisms can be divided into two main categories: (1) the chronic effects of pesticides upon the growth, physiology, and reproduction of organisms, and (2) the ecological effects of pesticides on populations and communities of living organisms. Usually, only when indirect effects are spectacular, such as (1) the thinning of the eggshells of important endangered raptor species of birds or (2) major fish kills, do we hear much of them.

Chronic Effects

If a pesticide does not kill an organism it can still have very significant sublethal effects on its functioning by influencing its length of life, growth, physiology, behavior, or reproduction.

Effects on the Growth of Organisms

There are relatively few data available on the effects of pesticides on the growth of organisms. A number of pesticides seem to affect the growth of earthworms (Edwards and Bohlen, 1991), marine organisms (Newell, 1979) such as mussels (Bayne, 1975) and fish (Warren and Davis, 1971, Benoit, 1975), birds (Stickel et al, 1984), pigeons (Jefferies, 1975), and deer (Edwards, 1973a). In the case of organochlorine insecticides there were good correlations between residues in the fat and weight gain (Stickel et al., 1984). Two herbicides, bromacil and diuron, were reported to affect the growth of fish (Call et al., 1987), and so was the pyrethroid insecticide permethrin (Kingsbury and Krcutzweiser, 1987). However, some of these changes in growth rates have been considered to result from the effects of the pesticides on feeding by the animal, and they are usually only short term.

Effects on the Behavior of Organisms

The diversity of known sublethal effects of pesticides on organisms is enormous (Moriarty, 1983). They range from a slowing down of activity to hyperactivity, as well as many other complex behavioral changes.

Data on the effects of pesticides on the behavior and physiology of the invertebrates that inhabit soil and water are extremely sparse. Many pesticides inhibit acetylcholinesterase activity in insects, mites, ticks, molluscs, earthworms, and nematodes (Edwards and Fisher, 1991). The most general symptoms include hyperactivity followed by a decrease in movement and often death. However, in some instances the invertebrates recover. Continual exposure of invertebrates to a low dose of a pesticide may result in a longer-term slowing down of their activity.

Insecticides such as dimethoate have been reported to affect the behavior of aquatic crustacea (Vogt, 1987). Various behavioral side effects have been reported for fish exposed to pesticides (Brown, 1978). For instance, fish poisoned sublethally may prefer warmer or even more saline water. Various other changes in behavior such as aversion to light, hyperactivity, or changes in turning behavior have also been reported in fish exposed to sublethal doses of pesticides (Brown, 1978).

There is an extensive literature on the effects of pesticides on the behavior and physiology of birds (Riseborough, 1986) and mammals (Moriarty, 1983). These effects are too diverse to allow a detailed consideration here.

Effects on the Reproduction of Organisms

It is extremely difficult to assess the effects of pesticides on the reproduction of aquatic or soil-inhabiting invertebrates in the laboratory. Hence, although we have field data on the effects of pesticides on reproduction of invertebrates we have only very sparse data from laboratory experiments. The only reliable data available are for key-indicator species such as the earthworm *Eisenia fetida* (Edwards and Bohlen, 1991), aquatic crustacea such as *Daphnia*, cladocerans, and copepods (Day, 1989). There are numerous reports of pesticides affecting the reproduction of fish through decreased production of eggs, or increased egg mortality, or decreased hatching (Brungs, 1969; Benoit, 1975). Pesticides have also been reported to produce deformations in the fry. For instance, White et al. (1983a) reported that large DDT and toxaphene residues impaired the reproduction of fish and caused the fry to be deformed. Call et al. (1987) reported decreased hatching and incidence of abnormal fry in fish exposed to the herbicides, bromacil, and diuron. Clearly some pesticides do inhibit reproduction of fish species. Some of the organochlorine insecticides had spectacular effects on the breeding of predatory birds, through the thinning of their eggshells, and this has been fully summarized by Cooper (1991). There is not much information on the effects of pesticides on the reproduction of mammals (Brown, 1978).

Uptake of Pesticides Into the Bodies of Organisms

Almost all pesticides are taken up into the bodies of organisms but many are usually metabolized quite rapidly. However, some pesticides, including most organochlorine insecticides, are bioconcentrated from the food or medium in which they live, so large residues accumulate in the organism's bodies. Even small invertebrates such as aquatic protozoa can bioconcentrate pesticides such as dieldrin, dimethoate, and permethrin into their tissues to levels that are 1,000–3,500 times greater than those in the water in which they live (Bhatagnar et al., 1988). Earthworms and molluscs accumulate residues of organochlorine insecticides at levels much higher than those in the soils in which they live (Edwards and Thompson, 1973). Many fish can take up pesticides into their tissues from the water in which they live. This can be via their food or directly from the water (Edwards, 1973a,b).

Many organochlorine insecticides are taken up by birds in their food, whether plant or animal. They may not kill the birds immediately. They can be stored in the fatty tissues with little immediate effect but may have drastic effects in winter when the birds are poorly fed and their fat reserves are mobilized. Moreover, it is clear that several of these insecticides, particularly DDT and dieldrin, when taken up into the bodies of birds of prey, can cause thinning of the egg shells and consequent breeding failures (Edwards, 1973a,b; Cooper, 1991).

Fortunately, with the phasing out of the use of organochlorine insecticides there has been a recovery in populations of many species of bird raptors, even though residues still occur in their tissues and eggs. Fleming et al. (1983) reported that although environmental organochlorine contamination was decreasing, some still existed and was affecting predatory bird species, and McEwen et al. (1984) presented data that showed that black-crowned night herons were still affected by DDT. Stickel et al. (1984) reported that there were still high levels of DDT residues in birds in certain parts of the United States. White et al. reported effects of DDT on herons and anhiugas 13 years after DDT was banned (White et al., 1988). The organophosphate, carbamate, and pyrethroid insecticides which have largely replaced the organochlorines, tend to have greater acute toxicities to birds, but they are not stored in the bird's tissues in such large quantities, and most have little effect on reproduction. However, White et al. (1983b) reported that organophosphate insecticides had considerable effects on the nesting ecology of gulls. Reports of field incidents of the effects of pesticides on bird populations have decreased dramatically since the 1970s. Although there have been many localized bird kills, these have mainly been due to acute poisoning rather than to storage of pesticides in tissues. However, although DDT use was phased out in 1973, very large residues still occur in some animals 10 years later. For instance, White and Krynitsky (1986) reported large residues of DDT in birds, lizards, and bats in New Mexico and Texas.

There is much less evidence of pesticides that are taken up into their bodies having effects on mammals than on birds. Although organochlorine insecticides are stored in tissues of mammals, there is little evidence that these residues are toxic or have major effects.

Indirect Ecological Effects

To have a significant indirect ecological effect a pesticide must influence the availability of food, the habitat, the predators and parasites of an organism, and an organism's interactions with other organisms. It is extremely difficult to assess the indirect effects of pesticides on populations or communities of organisms in the absence of detailed knowledge of the functioning of these systems.

Effects on the Availability of Food

Both soil and aquatic ecosystems involve complex interactions between the flora and fauna that live in them. These are illustrated best by models of food webs or trophic-level food chains. Clearly, if a pesticide has a major impact upon any important organism in such a web or chain these effects can impact upon a large number of other organisms in the system that prey or depend upon it, particularly if the organism affected is present in large numbers or biomass.

There are many examples of such interactions known and it is impossible to review these all here; indeed, only a few food webs are understood sufficiently for such indirect effects to be evaluated. However, when the direct effects on a key organism are considerable the indirect effects often become obvious. For instance, the fruit- tree red spider mite did not become a pest until its natural enemies, particularly predatory mites, were eliminated by the widespread use of DDT soon after the introduction of this insecticide. If the weeds are completely eliminated in agricultural fields, pest and disease problems often become worse because the weeds provide alternative food sources for the pests. In aquatic systems, pesticides which have drastic effects upon the phytoplankton can have major indirect effects on fish by removing much of their food. Populations of plant parasitic nematodes, kept under control by fungi, can increase very significantly when these fungi are killed by a fungicide such as captan (Kerry, 1988).

Fish can be affected drastically when pesticides kill the phytoplankton and other organisms that provide their food (Muirhead- Thomason, 1971). Bird populations can be significantly decreased when insecticides affect their food supply (Stromborg, 1982). Birds are affected very much by how pesticides influence the availability of food (Riseborough, 1986). Grain treated with insecticides such as dieldrin, chlorfenvinphos, carbofuran, phorate, fonofos, and fensulfothion has been reported to be toxic to birds in various incidents (Hardy, 1990).

Indirect Effects on the Habitats of Organisms

Pesticides can have dramatic indirect effects on the habitat of organisms. Probably the most significant effects of this kind on habitats are by herbicides. If a thorough herbicide program is used in an agricultural field the soil is usually kept relatively bare. Such a drastic change can have many indirect environmental effects. The bare soil is much more susceptible to wind and water erosion. There will be a considerable effect on the availability of organic matter for soil-inhabiting organisms resulting from the roots of weeds and dead weeds. Weeds provide shelter for many antagonists and predators and parasites of pests and diseases, so removal of the weeds can result in dramatically increased pest attack in many instances. Santillo et al. (1989) reported that the herbicide glyphosate had significant effects on populations of small mammals by affecting their habitats.

There are many other similar examples of how pesticides can have indirect environmental effects by changing the habitat in an ecosystem (Edwards, 1990).

Effects on Natural Enemies of Pests and Other Organisms

Few pesticides are highly selective. Many insecticides are toxic to the predators or parasites of pests as well as to the pest. Indeed, some insecticides like DDT are more toxic to the natural enemies such as lady-bird beetles than the pests themselves. When this occurs the response is usually an increase in numbers of the pest once the effects of the pesticide have disappeared or even creation of new pests by the elimination of the natural enemies of a previously relatively innocuous organism. There are many practical examples of such interactions in the literature—for instance, when heavy spray programs are applied to cotton.

The role of the natural enemies of pests in keeping them under control can be considerable and the importance of these parasites, predators, and antagonists is often underestimated (Pimentel, 1988). The total number of insect species so far described is about 1.5 million, of which about 5,000 are considered to be important as pests from the economic point of view, because they damage crops, domestic animals, or man either directly, or indirectly by transmission of disease. If these pests had no natural enemies, this figure would be much larger. It is unfortunate that parasites and predators of pests are often more susceptible to pesticides than their prey, and there are few types of pesticides selective enough to kill a particular pest without affecting its predators, although systemic insecticides do to a certain extent fulfill this ideal.

Furthermore, pesticides sometimes kill not only enemies of existing pests but also those of relatively innocuous plant-feeding species, which, released from predator/parasite pressure, may multiply rapidly in number and become new pests. An example of this is the fruit tree red spider mite (*Panonychus ulmi* Koch), which prior to the advent of DDT, was not a pest (Edwards, 1973b). However, this species was kept in check by various predatory mites which were very susceptible to the large quantities of DDT that were then used in orchards. When these main predators were killed the red spider mite became a major pest.

Effects on Biodiversity

Most agricultural practices tend to decrease the biodiversity of plants and animals. Pesticides have major impacts in decreasing the biodiversity of agricultural ecosystems. In soil, they have major effects on decreasing the diversity of soil-inhabiting organisms since they selectively kill particular groups of organisms (Edwards and Thompson, 1973). By using insecticides and fungicides to control pests and diseases that can carry over from one year to the next if the same crop is grown twice, they decrease the need for cropping rotations and diversity and encourage monoculture. Herbicides have a major impact in lessening plant diversity in agricultural ecosystems.

There is good evidence that pesticides act to increase dependence on biodiversity by suppressing the natural control mechanisms of pests and diseases, which is dependent in many cases on biodiversity (Dover and Talbot, 1987).

Pesticides also decrease the biodiversity of aquatic insects and fish, like trout, in aquatic systems that are treated or become contaminated with agricultural pesticides and have very similar effects to those on terrestrial systems (Edwards, 1977).

Development of Resistance

Continued and frequent uses of pesticides have led to the development of resistance to these chemicals in arthropods (504 species) (Georghiou, 1990), pathogens (150 species) (Eckert, 1988), and weeds (273 species) (LeBaron and McFarland, 1990) that they are used to control. This has developed most rapidly among insects but is developing at an increasing rate in disease organisms and weeds. Many of the genetic characters are cross-linked so resistance to one chemical can make an organism resistant to others. Resistance depends upon repeated exposure to a chemical and develops most rapidly in species with short life cycles. The environmental impact of the development of resistance is that when it occurs, increasing quantities of pesticides need to be used to maintain satisfactory chemical control, or new chemicals to which there is no resistance have to be used.

Effects of Pesticides on Ecosystems

Most of the direct and indirect effects of pesticides discussed so far are related to the effects of pesticides on populations of individual organisms and to how these effects affect populations of other organisms. However, sometimes the impact of a pesticide can be so drastic as to influence the functioning of the whole ecosystem in a major way.

Effects of Pesticides on the Breakdown of Organic Matter in Soil

Organic matter is fragmented by a wide range of soil-inhabiting invertebrates and broken down progressively by their action and that of microorganisms that grow on the fragmented material. During the breakdown process the nutrients N, P, and K are released from the organic matter and returned to the soil, where they are available to plants.

Many pesticides, which are toxic to key organisms in this process—for instance, earthworms—can slow the breakdown process considerably; for instance, a slowdown in organic matter breakdown has been reported for simazine (Edwards, 1989). Such effects are much more common than is reported in the literature because relatively few pesticides have been tested for their effects on organic matter breakdown. When organic matter breakdown is slowed down significantly, this in turn can have major effects on the primary productivity of the ecosystem.

Effects of Pesticides on Soil Respiration

Soil respiration is an index of overall microbial and invertebrate activity in soil, and soils that respire little or are anaerobic accumulate mats of undecomposed organic matter and have poor fertility. Microorganisms can also use some pesticides as substrates, so these pesticides cause a significant increase in soil respiration. However, most fertile soils typically have a high respiration rate, and long-term suppression of respiration usually represents an adverse effect of a pesticide. There have been relatively few studies of the effects of pesticides on soil respiration other than in assessments of the effects of pesticides on populations of microorganisms (Edwards, 1989).

Effects of Pesticides on Nutrient Cycles

Since mineral nitrogen is a major plant nutrient it is important to know whether pesticides influence dynamic soil processes that affect the availability of nitrogen. Key components of such processes include ammonification and nitrification as part of the N-mineralization process. Both of these processes are driven by nitrogen-fixing bacteria which may be sensitive to pesticides.

There is an extensive literature on the effects of pesticides on nitrogen fixation. There are many reports of depression of this process by pesticides, particularly herbicides (Edwards, 1989), but such adverse affects are seldom long-term or serious.

Effects on Eutrophication in Freshwater

Eutrophication results from enrichment with nutrients. Discharge of nutrients such as nitrate or phosphate into aquatic systems in runoff from agricultural land can cause excessive eutrophication and cause algae to multiply rapidly. Some pesticides, particularly herbicides, can reverse this excessive eutrophication by killing the algae that are responsible.

Contamination of the Environment by Pesticides

Enormous quantities of pesticides are currently used in developing countries and some tropical countries. These pesticides range in persistence from compounds that degrade rapidly, and are broken down in hours or days, to some of the most complex and persistent molecules known. We still do not know the full degradation pathways or ultimate fate of many pesticides in the field. Many pesticides do not reach their targets but instead end up on crops, trees, animals, soils, or water. Residues that reach crops, trees, or animals, if they are persistent, usually end up either in soils or aquatic sediments in freshwater or the sea.

Soils

By far the greater quantity of pesticides applied to crops end up in the soil, either through aerial drift, runoff from plants, or eventual death of the plants. Depending on the nature of the pesticide it may be broken down rapidly, usually by soil microorganisms, or become bound progressively onto soil fractions, such as organic matter or clay minerals, and persist weeks, months, or even many years (Edwards, 1966). This binding may be readily reversible or irreversible. Some of even the most nonvolatile pesticides volatize from the soil surface or from deeper soil by a "wick" process and reach the atmosphere, where they may be adsorbed onto atmospheric particles. They may be washed out from the atmosphere in precipitation to contaminate untreated soils. Often urban soils are more heavily contaminated by pesticides than agricultural soils in the same areas (Carey, 1979).

Pesticides are also lost from soils by wind and water soil erosion in quite large quantities. Some of the more persistent pesticides such as DDT may end up bound up in humic materials and persist for many years. For instance, Buck et al. (1983) reported residues of DDT in Arizona soils 12 years after this pesticide was withdrawn.

Water

Pesticides can reach water as a result of direct treatment to control pests, such as mosquito larvicides, or molluscides used to control disease-carrying snails. More commonly, pesticides contaminate aquatic systems by fallout from aerial sprays, through drainage from soil and water erosion, or through disposal of pesticide containers or effluent from pesticide factories. Runoff from agricultural land can carry between 0.5% and 15% of a pesticide treatment into an aquatic system (Wauchope, 1978). The fate and distribution of a pesticide once it has reached an aquatic system depends mainly on its solubility and persistence. Pesticides can volatize into the atmosphere from water surfaces, but for most pesticides this occurs mainly soon after the treatment is applied or the pesticide reaches the water. Any pesticide that persists in an aquatic system usually becomes adsorbed onto floating particles and eventually ends up in the bottom sediment. Some pesticides can persist in the sediment for many years and are periodically recycled into the water when the sediment is disturbed (Edwards, 1977). In general, the more persistent the pesticide the greater has been its effect on aquatic life (Edwards, 1973a,b). However, some of the less-persistent organophosphate and pyrethroid insecticides now used have also caused very serious kills of fish and their food organisms (Alfoldi, 1983). Pesticides in fast-flowing waterways become progressively carried down to the mouth of rivers, estuaries, or bays. Here they are also retained in the bottom sediment, where they can affect many bottom-living organisms. The more soluble a pesticide is the greater its potential for contaminating aquatic systems and groundwater. The most commonly reported pesticides are atrazine, alachlor, and aldicarb.

Air

There is good evidence that large quantities of even extremely nonvolatile pesticides such as DDT eventually volatize into the atmosphere, particularly in the humid tropics (Nash, 1983). The ultimate fate of these residues is still poorly understood, but there is good evidence of global transport of these residues over long distances; it seems probable that some residues fall back to the earth in precipitation, and some may be carried into the upper atmosphere. It is also possible that some residues of even the more persistent pesticides may be susceptible to photodegradation in the upper atmosphere (Edwards, 1973a,b)

Food

Many pesticides are systemic and are translocated into crops from soils. Even nonsystemic pesticides can be taken up into crops. For instance, Carey et al. (1979) reported that 40% of all crops contained detectable levels of organochlorine insecticides and 10% organophosphate insecticides.

More recently, it has been estimated that about 35% of the food consumed in the United States has detectable pesticide residues (F.D.A., 1990).

Minimization of the Environmental Effects of Pesticides

Much has been learned about minimizing environmental pollution by pesticides since the period after the Second World War, when many persistent pesticides were used in a relatively haphazard and indiscriminate way. At that time, large-scale aerial spraying of large areas of land was common. Now, such aerial spraying is under considerable constraints and legislation in many countries (see Chap. 10).

Clearly some of the environmental impacts of pesticides are serious. However, there are many ways in which potential environmental impacts of pesticides can be minimized or avoided, through choice of chemical method of application and use of supplementary cultural and biological methods of control.

Technical Methods

Improved Pesticides

Whenever possible, pesticides of low mammalian toxicity should be used. It seems certain that the more mammalian-toxic mercury-based compounds, arsenicals, organochlorines, and organophosphates still in current use will be phased out gradually. Many of the newer pesticides such as the pyrethroid insecticides not only have low mammalian toxicities but are effective at very low doses that minimize their environmental effects. Whenever possible, pesticides specific to particular pests or groups of pests that have the minimum of side effects should be used. There is considerable legislative pressure in most countries to restrict pesticides that have been demonstrated to be sources of actual or potential hazards to wildlife or to humans or that persist in the environment. There are economic problems in that the development of new pesticides is expensive, particularly in obtaining registration for use, which unfortunately mediates against the development of highly selective pesticides.

Improved Pesticide Application: Methods

There are many ways of improving the methods of applying pesticides. Spray equipment that should be used includes ultra-low-volume sprayers, charged electrodyne sprayers, aerosols, granule applicators, slow-release formulations such as granules in soil, and many other methods that minimize side effects by lowering the amount of chemical used and improving its placement. The development of such techniques is progressing rapidly and holds considerable promise for lessened environmental impact.

Biological Pest Control

Historically, biological pest control has held great promise, but its overall successes have been relatively limited. With the development of new organisms, novel methods, and biotechnology the potential for these methods is greatly increased.

- Use of microorganisms:
  Relatively few insect pathogens have been sufficiently successful for extensive commercial development. However, considerable advances are being made in the selection of different strains of pathogens that attack insects such as *Bacillus thuringiensis*, which are specific to different groups of insects. This, together with identification of associated toxins, and the development of genetic engineering, which could produce new strains which are much more host-specific, greatly increases the potential of such organisms as control agents. Recently, there have been developments in the identification of other suitable pathogens. Other advances have been in the commercial production of microbially produced pesticides such as the avermectins. There has even been progress in the biological control of diseases such as *Phytophthora*. The use of nematodes that attack insects holds considerable promise.
- Use of semiochemicals:
  Many pheromones and attractants for important pests have been identified and a number of the active ingredients that they contain have been isolated. Some of these have found their way into field-control programs, e.g., in control of caterpillar pests of cotton, a crop which currently uses the greatest amount of pesticides (see Chap. 10).
- Use of predators and parasites:
  There has been considerable development in the identification and rearing of predators and parasites of pests. This has been taken to commercial development, particularly for antagonists of greenhouse pests. There seems considerable potential for development of the use of entomophilic nematodes that attack and kill insects such as the Japanese beetles and codling moth (Edwards et al., 1990).

Integrated Pest Management

There have been considerable advances in recent years in the development of integrated control and pest-management programs for major crops, and this holds considerable promise for the future (Edwards et al., 1991; Pimentel, 1991). Such programs can include a combination of any suitable techniques to decrease pest populations and maintain them at levels below those causing economic injury. This differs from "supervised control," which is concerned with using pesticides only when necessary. Implicit in integrated control is minimizing harm by chemicals to beneficial natural enemies of pests. Combinations of practices that are acceptable to the farmer are usually combined with minimal pesticide use.

Sustainable Agriculture

In the 1980s, there has been a strongly growing movement in Europe and the United States to adopt agricultural systems that depend much less on synthetic-fossil-fuel-based chemicals and to maximize the use of cultural and biological practices (Edwards, 1988, 1990). The economic and environmental advantages of sustainable agriculture have been demonstrated very clearly (Edwards et al., 1990). Legislative programs have been developed in Sweden, Denmark, The Netherlands, and the province of Ontario, Canada, to reduce use of chemicals, including pesticides, in crop production by 50% in the next few years. Such targets seem to be achievable with very little loss of crop yields or overall productivity and are very important in decreasing the environmental impact of pesticides. Similar principles are being used in the development of programs that aim to increase sustainability in agriculture in the developing countries (N.R.C., 1991; Edwards, 1991) so that they can lessen dependence upon synthetic chemicals or improve productivity without the need for fossil-fuel-based pesticides that require hard currency.

Legislative and Political Activities

In recent years there has been steadily increasing anxiety about the potential human and environmental hazards associated with pesticide use. This has generated many activities by government organizations, politicians, and private-interest groups.

Pesticide Registration and Government Control

The requirements of registration authorities on pesticide manufacturers before the use of a pesticide is authorized have become increasingly stringent in the developed countries. Clearly, it is difficult to predict possible environmental effects accurately from data presented to registration authorities, but as manufacturers have gained experience, the numbers of reports of adverse environmental effects of pesticides have become fewer.

Intense research and administrative activities were initiated as a result of appreciation of the environmental impact of the organochlorine insecticides, and many national residue-monitoring schemes were set up in the 1970s. Unfortunately, once the use of the organochlorines was restricted or banned, many of these monitoring schemes were abandoned, so we have few data on the continuing residues of these insecticides in many parts of the environment.

There has been considerable work done to minimize the movement of pesticide-contaminated foods as exports across international borders. Each developed country has set stringent limits on the amounts of particular pesticides permitted in its food imports and exports. This has been codified by the Food and Agriculture Organization of the United Nations (FAO), which has set international limits on such residues.

The Organization for Economic Cooperation and Development (OECD) and the European Economic Community (EEC) have developed detailed protocols for testing the environmental impact of chemicals in soil, water, and the biota. All these governmental activities have had beneficial effects on the environmental impacts of pesticides.

Political Activities

The current registration requirements for new pesticides in developed countries have resulted in decreased numbers of reports of wildlife incidents, human poisoning, and environmental pollution by pesticides compared with those in the 1960s and 1970s. However, we are still a long way from developing or using pesticides that have little toxicity to most living organisms other than their target species.

There have been many political activities associated with the use of pesticides. Some of these, through promotion of monoculture or biculture by subsidizing crops such as corn and soybeans in the United States encouraged the use of pesticides. However, most political activity in recent years has been in response to constituents who have pressured politicians to try to promote legislation to restrict the use of many pesticides. In some countries this has led to legislative action to decrease pesticide use. One U.S. state, Iowa, passed legislation which taxed the use of certain herbicides. Several U.S. states have discouraged the use of pesticides by providing considerable funding for sustainable agriculture systems that minimize the use of pesticides. International organizations such as the FAO and UNESCO have had programs to assess the environmental impact of pesticides and promote methods using less pesticides.

Private-Interest Groups

During the last 20 years private-interest groups targeted at environmental conservation and protection have expanded in membership and increased greatly in numbers in both the United States and Europe. These groups have a multiplicity of aims, some wishing to protect wildlife or plants, others anxious about pollution of the environment or food. All of them wish to minimize the use of chemicals such as pesticides. They exert considerable pressure on national and local legislators and administrations to limit the use of pesticides and many have effective lobbying organizations that increase pressure on politicians.

Conclusions

It is clear from the discussion in the preceding sections that the environmental impacts of pesticides are extremely diverse; some of these are relatively obvious whereas others are extremely subtle and complex. Some are highly specific and others broad and not immediately obvious. It may be of value to review those impacts which may have the greatest importance. They fall into three main areas:

Effects on Wildlife

One of the most striking effects on wildlife has been that on *birds*, particularly those in the higher trophic levels of food chains such as bald eagles, hawks, and owls. These birds are rare, often endangered, and susceptible to poisonous residues in their food, such as occurred through the bioconcentration of organochlorine insecticides through food chains. Grain- and plant-feeding birds are also susceptible to pesticides contained on or in their food and there have been many incidents of kills of rare birds such as ducks and geese (Hardy, 1990). There have also been dramatic diminutions in populations of insect-eating birds such as partridges, grouse, and pheasants due to a loss of their food in agricultural fields through the use of insecticides. It is extremely difficult to protect birds from such hazards when persistent or highly poisonous pesticides are used. It is also difficult to put a commercial value on rare birds, but clearly they are very important to many conservation-minded people.

Probably the second-most-important impact on wildlife that has occurred has been the *fish and marine crustacean* kills that have occurred due to contamination of aquatic systems with pesticides. This has resulted not only from the agricultural contamination of waterways as fallout, drainage, or runoff erosion but also from the discharge of industrial effluents containing pesticides into waterways. For instance, most of the fish in the Rhine River have been killed by discharge of pesticides and other chemicals into this river and at one time fish populations in the Great Lakes became low mainly due to pesticide contamination. It is extremely difficult to assess accurately the global extent of the effects of pesticides on fish but it is certain that pesticides cause major losses in fish production because many fish are killed by pesticides that reach aquatic systems. Additionally, many of the organisms that provide food for fish are extremely susceptible to pesticides, so the indirect effects of pesticides on fish food supply may have an even greater effect on fish populations.

Bees are extremely important in the pollination of crops and wild plants. Although in most countries pesticides are screened for their toxicity to bees before they can be registered for use, and the use of pesticides toxic to bees is permitted only under stringent conditions, there are still extremely serious losses of bees due to pesticides. This results in considerable loss of yield of crops dependent on bee pollination (Pimentel et al., Chapter 3). The financial cost of the impacts of pesticides on bees is enormous.

The literature on pest control is replete with examples of the *development of new pest species* when their natural enemies are killed as a result of pesticide use. Disruption of food webs of this kind creates a further dependence on synthetic chemicals for pest control not dissimilar to drug dependence. It is well documented (Pimentel et al., Chapter 10) that in spite of the vast increase in use of synthetic pesticides over the last 50 years, losses to pests have actually increased. Clearly, chemical pesticides cannot alone be a sustainable solution to pest problems.

Finally, the effects of pesticides on the *biodiversity* of plants and animals in agricultural landscapes, whether caused directly or indirectly by pesticides, constitute a major adverse impact of pesticides that is extremely difficult to quantify or to assess a value of.

Effects on Soil and Water

Serious direct effects of pesticides on *soil structure and fertility* are probably rare but there are also indirect effects of pesticides which are much more difficult to assess or quantify. Probably such effects are important but they are not usually long term. However, the indirect effects of pesticides in accelerating soil erosion have been much more obvious and adverse to the environment.

The movement of pesticides into *surface and groundwater* is well-documented. Wildlife is affected, and drinking water is contaminated, sometimes beyond accepted safety levels.

The *sediments* dredged from U.S. waterways are often so heavily contaminated with pesticides that there may be problems in disposing of them on land.

Effects on Humans

Clearly, the environmental effects of pesticides on wildlife, soil, and water all impact strongly on the human quality of life. However, there is also increasing anxiety as to the importance of small residues of pesticides, often suspected of being carcinogens, in *drinking water and food*. In spite of stringent regulations by national and international regulatory agencies, there are many reports of small pesticide residues in various foods, both imported and home produced (Sachs et al., 1987). There is considerable pressure from environmental associations and other concerned groups to take actions to eliminate as many of these contaminants as possible.

Finally, over the last 50 years there have been many *human illnesses and deaths* due to pesticides, with up to 20,000 deaths per annum. Some of these incidents have been due to attempted or successful suicides, but the majority have involved some form of accidental exposure to pesticides. Such accidents are common among farmers and spray operators, who are careless in handling pesticides or wear insufficient protective clothing and equipment. However, there have been a number of major incidents which have led to the death or sickness of many thousands of human beings. Cases include emissions from chemical plants, such as the Bhopal disaster, where 2,500–5,000 deaths resulted from methyl isocyanate; the TCDD incident in Italy, where 32,000 people were affected; and the death of 459 people and illness of 6,070 from eating grain treated with pesticides (Hayes and Lawes, 1991).

Clearly, we have progressed a long way from the initial idea that pesticides control pests with low economic, environmental, and human costs. Even the direct economic benefits of pesticides are under question, since progressively increasing costs of pesticides are increasingly not correlated with increasing financial returns to farmers. There is little doubt that if the costs of the environmental impacts of pesticides were subtracted from the economic benefits, pesticide use would be much less attractive, particularly if the users or the producers of the pesticides were required to pay these costs. Examples of such costs include those of extracting pesticides from contaminated drinking water, provision of land for disposal of highly contaminated material dredged from rivers and waterways, loss of fish productivity in contaminated freshwater such as the Great Lakes, losses of crustacea that provide human food in contaminated estuaries, and effects on crop yields through decreased pollination.

In developed countries, there is a demand for fruits and vegetables that are "cosmetically" attractive and have no blemishes. However, there is increasing question as to whether the cost of achieving this in both financial terms as well as in the accompanying contamination of the food is worthwhile. There is an increasing pressure by consumers for "clean" and uncontaminated foods. This in turn is putting an increasing demand on the pesticide industry to produce chemicals with low mammalian toxicity that can be used at low doses with little environmental impact. There are increasing costs in the production of chemicals as the cost of oil increases and increasing requirements for data to prove their environmental safety. Paradoxically, as more environmental data are required for each pesticide, so the cost of producing a new pesticide goes up. This economic pressure on the agrochemical industry makes it uneconomical to develop chemicals that are highly specific for certain organisms as pesticides, because of the limited market for such chemicals.

Clearly, the overall environmental impact of pesticides has many unacceptable aspects, although there has been much progress in minimizing that impact in recent years. We must progressively explore alternatives to pesticides that are more ecologically acceptable and keep use of pesticides at levels which create no environmental or human problems.

References

Alfoldi, L. 1983. Movement and interaction of nitrates and pesticides in the vegetarian cover-soil groundwater-rock system. Environmental Geology, **5**, 19–25.

Anderson, R.L. 1989. Toxicity of synthetic pyrethroids to freshwater invertebrates. Environmental Toxicology and Chemistry, **8** (5), 403–410.

Benoit, D.A. 1975. Chronic effects of copper on the survival, growth and reproduction of the bluegill, *Lepomis macrchirus*. Transactions of the American Fisheries Society, **104** (2), 353–358.

Bhatagnar, P., Kumar, S. and Lal, R. 1988. Uptake and bioconcentration of dieldrin, dimethoate and permethrin by *Tetrahymena pyriformis*. Water, Air and Soil Pollution, **40** (3–4), 345–349.

Brown, A.W.A. 1978. Ecology of Pesticides. Wiley and Sons, New York, Chichester, Brisbane, Toronto, 525 pp.

Brungs, W.A. 1969. Chronic toxicity of the fathead minnow, *Pimephalus promelas*. Transactions of the American Fisheries Society, **98**, 272–279.

Buck, N.A., Estesen, B.J. and Warne, G.W. 1983. DDT moratorium in Arizona: Residues in soil and alfalfa after 12 years. Bulletin of Environmental Contamination and Toxicology, **31**, 66–72.

Butler, G.C. (ed) 1978. Principles of Ecotoxicology. Scope 12. Wiley and Sons, Chichester, New York, Brisbane, Toronto, 350 pp.

Call, D.J., Brooke, L.T., Kent, R.J., Knuth, M.L., Poirier, S.H., Huot, J.M. and Lima, A.R. 1987. Bromacil and diuron herbicides: Toxicity, uptake and elimination in freshwater fish. Archives of Environmental Contamination and Toxicology, **16** (5), 607–613.

Carey, A.E. 1979. Monitoring pesticides in agricultural and urban soils in the United States. Pesticides Monitoring Journal, **13** (1), 23–27.

Carey, A.E., Garven, J.A., Tai, H., Mitchell, W.G. and Wiersma, G.B. 1979. Pesticide residue levels in soils and crops from 37 states—National Soils Monitoring Program (IV). Pesticides Monitoring Journal, **12** (4), 209–233.

Carroll, C.R., Vandermeer, J.H. and Rosset, P.M. (ed) 1990. Agroecology. McGraw Hill Publishing Co., New York, 641 pp.

Carson, R. 1962. Silent Spring. Hamish Hamilton, London, 304 pp.

Cooper, K. 1991. Effects of pesticides on wildlife. In Handbook of Pesticide Toxicology, W.J. Hayes, Jr. and E.R. Lawes, Jr. (eds), **11**, 463–496. Academic Press, New York.

Day, K.E. 1989. Acute chronic and sublethal effects of synthetic pyrethroids in freshwater. Environmental Toxicology and Chemistry, **8** (5), 411–416.

Domsch, K.H. 1963. Influence of plant protection chemicals on the soil microflora: Review of the literature. Mitt. Biologische Bundesanstalt Land-Forstirwch. Berlin Dahlem, 105, 5.

Domsch, K.H. 1983. An ecological concept for the assessment of the side-effects of agrochemicals on soil microorganisms. Residue Reviews, **86**, 65.

Dover, M. and Talbot, L.M. 1987. To Feed the Earth: Agroecology for Sustainable Development. World Resources Institute, Washington D.C., 88 pp.

Duffus, J.H. 1980. Environmental Toxicology. Resource and Environmental Sciences Series, Halsted Press, John Wiley and Sons, New York, 164 pp.

Eckert, J.W. 1988. Historical development of fungicide resistance in plant pathogens. In Fungicide Resistance in N. America, C.J. Delp (ed). American Phytopathological Society Press, St. Paul, 1–3.

Edwards, C.A. 1966. Insecticide residues in soils. Residue Reviews, **13**, 83–132.

Edwards, C.A. 1973a. Persistent Pesticides in the Environment (2nd ed.). C.R.C. Press, Cleveland, Ohio, 170 pp.

Edwards, C.A. (ed) 1973b. Environmental Pollution by Pesticides. Plenum Press, London and New York, 542 pp.

Edwards, C.A. 1977. Nature and origins of pollution of aquatic systems by pesticides. In Pesticides in Aquatic Environments, M.A.Q. Khan (ed). Plenum Press, New York, 11–37.

Edwards, C.A. 1985. Agrochemicals as environmental pollutants. In Control of Pesticide Residues in Food. A Directory of National Authorities and International Organizations. Bengt V. Hofsten., 1–19.

Edwards, C.A. 1988. The concept of integrated sustainable farming systems. American Journal of Alternative Agriculture, Washington, **2** (4), 148–152.

Edwards, C.A. 1989. Impact of herbicides on soil ecosystems. Critical Reviews in Plant Science, **8** (3), 221–257.

Edwards, C.A. 1990. The importance of integration in lower input agricultural systems. Agriculture, Ecosystems and Environment, **27**, 25–35.

Edwards, C.A. 1991. A Strategy for Developing and Implementing Sustainable Agriculture in Developing Countries. Agriculture, Ecosystems and Environment (in press).

Edwards, C.A. and Thompson, A.R. 1973. Pesticides and the soil fauna. Residue Reviews, **45**, 1–79.

Edwards, C.A., Krueger, H.R. and Veeresh, G.K. (eds) 1980. Pesticide Residues in the Environment in India. University of Agricultural Science Press, Bangalore, India, 650 pp.

Edwards, C.A., Lal, R., Madden, P., Miller, R.H. and House, G. (eds) 1990. Sustainable Agricultural Systems. Soil and Water Conservation Society, Ankeny, Iowa, 696 pp.

Edwards, C.A. and Bohlen, P. 1991. The assessment of the effects of toxic chemicals on earthworms. Review of Environmental Contamination and Toxicology. Vol 125, 23–99.

Edwards, C.A. and Fisher, S. 1991. The use of cholinesterase measurements in assessing the impacts of pesticides on terrestrial and aquatic invertebrates. In Cholinesterase-Inhibiting Insecticides: Their Impact on Wildlife and The Environment, P. Mineau (ed). Elsevier Science Publishers, B.V., Amsterdam, Holland, 255–276.

Edwards, C.A., Thurston, H.D. and Janke, R., 1991. Integrated Pest Management for Sustainability in Developing Countries. In Toward Sustainability: A Plan for Collaborative Research on Agricultural and Natural Resource Management. National Research Council, National Academy Press, Washington, D.C., 109–133.

Environmental Protection Agency 1989. Pesticide Industry Sales and Usage: 1988 Market Estimates. E.P.A. Economic Analysis Branch, Washington, D.C.

Fleming, W.J., Clark, D.R. Jr. and Henny, C.J. 1983. Organochlorine Pesticides and P.C.B.'s: A Continuing Problem for the 1980s. Transactions of the 48th N. American Wildlife and National Resources Conference. Wildlife Management Institute, Washington, D.C., 186–199.

Food and Drug Administration 1990. F.D.A. Pesticide program on residues in foods. Journal of Association of Official Analytical Chemists, **73**, 127–146.

Georghion, G.P. 1990. Overview of insecticide resistance. In Managing Resistance to Agrochemicals: From Fundamental Research to Practical Strategies, M.B. Green, H.M. LeBaron, and W.K. Moberg (eds). American Chemical Society, Washington D.C., 18–41.

Graham, F. Jr. 1970. Since Silent Spring. Hamish Hamilton, London, 297 pp.

Guenzi, W.D. (ed) 1974. Pesticides in Soil and Water: Soil Science Society of America. Madison, WI, 315 pp.

Gurney, S.E. and Robinson G.G.C. 1989. The influence of two herbicides on the productivity, biomass and community composition of freshwater marsh periphyton. Aquatic Botany, **36** (1), 1–22.

Hardy, A.R. 1990. Estimating exposure: the identification of species at risk and routes of exposure. In Effects of Pesticides on Terrestrial Wildlife, L. Somerville and C.H. Walker (eds). Taylor and Francis, London, 81–97.

Hayes, W.J. Jr. and Lawes, E.R. 1991. Handbook of Pesticide Toxicology (3 volumes). Academic Press, Inc. San Diego, New York, Boston, London, Sydney, Tokyo, Toronto, 1576 pp.

Irvine, D.E.G. and Knights, B. 1974. Pollution and the Use of Chemicals in Agriculture. Butterworths, London, 136 pp.

Kerry, B. 1988. Fungal parasites of cyst nematodes. In Biological Interactions in Soil, C.A. Edwards, B.R. Stinner, D. Stinner, and S. Rabatin (eds). Elsevier, Amsterdam, 293–306.

Kingsbury, P.D. and Kreutzweiser, D.P. 1987. Permethrin treatments in Canadian forests, part 1. Impact on stream fish. Pesticide Science, **19** (1), 35–48.

LeBaron, H.M. and McFarland, J. 1990. Herbicide resistance in weeds and crops. In Managing Resistance: From Fundamental Research to Practical Strategies, M.B. Green, H.M. LeBaron and W.K. Moberg (eds). Moberg American Chemical Society, Washington, D.C., 336–352.

McEwen, S.L. and Stevenson, G.R. 1979. Use and Significance of Pesticides in the Environment. Wiley and Sons, New York, Chichester, Brisbane, Toronto.

McEwen, L.C., Stafford, C.J. and Henslev, G.L. 1984. Organochlorine residues in eggs of black crowned night herons. Environmental Toxicology and Chemistry, **3**, 367–376.

Mellanby, K. 1967. Pesticides and Pollution. New Naturalist Series No. 50. Collins, London, 221 pp.

Moore, N.W. 1966. Pesticides in the Environment and Their Effects on Wildlife. Blackwell, Oxford, 311 pp.

Moriarty, F. 1983. Ecotoxicology: The Studies of Pollutants in Ecosystems. Academic Press, London and New York, 233 pp.

Muirhead-Thomson, R.C. 1971. Pesticides and Freshwater Fauna. Academic Press, London and New York, 248 pp.

Nash, R.G. 1983. Comparative volatilization and dissipation rates of several pesticides from soil. Journal of Agricultural and Food Chemistry, **31**, 210–217.

National Research Council 1991. Toward Sustainability: A Plan for Collaborative Research on Agriculture and Natural Resource Management. National Academy Press, Washington D.C., 147 pp.

Neely, W.B. 1980. Chemicals in the Environment: Distribution, Transport, Fate and Analysis. Dekker, New York.

Norton, G.A. and Holling, C.S. 1979. Proceedings of an International Conference on Pest Management, October 25-29, 1976. Pergamon Press, New York.

Parr, J.F. 1974. Effects of pesticides on microorganisms in soil and water. In Pesticides in Soil and Water, W.D. Guenzi (ed). Soil Science Society of America, Madison, WI, 315 pp.

Pimentel, D. 1988. Herbivore population feeding pressure on plant host: feedback evolution and host conservation. Oikos, **53**, 185–238.

Pimentel, D. (ed) 1991. C.R.C. Handbook of Pest Management in Agriculture (2nd ed., 3 volumes). C.R.C. Press, Boca Raton, Ann Arbor, Boston.

Poirier, D.G. and Surgeoner, G.A. 1988. Evaluation of a field bioassay technique to predict the impact of aerial applications of forestry insecticides on stream invertebrates. Canadian Entomologist, **120** (7), 627–637.

Riseborough, R.W. 1986. Pesticides and bird populations. In Current Ornithology, R.F. Johnston (ed). Plenum Press, London and New York, 397–427.

Rosen, A.A. and Kraybill, H.F. 1966. Organic Pesticides in the Environment. Advances in Chemistry. Series 60. American Chemical Society, Washington, D.C., 309 pp.

Rudd, R.L. 1964. Pesticides and the Living Landscape. Faber and Faber, London, 320 pp.

Sachs, C., Blair, D. and Richter, C. 1987. Consumer pesticide concerns. Journal of Consumer Affairs, **21**, 96–107.

Santillo, D.J., Leslie, D.M. and Brown, P.W. 1989. Responses of small mammals and habitat to glyphosate application on clearcuts. Journal of Wildlife Management, **53** (1), 164–172.

Smith, R.F. and Reynolds, H.T. 1965. Principles, definitions and scope of integrated pest control. Proceedings of F.A.O. Symposium on Integrated Pest Control **1**, 11–17.

Stern, V.M., Smith, R.F., van der Bosch, R. and Hagen, K.S. 1959. The integrated control concept. Hilgardia **29**, 81–101.

Stickel, L.R. 1968. Organochlorine Pesticides in the Environment. Report of U.S,D.I. Bureau of Sports Fisheries and Wildlife, 1, 119 pp.

Stromborg, K.L. 1982. Modern Pesticide and Bobwhite Populations. Proceedings 2nd National Bobwhite Quail Symposium, F. Schitoskey, E.C. Schitoskey, and L.G. Talent (eds). Oklahoma State University, Stillwater, OK, 69–73.

Thompson, A.R. and Edwards, C.A. 1974. Effects of pesticides on non-target invertebrates in freshwater and soil. In Pesticides in Soil and Water, W.D. Guenzi (ed), **13**, 341–386. Soil Science Society of America, Madison, WI.

Truhart, R. 1975. Ecotoxicology—A new branch of toxicology. In Ecological Toxicology Research, A.D. McIntrye and C.F. Mills (eds). Proceedings of NATO Science Communications Conference, Quebec. Plenum Press, New York, 323 pp.

Vogt, G. 1987. Monitoring of environmental pollutants such as pesticides in prawn aquaculture by histological diagnosis. Aquaculture, **67** (1-2), 157–164.

Warren, C.E. and Davis, G.E. 1971. Laboratory stream research: objectives, possibilities and constraints. American Review of Ecological Systems, **2**, 111–144.

Wauchope, R.D. 1978. The Pesticide Content of Water Draining from Agricultural Fields—A Review. Journal of Environmental Quality, **7**, 459–472.

White, D.H., Mitchell, C.A., Kennedy, H.D., Krynitsky, A.J. and Ribick, M.A. 1983a. Elevated DDT and toxaphene residues in fishes and birds reflect local contamination in the lower Rio Grande Valley, Texas. Southwestern Naturalist, **28** (3), 325–333.

White, D.H., Mitchell, C.A. and Prouty, R.M. 1983b. Nesting biology of laughing gulls and relation to agricultural chemicals. Wilson Bulletin, **95** (4), 540–551.

White, D.H. and Krynitsky, A.J. 1986. Wildlife in Some Areas of New Mexico and Texas Accumulate Elevated DDE Residues, 1983. Archives of Environmental Contamination and Toxicology, **15**, 149–157.

White, D.H., Fleming, W.J. and Eusor, K.L. 1988. Pesticide contamination and hatching success of waterbirds in Mississippi. Journal of Wildlife Management, **52** (4), 724–729.

Willis, G.H., McDowell, L.L., Harper, L.A., Southwick, L.M. and Smith, S. 1983. Journal of Environmental Quality, **12** (1), 80–85.

3

Assessment of Environmental and Economic Impacts of Pesticide Use

David Pimentel, H. Acquay, M. Biltonen, P. Rice, M. Silva, J. Nelson, V. Lipner, S. Giordano, A. Horowitz, and M. D'Amore

Introduction

Worldwide, about 2.5 million tons of pesticides are applied each year with a purchase price of $20 billion (PN, 1990). In the United States approximately 500,000 tons of 600 different types of pesticides are used annually at a cost of $4.1 billion (including application costs) (Pimentel et al., 1991).

Despite the widespread use of pesticides in the United States, pests (principally insects, plant pathogens, and weeds) destroy 37% of all potential food and fiber crops (Pimentel, 1990). Estimates are that losses to pests would increase 10% if no pesticides were used at all; specific crop losses would range from zero to nearly 100% (Pimentel et al., 1978). Thus, pesticides make a significant contribution to maintaining world food production. In general, each dollar invested in pesticidal control returns about $4 in crops saved (Carrasco-Tauber, 1989; Pimentel et al., 1991).

Although pesticides are generally profitable, their use does not always decrease crop losses. For example, even with the 10-fold increase in insecticide use in the United States from 1945 to 1989, total crop losses from insect damage have nearly doubled from 7% to 13% (Pimentel et al., 1991). This rise in crop losses to insects is, in part, caused by changes in agricultural practices. For instance, the replacement of corn-crop rotations with the continuous production of corn on about half of the original hectarage has resulted in nearly a fourfold increase in corn losses to insects despite approximately a 1,000-fold increase in insecticide use in corn production (Pimentel et al., 1991).

Most benefits of pesticides are based only on direct crop returns. Such assessments do not include the indirect environmental and economic costs associated with pesticides. To facilitate the development and implementation of a balanced, sound policy of pesticide use, these costs must be examined. Over a decade ago the U.S. Environmental Protection Agency pointed out the need for such a risk investigation (EPA, 1977). Thus far, only a few scientific papers on this complex and difficult subject have been published.

The obvious need for an updated and comprehensive study prompted this investigation of the complex of environmental and economic costs resulting from the nation's dependence on pesticides. Included in the assessment are analyses of pesticide impacts on human health; livestock and livestock product losses; increased control expenses resulting from pesticide-related destruction of natural enemies and from the development of pesticide resistance; crop pollination problems and honeybee losses; crop and crop product losses; fish, wildlife, and microorganism losses; and governmental expenditures to reduce the environmental and social costs of pesticide use.

Public Health Effects

Human pesticide poisonings and illnesses are clearly the highest price paid for pesticide use. A recent World Health Organization and United Nations Environmental Programme report (WHO/UNEP, 1989) estimated there are 1 million human pesticide poisonings each year in the world with about 20,000 deaths. In the United States, pesticide poisonings reported by the American Association of Poison Control Centers total about 67,000 each year (Litovitz et al., 1990). J. Blondell (EPA, PC [personal communication], 1990) has indicated that because of demographic gaps, this figure represents only 73% of the total. The number of accidental fatalities is about 27 per year (J. Blondell, EPA, PC, 1990).

While the developed countries annually use approximately 80% of all the pesticides produced in the world (Pimentel, 1990), less than half of the pesticide-induced deaths occur in these countries (Committee, House of Commons Agriculture, 1987). Clearly, a higher proportion of pesticide poisonings and deaths occurs in developing countries where there are inadequate occupational and other safety standards, insufficient enforcement, poor labeling of pesticides, illiteracy, inadequate protective clothing and washing facilities, and users' lack of knowledge of pesticide hazards (Bull, 1982).

Both the acute and chronic health effects of pesticides warrant concern. Unfortunately, while the acute toxicity of most pesticides is well documented (Ecobichon et al., 1990), information on chronic human illnesses resulting from pesticide exposure is not as sound (Wilkinson, 1990). Regarding cancer, the International Agency for Research on Cancer found "sufficient" evidence of carcinogenicity for 18 pesticides, and "limited" evidence of carcinogenicity for an additional 16 pesticides based on animal studies (Lijinsky, 1989; WHO/UNEP, 1989). With humans the evidence concerning cancer is mixed. For example, a recent study in Saskatchewan indicated no significant difference in non-Hodgkin's lymphoma mortality between farmers and nonfarmers (Wigle et al., 1990), whereas others have reported some human cancer difference (WHO/UNEP, 1989). A realistic estimate of the number of U.S. cases of cancer in humans due to pesticides is given by D. Schottenfeld (University of Michigan, PC, 1991), who estimated that less than 1% of the nation's cancer cases are caused by exposure to pesticides. Considering that there are approximately 1 million cancer cases/year (USBC, 1990), Schottenfeld's assessment suggests less than 10,000 cases of cancer due to pesticides per year.

Many other acute and chronic maladies are beginning to be associated with pesticide use. For example, the recently banned pesticide dibromochloropropane (DBCP) caused testicular dysfunction in animal studies (Foote et al., 1986; Sharp et al., 1986; Shaked et al., 1988) and was linked with infertility among human workers exposed to DBCP (Whorton et al., 1977; Potashnik and Yanai-Inbar, 1987). Also, a large body of evidence suggesting pesticides can produce immune dysfunction has been accumulated over recent years from animal studies (Devens et al., 1985; Olson et al., 1987; Luster et al., 1987; Thomas and House, 1989). In a study of women who had chronically ingested groundwater contaminated with low levels of aldicarb (mean = 16.6 ppb), Fiore et al. (1986) reported evidence of significantly reduced immune response, although these women did not exhibit any other overt health problems.

Of particular concern are the chronic health problems associated with effects of organophosphorous pesticides which have largely replaced the banned organochlorines (Ecobichon et al., 1990). The malady OPIDP (organophosphate-induced delayed polyneuropathy) is well documented and includes irreversible neurological defects (Lotti, 1984). Other defects in memory, mood, and abstraction have been documented by Savage et al. (1988). Well-documented cases indicate that persistent neurotoxic effects may be present even after the termination of an acute poisoning incident (Ecobichon et al., 1990).

Such chronic health problems are a public health issue, because everyone, everywhere, is exposed to some pesticide residues in food, water, and the atmosphere. About 35% of the foods purchased by U.S. consumers have detectable levels of pesticide residues (FDA, 1990). Of this from 1% to 3% of the foods have pesticide residue levels above the legal tolerance level (Hundley et al., 1988; FDA, 1990). Residue levels may well be higher than has been recorded because the U.S. analytical methods now employed detect only about one-third of the more than 600 pesticides in use (OTA,

1988). Certainly the contamination rate is higher for fruits and vegetables because these foods receive the highest dosages of pesticides. Therefore, there are many good reasons why 97% of the public is genuinely concerned about pesticide residues in their food (FDA, 1989).

Food residue levels in developing nations often average higher than those found in developed nations, either because there are no laws governing pesticide use or because the numbers of skilled technicians available to enforce laws concerning pesticide tolerance levels in foods are inadequate or because other resources are lacking. For instance, most milk samples assayed in a study in Egypt had high residue levels (60% to 80%) of 15 pesticides included in the investigation (Dogheim et al., 1990).

In all countries, the highest levels of pesticide exposures occur in pesticide applicators, farm workers, and people who live adjacent to heavily treated agricultural land (L.W. Davis, Com. of Agr. and Hort., Agr. Chem. and Environ. Services Div., Arizona, PC, 1990). Because farmers and farm workers directly handle 70% to 80% of all pesticides used, the health of these population groups is at the greatest risk of being seriously affected by pesticides. The epidemiological evidence suggests significantly higher cancer incidence among farmers and farm workers in the United States and Europe than among non–farm workers in some areas (e.g., Sharp et al., 1986; Blair et al., 1985; Brown et al., 1990). A consistent association has been documented between lung cancer and exposure to organochlorine insecticides (Blair et al., 1990). Evidence is strong for an association between lymphomas and soft-tissue sarcomas and certain herbicides (Hoar et al., 1986; Blair and Zahm, 1990; Zahm et al., 1990).

Medical specialists are concerned about the lack of public health data about pesticide usage in the United States (GAO, 1986). Based on an investigation of 92 pesticides used on food, the GAO (1986) estimates that 62% of the data on health problems associated with registered pesticides contain little or no information on tumors and even less on birth defects.

Although no one can place a precise monetary value on a human life, the "costs" of human pesticide poisonings have been estimated. Studies done for the insurance industry have computed monetary ranges for the value of a "statistical life" at between $1.6 and $8.5 million (Fisher et al., 1989). For our assessment, we use the conservative estimate of $2 million per human life. Based on the available data, estimates indicate that human pesticide poisonings and related illnesses in the United States total about $787 million each year (Table 3.1).

Domestic Animal Poisonings and Contaminated Products

In addition to pesticide problems that affect humans, several thousand domestic animals are poisoned by pesticides each year, with dogs and cats

Table 3.1. Estimated economic costs of human pesticide poisonings and other pesticide-related illnesses in the United States each year.

Human health effects from pesticides	Total costs ($)
Cost of hospitalized poisonings	
2380[a] × 2.84 days @ $1,000/day	6,759,000
Cost of outpatient treated poisonings	
27,000[c] × $630[b]	17,010,000
Lost work due to poisonings	
4680[a] workers × 4.7 days × $80/day	1,760,000
Pesticide cancers	
<10,000[d] cases × $70,700[c]/case	707,000,000
Cost of fatalities	
27 accidental fatalities[c] × $2 million	54,000,000
TOTAL	786,529,000

[a]Keefe et al. (1990).

[b]Includes hospitalization, foregone earnings, and transportation.

[c]J. Blondell, EPA, PC (1991).

[d]See text for details.

representing the largest number (Table 3.2). For example, of 25,000 calls made to the Illinois Animal Poison Control Center in 1987, nearly 40% of all calls concerned pesticide poisonings in dogs and cats (Beasley and Trammel, 1989). Similarly, Kansas State University reported that 67% of all animal pesticide poisonings involved dogs and cats (Barton and Oehme, 1981). This is not surprising because dogs and cats usually wander freely about the home and farm and therefore have greater opportunity to come into contact with pesticides than other domesticated animals.

The best estimates indicate that about 20% of the total monetary value of animal production, or about $4.2 billion, is lost to all animal illnesses, including pesticide poisonings (Gaafar et al., 1985). Colvin (1987) reported that 0.5% of animal illnesses and 0.04% of all animal deaths reported to a veterinary diagnostic laboratory were due to pesticide toxicosis. Thus, $21.3 and $8.8 million, respectively, are lost to pesticide poisonings (Table 3.2).

This estimate is considered low because it is based only on poisonings reported to veterinarians. Many animal deaths that occur in the home and on farms go undiagnosed and are attributed to factors other than pesticides. In addition, when a farm animal poisoning occurs and little can be done for the animal, the farmer seldom calls a veterinarian but, rather either waits for the animal to recover or destroys it (G. Maylin, Cornell University, PC, 1977). Such cases are usually unreported.

Table 3.2. Estimated domestic-animal pesticide poisonings in the United States.

Livestock	Number × 1000	$ per head	Number Ill[e]	$ cost per poisoning[f]	$ cost of poisonings	Number Deaths[d]	$ cost of Deaths × 1,000[g]	Total $ × 1,000
Cattle	99,484[a]	607[a]	100	121.40	12,140	8	4,856	16,996
Dairy cattle	10,298[a]	900[a]	10	180.00	1,800	1	900	2,700
Dogs	52,000[c]	125[h]	50	25.00	1,250	4	500	1,750
Horses	10,600[b]	1,000[c]	11	200.00	2,200	1	1,000	3,200
Cats	55,000[c]	20[h]	50	4.00	200	4	80	280
Swine	52,485[a]	66.30[a]	53	13.26	703	4	265	968
Chickens	5,700,000[a]	2.04[a]	5,700	0.40	2,280	456	912	3,192
Turkeys	260,000[a]	10[c]	260	2.00	520	21	210	730
Sheep	10,800[a]	82.40[a]	11	16.48	181	1	82	263
TOTAL	6,250,667				21,274		8,805	30,079

[a]USDA (1989a).

[b]FAO (1986).

[c]USBC (1990).

[d]Based on a 0.008% mortality rate. (See text.)

[e]Based on a 0.1% illness rate. (See text.)

[f]Based on each animal illness costing 20% of total production value of that animal.

[g]The death of the animal equals the total value for that animal.

[h]Estimated.

Additional economic losses occur when meat, milk, and eggs are contaminated with pesticides. In the United States, all animals slaughtered for human consumption, if shipped interstate, and all imported meat and poultry, must be inspected by the U.S. Department of Agriculture. This is to insure that the meat and products are wholesome, properly labeled, and do not present a health hazard. One part of this inspection, which involves monitoring meat for pesticide and other chemical residues, is the responsibility of the National Residue Program (NRP). The samples taken are intended to insure that if a chemical is present in 1% of the animals slaughtered it will be detected (NAS, 1985).

However, of more than 600 pesticides now in use, NRP tests are made for only 41 different pesticides (D. Beermann, Cornell University, PC, 1991), which have been determined by FDA, EPA, and FSIS to be of public health concern (NAS, 1985). While the monitoring program records the number and type of violations, there is no significant cost to the animal industry because the meat is generally *sold and consumed* before the test results are available. About 3% of the chickens with illegal pesticide residues are sold in the market (NAS, 1987).

Compliance sampling is designed to prevent meat and milk contamination with pesticides. When a producer is suspected of marketing contaminated livestock, the carcasses are detained until the residue analyses are reported. If there are illegal residues present, the carcasses or products are condemned and the producer is prohibited from marketing other animals until it is confirmed that all the livestock are safe (NAS, 1985). If carcasses are not suspected of being contaminated, then by the time the results of the residue tests are reported the carcasses have been sold to consumers. This points to a major deficiency in the surveillance program.

In addition to animal carcasses, pesticide-contaminated milk cannot be sold and must be disposed of. In certain incidents these losses are substantial. For example, in Oahu, Hawaii, in 1982, 80% of the milk supply, worth more than $8.5 million, was condemned by public health officials because it had been contaminated with the insecticide heptachlor (van Ravenswaay and Smith, 1986). This incident had immediate and far-reaching effects on the entire milk industry on the island. Initially, reduced milk sales due to the contaminated milk alone were estimated to cost each dairy farmer $39,000. Subsequently, the structure of the island milk industry has changed. Because island milk was considered by consumers to be unsafe, most of the milk supply is now imported. The $500 million lawsuit against the producers brought by consumers is still pending (van Ravenswaay and Smith, 1986).

When the costs attributable to domestic animal poisonings and contaminated meat, milk, and eggs are combined, the economic value of all livestock products in the United States lost to pesticide contamination is estimated to be at least $29.6 million annually (Table 3.2).

Similarly, other nations lose significant numbers of livestock and large amounts of animal products each year due to pesticide-induced illness or death. Exact data concerning livestock losses do not exist and the available information comes only from reports of the incidence of mass destruction of livestock. For example, when the pesticide leptophos was used by Egyptian farmers on rice and other crops, 1,300 draft animals were poisoned and lost (Sebae, 1977, in Bull, 1982). The estimated economic losses were significant but exact figures are not available.

In addition, countries exporting meat to the United States can experience tremendous economic losses if the meat is found contaminated with pesticides. In a 15-year period, the beef industries in Guatemala, Honduras, and Nicaragua lost more than $1.7 million due to pesticide contamination of exported meat (ICAITI, 1977). In these countries, meat which is too contaminated for export is sold in local markets. Obviously such policies contribute to public health problems.

Destruction of Beneficial Natural Predators and Parasites

In both natural and agroecosystems, many species, especially predators and parasites, control or help control herbivorous populations. Indeed, these natural beneficial species make it possible for ecosystems to remain "green." With the parasites and predators keeping herbivore populations at low levels, only a relatively small amount of plant biomass is removed each growing season (Hairston et al., 1960). Natural enemies play a major role in keeping populations of many insect and mite pests under control (DeBach, 1964; Huffaker, 1977; Pimentel, 1988).

Like pest populations, beneficial natural enemies are adversely affected by pesticides (van den Bosch and Messenger, 1973; Adkisson, 1977; Ferro, 1987; Croft, 1990). For example, the following pests have reached outbreak levels in cotton and apple crops following the destruction of natural enemies by pesticides: *cotton*—cotton bollworm, tobacco budworm, cotton aphid, spider mites, and cotton loopers (Adkisson, 1977; OTA, 1979); and *apple*—European red mite, red-banded leafroller, San Jose scale, oystershell scale, rosy apple aphid, wooly apple aphid, white apple aphid, two-spotted spider mite, and apple rust mite (Tabashnik and Croft, 1985; Messing et al., 1989; Croft, 1990; Kovach and Agnello, 1991). Significant pest outbreaks also have occurred in other crops (Huffaker and Kennett, 1956; Huffaker, 1977; OTA, 1979; Croft, 1990; Pimentel, 1991). Also, because parasitic and predaceous insects often have complex searching and attack behaviors, sublethal insecticide dosages may alter this behavior and in this way disrupt effective biological controls (L.E. Ehler, University of California, PC, 1991).

Fungicides also can contribute to pest outbreaks when they reduce fungal pathogens that are naturally parasitic on many insects. For example, the use of benomyl reduces populations of entomopathogenic fungi, resulting in increased survival of velvet bean caterpillars and cabbage loopers in soybeans. This eventually leads to reduced soybean yields (Ignoffo et al., 1975; Johnson et al., 1976).

When outbreaks of secondary pests occur because their natural enemies are destroyed by pesticides, additional and sometimes more expensive pesticide treatments have to be applied in efforts to sustain crop yields. This raises overall costs and contributes to pesticide-related problems.

An estimated $520 million can be attributed to costs of additional pesticide applications and increased crop losses, both of which follow the destruction of natural enemies by pesticides applied to crops (Table 3.3).

Table 3.3. Losses due to the destruction of beneficial natural enemies in U.S. crops ($ millions).

Crops	Total expenditures for insect control with pesticides ($)[a]	Amount of added control costs ($)
Cotton	320	160
Tobacco	5	1
Potatoes	31	8
Peanuts	18	2
Tomatoes	11	2
Onions	1	0.2
Apples	43	11
Cherries	2	1
Peaches	12	2
Grapes	3	1
Oranges	8	2
Grapefruit	5	1
Lemons	1	0.2
Nuts	160	16
Other	500	50
TOTAL		$257.4 ($520)[b]

[a]Pimentel et al. (1991).

[b]Because the added pesticide treatments do not provide as effective control as the natural enemies, we estimate that at least an additional $260 million in crops are lost to pests. Thus, the total loss due to the destruction of natural enemies is estimated to be at least $520 million/year.

As in the United States, natural enemies are being adversely affected by pesticides worldwide. Although no reliable estimate is available concerning the impact of this in terms of increased pesticide use and/or reduced yields, general observations by entomologists indicate the impact of loss of natural enemies is severe in many parts of the world. For example, from 1980 to 1985 insecticide use in rice production in Indonesia drastically increased (Oka, 1991). This caused the destruction of beneficial natural enemies of the brown planthopper and the pest populations exploded. Rice yields dropped to the extent that rice had to be imported into Indonesia. The estimated loss in rice in just a 2-year period was $1.5 billion (FAO, 1988).

Following that incident, Dr. I. N. Oka and his associates, who previously had developed a successful low-insecticide program for rice pests in Indonesia, were consulted by Indonesian President Suharto's staff to determine what should be done to rectify the situation (I. N. Oka, Bogor Food Research Institute, Indonesia, PC, 1990). Their advice was to substantially reduce insecticide use and return to a sound "treat-when-necessary" program that protected the natural enemies. Following Oka's advice, President Suharto mandated in 1986 that 57 of 64 pesticides would be withdrawn from use on rice and pest management practices would be improved. Pesticide subsidies also were reduced to zero. Subsequently, rice yields increased to levels well above those recorded during the period of heavy pesticide use (FAO, 1988).

D. Rosen (Hebrew University of Jerusalem, PC, 1991) estimates that natural enemies account for up to 90% of the control of pest species achieved in agroecosystems and natural systems; we estimate that at least 50% of control of pest species is due to natural enemies. Pesticides give an additional control of 10% (Pimentel et al., 1978), while the remaining 40% is due to host-plant resistance and other limiting factors present in the agroecosystem (Pimentel, 1988).

Parasites, predators, and host-plant resistance are estimated to account for about 80% of the nonchemical control of pest insects and plant pathogens in crops (Pimentel et al., 1991). Many cultural controls such as crop rotations, soil and water management, fertilizer management, planting time, crop-plant density, trap crops, polyculture, and others provide additional pest control. Together these nonchemical controls can be used effectively to reduce U.S. pesticide use by as much as one-half without any reduction in crop yields (Pimentel et al., 1991).

Pesticide Resistance in Pests

In addition to destroying natural enemy populations, the extensive use of pesticides has often resulted in the development of pesticide resistance in insect pests, plant pathogens, and weeds. In a report of the United Nations Environment Programme pesticide resistance was ranked as one of the top four environmental problems in the world (UNEP, 1979). About 504 insect and mite species (Georghiou, 1990), a total of nearly 150 plant pathogen species (Georghiou, 1986; Eckert, 1988), and about 273 weed species are now resistant to pesticides (LeBaron and McFarland, 1990).

Increased pesticide resistance in pest populations frequently results in the need for several additional applications of the commonly used pesticides to maintain expected crop yields. These additional pesticide applications compound the problem by increasing environmental selection for resistance traits. Despite attempts to deal with it, pesticide resistance continues to develop (Dennehy et al., 1987).

The impact of pesticide resistance, which develops gradually over time, is felt in the economics of agricultural production. A striking example of this occurred in northeastern Mexico and the Lower Rio Grande of Texas (Adkisson, 1972; NAS, 1975). Over time extremely high pesticide resistance had developed in the tobacco budworm population on cotton. Finally in early 1970 approximately 285,000 ha of cotton had to be abandoned, because pesticides were ineffective and there was no way to protect the crop from the budworm. The economic and social impact on these Texan and Mexican farming communities dependent upon cotton was devastating.

The study by Carrasco-Tauber (1989) indicates the extent of costs attributed to pesticide resistance. They reported a yearly loss of $45 to $120/ha to pesticide resistance in California cotton. A total of 4.2 million hectares of cotton were harvested in 1984; thus, assuming a loss of $82.50/ha, approximately $348 million of the California cotton crop was lost to resistance. Since $3.6 billion of U.S. cotton were harvested in 1984 (USBC, 1990), the loss due to resistance for that year was approximately 10%. Assuming a 10% loss in other major crops that receive heavy pesticide treatments in the United States, crop losses due to pesticide resistance are estimated to be $1.4 billion/year.

A detailed study by Archibald (1984) further demonstrated the hidden costs of pesticide resistance in California cotton. She reported that 74% more organophosphorus insecticides were required in 1981 to achieve the same kill of pests, like *Heliothis* spp., than in 1979. Her analysis demonstrated that the diminishing effect of pesticides plus the intensified pest control reduced the economic return per dollar of pesticide invested to only $1.14.

Furthermore, efforts to control resistant *Heliothis* spp. exact a cost on other crops when large, uncontrolled populations of *Heliothis* and other pests disperse onto other crops. In addition, the cotton aphid and the whitefly exploded as secondary cotton pests because of their resistance and their natural enemies' exposure to the high concentrations of insecticides.

The total external cost attributed to the development of pesticide resistance is estimated to range between 10% to 25% of current pesticide treatment costs (La Farge, 1985; Harper and Zilberman, 1990), or approximately $400 million each year in the United States alone. In other words, at least 10% of pesticide used in the United States is applied just to combat increased resistance that has developed in various pest species.

In addition to plant pests, a large number of insect and mite pests of both livestock and humans have become resistant to pesticides (Georghiou, 1986). Although a relatively small quantity of pesticide is applied for control of livestock and human pests, the cost of resistance has become significant. Based on available data, the yearly cost of resistance in insect and mite pests of livestock and humans we estimated to be about $30 million for the United States.

Although the costs of pesticide resistance are high in the United States, the costs in tropical developing countries are significantly greater, because pesticides are not only used to control agricultural pests but are also vital for the control of disease vectors. One of the major costs of resistance in tropical countries is associated with malaria control. By 1961, the incidence of malaria in India after early pesticide use declined to only 41,000 cases. However, because mosquitoes developed resistance to pesticides, as did malarial parasites to drugs, the incidence of malaria in India now has exploded to about 59 million cases per year (Reuben, 1989). Similar problems are occurring not only in India but also in the rest of Asia, Africa, and South America: the total incidence of malaria is estimated to be 270 million cases (WHO, 1990; NAS, 1991).

Honeybee and Wild Bee Poisonings and Reduced Pollination

Honeybees and wild bees are absolutely vital for pollination of fruits, vegetables, and other crops. Their direct and indirect benefits to agricultural production range from $10 to $33 billion each year in the United States (Robinson et al., 1989; E.L. Atkins, University of California, PC, 1990). Because most insecticides used in agriculture are toxic to bees, pesticides have a major impact on both honeybee and wild bee populations. D. Mayer (Washington State University, PC, 1990) estimates that approximately 20% of all honeybee colonies are adversely affected by pesticides. He includes the approximately 5% of U.S. bee colonies that are killed outright or die during winter because of pesticide exposure. Mayer calculates that the direct annual loss reaches $13.3 million (Table 3.4). Another 15% of the bee colonies either are seriously weakened by pesticides or suffer losses when apiculturists have to move colonies to avoid pesticide damage.

Table 3.4. Estimated honeybee losses and pollination losses from honeybees and wild bees.

Colony losses from pesticides	$13.3 million/year
Honey and wax losses	25.3 million/year
Loss of potential honey production	27.0 million/year
Bee rental for pollination	4 million/year
Pollination losses	200 million/year
TOTAL	$319.6 million/year

According to Mayer, the yearly estimated loss from partial bee kills, reduced honey production, plus the cost of moving colonies totals about $25.3 million. Also, as a result of heavy pesticide use on certain crops, beekeepers are excluded from 4 to 6 million ha of otherwise suitable apiary locations (D. Mayer, Washington State University, PC, 1990). He estimates the yearly loss in potential honey production in these regions is about $27 million.

In addition to these direct losses caused by the damage to bees and honey production, many crops are lost because of the lack of pollination. In California, for example, approximately 1 million colonies of honeybees are rented annually at $20 per colony to augment the natural pollination of almonds, alfalfa, melons, and other fruits and vegetables (R.A. Morse, Cornell University, PC, 1990). Since California produces nearly 50% of our bee-pollinated crops, the total cost for bee rental for the entire country is estimated at $40 million. Of this cost, we estimate at least one tenth or $4 million is attributed to the effects of pesticides (Table 3.4).

Estimates of annual agricultural losses due to the reduction in pollination by pesticides may range as high as $4 billion/year (J. Lockwood, University of Wyoming, PC, 1990). For most crops both crop yield and quality are enhanced by effective pollination. For example, McGregor et al. (1955) and Mahadevan and Chandy (1959) demonstrated that for several cotton varieties, effective pollination by bees resulted in yield increases from 20% to 30%. Assuming that a conservative 10% increase in cotton yield would result from more efficient pollination, and subtracting charges for bee rental, the net annual gain for cotton alone could be as high as $400 million. However, using bees to enhance cotton pollination is impossible at present because of the intensive use of insecticides on cotton (McGregor, 1976).

Mussen (1990) emphasizes that poor pollination will not only reduce crop yields, but more importantly, it will reduce the quality of crops such as melons and other fruits. In experiments with melons, E. L. Atkins (University of California, PC, 1990) reported that with adequate pollination melon yields were increased 10% and quality was raised 25% as measured by the dollar value of the crop.

Based on the analysis of honeybee and related pollination losses caused by pesticides, pollination losses attributed to pesticides are estimated to represent about 10% of pollinated crops and have a yearly cost of about $200 million. Adding the cost of reduced pollination to the other environmental costs of pesticides on honeybees and wild bees, the total annual loss is calculated to be about $320 million (Table 3.4). Clearly, the available evidence confirms that the yearly cost of direct honeybee losses, together with reduced yields resulting from poor pollination, are significant.

Crop and Crop Product Losses

Basically, pesticides are applied to protect crops from pests in order to increase yields, but sometimes the crops are damaged by pesticide treatments. This occurs when (1) the recommended dosages suppress crop growth, development, and yield; (2) pesticides drift from the targeted crop to damage adjacent nearby crops; (3) residual herbicides either prevent chemical-sensitive crops from being planted in rotation or inhibit the growth of crops that are planted; and/or (4) excessive pesticide residues accumulate on crops, necessitating the destruction of the harvest. Crop losses translate into financial losses for growers, distributors, wholesalers, transporters, retailers, food processors, and others. Potential profits as well as investments are lost. The costs of crop losses increase when the related costs of investigations, regulation, insurance, and litigation are added to the equation. Ultimately the consumer pays for these losses in higher marketplace prices.

Data on crop losses due to pesticide use are difficult to obtain. Many losses are never reported to the state and federal agencies because the parties often settle privately (B. D. Berver, Office of Agronomy Services, South Dakota, PC, 1990; R. Batteese, Board of Pesticide Control, Maine Dept. of Agriculture, PC, 1990; J. Peterson, Pesticide/Noxious Weed Division, Dept. of Agr., North Dakota, PC, 1990; E. Streams, EPA, region VII, PC, 1990). For example, in North Dakota, only an estimated one-third of the pesticide-induced crop losses are reported to the State Department of Agriculture (Peterson, PC, 1990). Furthermore, according to the Federal Crop Insurance Corporation, losses due to pesticide use are not insurable because of the difficulty of determining pesticide damage (E. Edgeton, Federal Crop Insurance Corp., Washington, D.C., PC, 1990).

Damage to crops may occur even when recommended dosages of herbicides and insecticides are applied to crops under normal environmental conditions (Chang, 1965; J. Neal, Chemical Pesticides Program, Cornell

University, PC, 1990). Heavy dosages of insecticides used on crops have been reported to suppress growth and yield in both cotton and strawberry crops (ICAITI, 1977; Reddy et al., 1987; Trumbel et al., 1988). The increased susceptibility of some crops to insects and diseases following normal use of 2,4-D and other herbicides was demonstrated by Oka and Pimentel (1976), Altman (1985), and Rovira and McDonald (1986). Furthermore, when weather and/or soil conditions are inappropriate for pesticide application, herbicide treatments may cause yield reductions ranging from 2% to 50% (von Rumker and Horay, 1974; Elliot et al., 1975; Akins et al., 1976).

Crops are lost when pesticides drift from the target crops to non–target crops located as much as several miles downwind (Henderson, 1968; Barnes et al., 1987). Drift occurs with almost all methods of pesticide application including both ground and aerial equipment; the potential problem is greatest when pesticides are applied by aircraft (Ware et al., 1969). With aircraft 50% to 75% of pesticides applied miss the target area (Ware et al., 1970; ICAITI, 1977; Ware, 1983; Akesson and Yates, 1984; Mazariegos, 1985). In contrast, 10% to 35% of the pesticide applied with ground application equipment misses the target area (Ware et al., 1975; Hall, 1991). The most serious drift problems are caused by "speed sprayers" and "mist-blower sprayers," because large amounts of pesticide are applied by these sprayers and with these application technologies about 35% of the pesticide drifts away from the target area (E. L. Atkins, University of California, PC, 1990).

Crop injury and subsequent loss due to drift are particularly common in areas planted with diverse crops. For example, in southwest Texas in 1983 and 1984, nearly $20 million of cotton was destroyed from drifting 2,4-D herbicide when adjacent wheat fields were aerially sprayed with the herbicide (Hanner, 1984).

Clearly, drift damage, human exposure, and widespread environmental contamination are inherent in the process of pesticide application and add to the cost of using pesticides. As a result, commercial applicators are frequently sued for damage inflicted during or after treatment. Therefore, most U.S. applicators now carry liability insurance at an estimated cost of about $245 million/year (FAA, 1988; D. Witzman, U.S. Aviation Underwriters, Tennessee, PC, 1990; H. Collins, Nat. Agr. Aviation Assoc., Washington, D.C., PC, 1990).

When residues of some herbicides persist in the soil, crops planted in rotation may be injured (Nanjappa and Hosmani, 1983; Rogers, 1985; Keeling, et al., 1989). In 1988/1989, an estimated $25 to $30 million of Iowa's soybean crop was lost due to the persistence of the herbicide Sceptor in the soil (R.G. Hartzler, Cooperative Extension Serv., Iowa State University, PC, 1990).

Herbicide persistence can sometimes prevent growers from rotating their crops and this situation may force them to continue planting the same crop (Altman, 1985; T. Tomas, Nebraska Sustainable Agriculture Society, Hartington, NE, PC, 1990). For example, the use of Sceptor in Iowa, as mentioned, has prevented farmers from implementing their plan to plant soybeans after corn (R. G. Hartzler, PC, 1990). Unfortunately, the continuous planting of some crops in the same field often intensifies insect, weed, and pathogen problems (PSAC, 1965; NAS, 1975; Pimentel et al., 1991). Such pest problems not only reduce crop yields but often require added pesticide applications.

Although crop losses caused by pesticides seem to be a small percentage of total U.S. crop production, their total value is significant. For instance, an average of 0.14% of San Joaquin County's (California) total crop production was lost to pesticides from 1986 to 1987 (OACSJC, 1990; OACSJC, Agricultural Commissioner, San Joaquin County, CA, PC, 1990). Similarly, in Yolo County, CA, approximately 0.18% of its total crop production was lost in 1989 (OACYC, Agricultural Commissioner, Yolo County, CA, PC, 1990; OACYC, 1990). Estimates from Iowa indicate that less than 0.05% of its annual soybean crop is lost to pesticides (R. G. Hartzler, PC, 1990).

An average 0.1% loss in the annual U.S. production of corn, soybeans, cotton, and wheat, which together account for about 90% of the herbicides and insecticides used in U.S. agriculture, was valued at $35.3 million in 1987 (USDA, 1989a; NAS, 1989). Assuming that only one-third of the incidents involving crop losses due to pesticides are reported to authorities, the total value of all crops lost because of pesticides could be as high as three times this amount, or $106 million annually.

However, this $106 million does not take into account other crop losses, nor does it include major but recurrent events such as the large-scale losses that occurred in Iowa in 1988–1989 ($25–$30 million), Texas in 1983–1984 ($20 million), and in California's aldicarb/watermelon crisis in 1985 ($8 million, see below). These recurrent losses alone represent an average of $30 million each year, raising the estimated average crop loss value from the use of pesticides to approximately $136 million.

Additional losses are incurred when food crops are disposed of because they exceed the EPA regulatory tolerances for pesticide residue levels. Assuming that all the crops and crop products that exceed the EPA regulatory tolerances (reported to be at least 1%) were disposed of as required by law, then about $550 million in crops annually would be destroyed because of excessive pesticide contamination (calculated based on data from FDA [1990] and USDA [1989a]). Because most of the crops with pesticides above the tolerance levels are neither detected nor destroyed, they are consumed by the public, avoiding financial loss but creating public health risks.

A well-publicized incident in California during 1985 illustrates this problem. In general, excess pesticides in the food go undetected unless a large number of people become ill after the food is consumed. Thus when more than 1,000 persons became ill from eating the contaminated watermelons, approximately $1.5 million dollars worth of watermelons were ordered destroyed (R. Magee, State of California Dept. of Food and Agriculture, Sacramento, PC, 1990). After the public became ill, it was learned that several California farmers had treated watermelons with the insecticide aldicarb (Temik), which is not registered for use on watermelons (Taylor, 1986; Kizer, 1986). Following this crisis the California State Assembly appropriated $6.2 million to be awarded to claimants affected by state seizure and freeze orders (*Legislative Counsel's Digest*, 1986). According to the California Department of Food and Agriculture an estimated $800,000 in investigative costs and litigation fees resulted from this one incident (R. Magee, CDFA, PC, 1990). The California Department of Health Services was assumed to have incurred similar expenses, putting the total cost of the incident at nearly $8 million.

To avoid other dangerous and costly incidents like the California watermelon crisis, many private distributors and grocers are testing their produce for the presence of pesticides to reassure themselves and consumers of the safety of the food they handle (C. Merrilees, Consumer Pesticide Project, Nat. Toxics Campaign, San Francisco, PC, 1990). Nationally, this testing is presently estimated to cost $1 million per year (C. Merrilees, PC, 1990), but if all the retail grocers nationwide were to undertake such testing, the calculated cost would be approximately $66 million per year based on data from California.

Special investigations of crop losses due to pesticide use, conducted by state and federal agencies, are also costly. From 1987 through 1989, the State of Montana Department of Agriculture conducted an average of 80 pesticide-related investigations per year at an average cost of $3,500 per investigation (S. F. Baril, State of Montana, Dept. of Agr., PC, 1990). Also, the State of Hawaii conducts approximately five investigations a year and these cost nearly $10,000 each (R. Boesch, Pesticide Programs, Dept. of Agriculture, State of Hawaii, PC, 1990). Averaging the number of investigations from seven states (Arkansas, Hawaii, Idaho, Iowa, Louisiana, Mississippi, and Texas) and using the low Montana figure of $3,500 per investigation, the average state conducts 70 investigations a year at a cost of $246,000 annually. Using these data, the investigations are estimated to total $10 million annually. This figure does not include investigation costs at the federal level.

When crop seizures, insurance, and investigation costs are added to the costs of direct crop losses due to the use of pesticides in commercial crop production, the total monetary loss is estimated to be about $942 million annually in the United States (Table 3.5).

Ground- and Surface Water Contamination

Certain pesticides applied to crops eventually end up in ground- and surface water. The three most common pesticides found in groundwater are aldicarb (an insecticide), alachlor, and atrazine (two herbicides) (Osteen and Szmedra, 1989). Estimates are that nearly one-half of the groundwater and well water in the United States is or has the potential to be contaminated (Holmes et al., 1988). The EPA (1990a) reported that 10.4% of community wells and 4.2% of rural domestic wells have detectable levels of at least one pesticide of the 127 pesticides tested in a national survey. It would cost an estimated $1.3 billion annually in the United States if well and groundwater were monitored for pesticide residues (Nielsen and Lee, 1987).

Two major concerns about groundwater contamination with pesticides are that about one-half of the population obtains its water from wells and that once groundwater is contaminated, the pesticide residues remain for long periods of time. Not only are there just a few microorganisms that have the potential to degrade pesticides (Larson and Ventullo, 1983; Pye and Kelley, 1984) but the groundwater recharge rate averages less than 1% per year (CEQ, 1980).

Monitoring pesticides in groundwater is only a portion of the total cost of U.S. groundwater contamination. There is also the high cost of cleanup. For instance, at the Rocky Mountain Arsenal near Denver, Colorado, the removal of pesticides from the groundwater and soil was estimated to cost

Table 3.5. Estimated loss of crops and trees due to the use of pesticides.

Impacts	Total costs (in millions of $)
Crop losses	136
Crop applicator insurance	245
Crops destroyed because of excess pesticide contamination	550
Investigations and testing	
Governmental	10
Private	1
TOTAL	$942 million

approximately $2 billion (NYT, 1988). If all pesticide-contaminated groundwater were cleared of pesticides before human consumption, the cost would be about $500 million (based on the costs of cleaning water [Clark, 1979; van der Leeden et al., 1990]). Note the cleanup process requires a water survey to target the contaminated water for cleanup. Thus, adding monitoring and cleaning costs, the total cost regarding pesticide-polluted groundwater is estimated to be about $1.8 billion annually.

Fishery Losses

Pesticides are washed into aquatic ecosystems by water runoff and soil erosion. About 18 tons/ha/year of soil are washed and/or blown from pesticide-treated cropland into adjacent locations including streams and lakes (USDA, 1989b). Pesticides also drift into streams and lakes and contaminate these aquatic systems (Clark, 1989). Some soluble pesticides are easily leached into streams and lakes (Nielsen and Lee, 1987).

Once in aquatic systems, pesticides cause fishery losses in several ways. These include high pesticide concentrations in water that directly kill fish; low-level doses that may kill highly susceptible fish fry; or the elimination of essential fish foods like insects and other invertebrates. In addition, because government safety restrictions ban the catching or sale of fish contaminated with pesticide residues, such unmarketable fish are considered an economic loss (EPA, 1990b; Knuth, 1989; ME & MNR, 1990).

Each year large numbers of fish are killed by pesticides. Based on EPA (1990b) data we calculate that from 1977 to 1987 the cost of fish kills due to all factors has been 141 million fish/year; from 6 to 14 million fish/year are killed by pesticides. These estimates of fish kills are considered to be low for the following reasons. First, 20% of the reported fish kills do not estimate the number of fish killed, and second, fish kills frequently cannot be investigated quickly enough to determine accurately the primary cause (Pimentel et al., 1980). In addition, fast-moving waters in rivers dilute pollutants so that these causes of kills frequently cannot be identified, and also wash away the poisoned fish, while other poisoned fish sink to the bottom and cannot be counted (EPA, 1990b). Perhaps most important is the fact that, unlike direct kills, few, if any, of the widespread and more frequent low-level pesticide poisonings are dramatic enough to be observed and therefore go unrecognized and unreported (EPA, 1990b).

The average value of a fish has been estimated to be about $1.70, using the guidelines of the American Fisheries Society (AFS, 1982); however, it was reported that Coors Beer might be "fined up to $10 per dead fish, plus other penalties" for an accidental beer spill in a creek (Barometer, 1991). At $1.70, the value of the estimated 6–14 million fish killed per year ranges from $10 to $24 million. For reasons mentioned earlier, this is considered an extremely low estimate; the actual loss is probably several times this amount.

Wild Birds and Mammals

Wild birds and mammals are also damaged by pesticides, but these animals make excellent "indicator species." Deleterious effects on wildlife include death from direct exposure to pesticides or secondary poisonings from consuming contaminated prey; reduced survival, growth, and reproductive rates from exposure to sublethal dosages; and habitat reduction through elimination of food sources and refuges (McEwen and Stephenson, 1979; Grue et al., 1983; Risebrough, 1986; Smith, 1987). In the United States, approximately 3 kg of pesticide per ha is applied on about 160 million ha/year of land (Pimentel et al., 1991). With such a large portion of the land area treated with heavy dosages of pesticide, it is to be expected that the impact on wildlife is significant.

The full extent of bird and mammal destruction is difficult to determine because these animals are often secretive, camouflaged, highly mobile, and live in dense grass, shrubs, and trees. Typical field studies of the effects of pesticides often obtain extremely low estimates of bird and mammal mortality (Mineau and Collins, 1988). This is because bird carcasses disappear quickly, well before the dead birds can be found and counted. Studies show only 50% of birds are recovered even when the bird's location is known (Mineau, 1988). Furthermore, where known numbers of bird carcasses were placed in identified locations in the field, 62% to 92% disappeared overnight due to vertebrate and invertebrate scavengers (Balcomb, 1986). Then, too, field studies seldom account for birds that die a distance from the treated areas. Finally, birds often hide and die in inconspicuous locations.

Nevertheless, many bird casualties caused by pesticides have been reported. For instance, White et al. (1982) reported that 1,200 Canada geese were killed in one wheat field that was sprayed with a 2:1 mixture of parathion and methyl parathion at a rate of 0.8 kg/ha. Carbofuran applied to alfalfa killed more than 5,000 ducks and geese in five incidents, while the same chemical applied to vegetable crops killed 1,400 ducks in a single incident (Flickinger et al., 1980, 1991). Carbofuran is estimated to kill 1–2 million birds each year (EPA, 1989). Another pesticide, diazinon, applied on just three golf courses killed 700 Atlantic brant geese of the wintering population of 2,500 geese (Stone and Gradoni, 1985).

Several studies report that the use of herbicides in crop production result in the total elimination of weeds that harbor some insects (Potts, 1986; R.

Beiswenger, University of Wyoming, PC, 1990). This has led to significant reductions in the grey partridge in the United Kingdom and in the common pheasant in the United States. In the case of the partridge, population levels have decreased more than 77%, because partridge chicks (also pheasant chicks) depend on insects to supply them with needed protein for their development and survival (Potts, 1986; R. Beiswenger, University of Wyoming, PC, 1990).

Frequently the form of a pesticide influences its toxicity to wildlife (Hardy, 1990). For example, treated seed and insecticide granules, including carbofuran, fensulfothion, fonofos, and phorate, are particularly toxic to birds when consumed. Estimates are that from 0.23 to 1.5 birds/ha were killed in Canada, while in the United States the estimates ranged from 0.25 to 8.9 birds/ha killed per year by the pesticides (Mineau, 1988).

Pesticides also adversely affect the reproductive potential of many birds and mammals. Exposure of birds, especially predatory birds, to chlorinated insecticides has caused reproductive failure, sometimes attributed to eggshell thinning (Stickel et al., 1984; Risebrough, 1986; Gonzalez and Hiraldo, 1988; Elliot et al., 1988). Most of the affected populations have recovered after the ban of DDT in the United States (Bednarz et al., 1990). However, DDT and its metabolite DDE remain a concern, because DDT continues to be used in some South American countries, which are the wintering areas for numerous bird species (Stickel et al., 1984).

Several pesticides, especially DBCP, dimethoate, and deltamethrinare, have been reported to reduce sperm production in certain mammals (Salem et al., 1988; Foote et al., 1986). Clearly, when this occurs the capacity of certain wild mammals to survive is reduced.

Habitat alteration and destruction can be expected to reduce mammal populations. For example, when glyphosphate was applied to forest clearcuts to eliminate low-growing vegetation, the southern red-backed vole population was greatly reduced because its food source and cover were practically eliminated (D'Anieri et al., 1987). Similar effects from herbicides on other mammals have been reported (Pimentel, 1971). However, overall, the impacts of pesticides on mammals have been inadequately investigated.

Although the gross values for wildlife are not available, expenditures involving wildlife made by humans are one measure of the monetary value. Nonconsumptive users of wildlife spent an estimated $14.3 billion on their sport in 1985 (USFWS, 1988). Yearly, U.S. bird-watchers spend an estimated $600 million on their sport and an additional $500 million on birdseed, or a total of $1.1 billion (USFWS, 1988). The money spent by bird hunters to harvest 5 million game birds was $1.1 billion, or approximately $216/bird (USFWS, 1988). In addition, estimates of the value of all types of birds ranged from $0.40 to more than $800/bird. The $0.40/bird was based on the costs of bird-watching and the $800/bird was based on the replacement costs of the affected species (Walgenbach, 1979; Tinney, 1981; Dobbins, 1986).

If it is assumed that the damages pesticides inflict on birds occur primarily on the 160 million ha of cropland that receives most of the pesticide, and the bird population is estimated to be 4.2 birds per ha of cropland (Blew, 1990), then 672 million birds are directly exposed to pesticides. If it is conservatively estimated that only 10% of the bird population is killed, then the total number killed is 67 million birds. Note this estimate is at the lower end of the range of 0.25 to 8.9 birds/ha killed per year by pesticides mentioned earlier in this section. Also, this is considered a conservative estimate because secondary losses due to reductions in invertebrate prey were not included in the assessment. Assuming the average value of a bird is $30, an estimated $2 billion in birds are destroyed annually.

Also, a total of $102 million is spent yearly by the U.S. Fish and Wildlife Service on their Endangered Species Program, which aims to reestablish species such as the bald eagle, peregrine falcon, osprey, and brown pelican that in some cases were reduced by pesticides (USFWS, 1991).

When all the above costs are combined, we estimate that the U.S. bird losses associated with pesticide use represent a cost of about $2.1 billion/year.

Microorganisms and Invertebrates

Pesticides easily find their way into soils, where they may be toxic to the arthropods, earthworms, fungi, bacteria, and protozoa found there. Small organisms are vital to ecosystems because they dominate both the structure and function of natural systems.

For example, an estimated 4.5 tons/ha of fungi and bacteria exist in the upper 15 cm of soil (Stanier et al., 1970). They with the arthropods make up 95% of all species and 98% of the biomass (excluding vascular plants). The microorganisms are essential to proper functioning in the ecosystem, because they break down organic matter, enabling the vital chemical elements to be recycled (Atlas and Bartha, 1987). Equally important is their ability to "fix" nitrogen, making it available for plants (Brock and Madigan, 1988). The role of microorganisms cannot be overemphasized, because in nature, agriculture, and forestry they are essential agents in biogeochemical recycling of the vital elements in all ecosystems (Brock and Madigan, 1988).

Earthworms and insects aid in bringing new soil to the surface at a rate of up to 200 tons/ha/year (Kevan, 1962; Satchel, 1967). This action improves soil formation and structure for plant growth and makes various nutrients more available for absorption by plants. The holes (up to 10,000 holes per square meter) in the soil made by earthworms and insects also facilitate the percolation of water into the soil (Hole, 1981; Edwards and Lofty, 1982), thereby slowing rapid water runoff from the land and preventing soil erosion.

Insecticides, fungicides, and herbicides reduce species diversity in the soil as well as the total biomass of these biota (Pimentel, 1971). Stringer and Lyons (1974) reported that where earthworms had been killed by pesticides, the leaves of apple trees accumulated on the surface of the soil. Apple scab, a disease carried over from season to season on fallen leaves, is commonly treated with fungicides. Some fungicides, insecticides, and herbicides can be toxic to earthworms, which would otherwise remove and recycle the surface leaves (Edwards and Lofty, 1977).

On golf courses and other lawns the destruction of earthworms by pesticides results in the accumulation of dead grass or thatch in the turf (Potter and Braman, 1991). To remove this thatch special equipment must be used at considerable expense.

Although these invertebrates and microorganisms are essential to the vital structure and function of all ecosystems, it is impossible to place a dollar value on the damage caused by pesticides to this large group of organisms. To date no relevant quantitative data on the value of microorganism destruction by pesticides has been collected.

Government Funds for Pesticide Pollution Control

A major environmental cost associated with all pesticide use is the cost of carrying out state and federal regulatory actions, as well as pesticide-monitoring programs needed to control pesticide pollution. Specifically, these funds are spent to reduce the hazards of pesticides and to protect the integrity of the environment and public health.

At least $1 million is spent each year by the state and federal government to train and register pesticide applicators (D. Rutz, Cornell University, PC, 1991). Also, more than $40 million is spent each year by the EPA just to register and reregister pesticides (GAO, 1986). Based on these known expenditures, estimates are that the federal and state governments spend approximately $200 million/year for pesticide pollution control (USBC, 1990) (Table 3.6).

Although enormous amounts of government money are being spent to reduce pesticide pollution, many costs of pesticides are not taken into account. Also, many serious environmental and social problems remain to be corrected by improved government policies.

Table 3.6. Total estimated environmental and social costs from pesticides in the United States.

Costs	Millions of $/year
Public health impacts	787
Domestic animals deaths and contamination	30
Loss of natural enemies	520
Cost of pesticide resistance	1,400
Honeybee and pollination losses	320
Crop losses	942
Fishery losses	24
Bird losses	2,100
Groundwater contamination	1,800
Government regulations to prevent damage	200
TOTAL	8,123

Ethical and Moral Issues

Although pesticides provide about $16 billion/year in saved U.S. crops, the data of this analysis suggest that the environmental and social costs of pesticides to the nation total approximately $8 billion. From a strictly cost/benefit approach, it appears that pesticide use is beneficial. However, the nature of environmental costs of pesticides has other trade-offs involving environmental quality and human health.

One of these issues concerns the importance of public health vs. pest control. For example, assuming that pesticide-induced cancers number 10,000 cases per year and that pesticides return a net agricultural benefit of $12 billion/year, each case of cancer is "worth" $1.2 million in pest control. In other words, for every $1.2 million in pesticide benefits, one person falls victim to cancer. Social mechanisms and market economics provide these ratios, but they ignore basic ethics and values.

In addition, pesticide pollution of the global environment raises numerous other ethical questions. The environmental insult of pesticides has the potential to demonstrably disrupt entire ecosystems. All through history, humans have felt justified in removing forests, draining wetlands, and constructing highways and housing everywhere. L. White (1967) has blamed the environmental crisis on religious teachings of mastery over nature. Whatever the origin, pesticides exemplify this attempt at mastery, and even a noneconomic analysis would question its justification. There is a clear need for a careful and comprehensive assessment of the environmental impacts of pesticides on agriculture and the natural ecosystem.

In addition to the ethical status of ecological concerns are questions of economic distribution of costs. Although farmers spend about $4 billion/year for pesticides, little of the pollution costs that result are borne by them or the pesticide chemical companies. Rather, most of the costs are borne off-site by public illnesses and environmental destruction. Standards of social justice suggest that a more equitable allocation of responsibility is desirable.

These ethical issues do not have easy answers. Strong arguments can be made to support pesticide use based on its definite social and economic benefits. However, evidence of these benefits should not cover up the public health and environmental problems. One goal should be to maximize the benefits while at the same time minimizing the health, environmental, and social costs. A recent investigation pointed out that U.S. pesticide use could be reduced by one-half without any reduction in crop yields or cosmetic standards and would increase food costs less than 0.6% (Pimentel et al., 1991). Judicious use of pesticides could reduce the environmental and social costs, while benefiting farmers economically in the short-term and supporting sustainability of agriculture in the long-term. That pesticide use be discontinued is not suggested, but that current pesticide policies be reevaluated to determine safer ways to employ them in pest control is suggested.

The major environmental and public health problems associated with pesticides are in large measure responsible for the loss of public confidence in state and federal regulatory agencies as well as in institutions that conduct agricultural research. A recent survey by Sachs et al. (1987) confirmed that confidence in the ability of the U.S. government to regulate pesticides declined from 98% in 1965 to only 46% in 1985. Another survey conducted by the FDA (1989) found that 97% of the public were genuinely concerned that pesticides contaminate their food.

Public concern over pesticide pollution confirms a national trend in the country toward environmental values. Media emphasis on the issues and problems caused by pesticides has contributed to a heightened public awareness of ecological concerns. This awareness is encouraging research in sustainable agriculture and nonchemical pest management.

Granted, substituting nonchemical pest controls in U.S. agriculture would be a major undertaking and would not be without its costs. The direct and indirect benefits and costs of implementation of a policy to reduce pesticide use should be researched in detail. Ideally, such a program would both enhance social equitability and promote public understanding of how to better protect human health and the environment while abundant, safe food is supplied. Clearly, it is essential that the environmental and social costs and benefits of pesticide use be considered when future pest control programs are being developed and evaluated. Such costs and benefits should be given ethical and moral scrutiny before policies are implemented, so that sound, sustainable pest management practices are available to benefit farmers, society, and the environment.

Conclusion

An investment of about $4 billion dollars in pesticide control saves approximately $16 billion in U.S. crops, based on direct costs and benefits (Pimentel et al., 1991). However, the indirect costs of pesticide use to the environment and public health need to be balanced against these benefits. Based on the available data, the environmental and social costs of pesticide use total approximately $8 billion each year (Table 3.6). Users of pesticides in agriculture pay directly for only about $3 billion of this cost, which includes problems arising from pesticide resistance and destruction of natural enemies. Society eventually pays this $3 billion plus the remaining $5 billion in environmental and public health costs (Table 3.6).

Our assessment of the environmental and health problems associated with pesticides faced problems of scarce data that made this assessment of the complex pesticide situation incomplete. For example, what is an acceptable monetary value for a human life lost or a cancer illness due to pesticides? Also, equally difficult is placing a monetary value on killed wild birds and other wildlife; on the death of invertebrates, or microbes; or on the price of contaminated food and groundwater.

In addition to the costs that cannot be accurately measured, there are additional costs that have not been included in the $8 billion/year figure. A complete accounting of the indirect costs should include accidental poisonings like the "aldicarb/watermelon" crisis; domestic animal poisonings; unrecorded losses of fish, wildlife, crops, trees and other plants; losses resulting from the destruction of soil invertebrates, microflora, and microfauna; true costs of human pesticide poisonings; water and soil pollution; and human health effects like cancer and sterility. If the full environmental and social costs could be measured as a whole, the total cost would be significantly greater than the estimate of $8 billion/year. Such a complete long-term cost/benefit analysis of pesticide use would reduce the perceived profitability of pesticides.

Human pesticide poisonings, reduced natural enemy populations, increased pesticide resistance, and honeybee poisonings account for a substantial portion of the calculated environmental and social costs of pesticide use in the United States. Fortunately some losses of natural enemies and pesticide resistance problems are being alleviated through carefully planned use of integrated pest management (IPM) practices. But a great deal remains to be done to reduce these important environmental costs (Pimentel et al., 1991).

This investigation not only underscores the serious nature of the environmental and social costs of pesticides but emphasizes the great need for more detailed investigation of the environmental and economic impacts of pesticides. Pesticides are and will continue to be a valuable pest control tool. Meanwhile, with more accurate, realistic cost/benefit analyses, we will be able to work to minimize the risks and develop and increase the use of nonchemical pest controls to maximize the benefits of pest control strategies for all society.

Acknowledgments

We thank the following people for reading an earlier draft of this manuscript, for their many helpful suggestions, and, in some cases, for providing additional information: A. Blair, National Institutes of Health; J. Blondell, U.S. Environmental Protection Agency; S. A. Briggs, Rachel Carson Council; L. E. Ehler, University of California (Davis); E. L. Flickinger, U.S. Fish and Wildlife Service; T. Frisch, NYCAP; E. L. Gunderson, U.S. Food and Drug Administration; R. G. Hartzler, Iowa State University; H. Lehman and G. A. Surgeoner, University of Guelph; P. Mineau, Environment Canada; I. N. Oka, Bogor Food and Agriculture Institute, Indonesia; C. Osteen, U.S. Department of Agriculture; O. Pettersson, The Swedish University of Agricultural Sciences; D. Rosen, Hebrew University of Jerusalem; J. Q. Rowley, Oxfam, United Kingdom; P. A. Thomson and J. J. Jenkins, Oregon State University; C. Walters, Acres, U.S.A.; G. W. Ware, University of Arizona; and D. H. Beerman, T. Brown, E. L. Madsen, R. Roush, C. R. Smith, Cornell University.

References

Adkisson, P.L. 1972. The integrated control of the insect pests of cotton. Tall Timbers Conf. Ecol. Control Habitat Mgmt. **4**:175–188.

Adkisson, P.L. 1977. Alternatives to the unilateral use of insecticides for insect pest control in certain field crops. Edited by L.F. Seatz. Symposium on Ecology and Agricultural Production. Knoxville: Univ. of Tennessee, pp. 129–144.

AFS. 1982. Monetary values of freshwater fish and fish-kill counting guidelines. Bethesda, MD: Amer. Fisheries Soc. Special Pub. No. 13.

Akesson, N.B., and W.E. Yates. 1984. Physical parameters affecting aircraft spray application. Edited by W.Y. Garner, and J. Harvey. Chemical and Biological Controls in Forestry. Washington, D.C.: Amer. Chem. Soc. Ser. **238**, pp. 95–111.

Akins, M.B., L.S. Jeffery, J.R. Overton, and T.H. Morgan. 1976. Soybean response to preemergence herbicides. Proc. S. Weed Sci. Soc. **29**:50.

Altman, J. 1985. Impact of herbicides on plant diseases. Edited by C.A. Parker, A.D. Rovia, K.J. Moore, and P.T.W. Wong. Ecology and Management of Soilborne Plant Pathogens. St. Paul: Amer. Phytopathological Soc., pp. 227–231.

Archibald, S.O. 1984. A Dynamic Analysis of Production Externalities: Pesticide Resistance in California. Davis, CA: Ph.D. Thesis. University of California (Davis).

Atlas, R.M., and R. Bartha. 1987. Microbial Ecology: Fundamentals and Applications. 2nd ed. Menlo Park, CA: Benjamin Cummings Co.

Balcomb, R. 1986. Songbird carcasses disappear rapidly from agricultural fields. Auk **103**:817–821.

Barnes, C.J., T.L. Lavy, and J.D. Mattice. 1987. Exposure of non-applicator personnel and adjacent areas to aerially applied propanil. Bul. Environ. Contam. and Tox. **39**:126–133.

Barometer. 1991. Too Much Beer Kills Thousands. Oregon State University Barometer, May 14, 1991.

Barton, J., and F. Oehme. 1981. Incidence and characteristics of animal poisonings seen at Kansas State University from 1975 to 1980. Vet. and Human Tox. **23**:101–102.

Beasley, V.R., and H. Trammel. 1989. Insecticide. Edited by R.W. Kirk. Current Veterinary Therapy: Small Animal Practice. Philadelphia: W.B. Saunders, pp. 97–107.

Bednarz, J.C., D. Klem, L.J. Goodrich, and S.E. Senner. 1990. Migration counts of raptors at Hawk Mountain, Pennsylvania, as indicators of population trends, 1934–1986. Auk **107**:96–109.

Blair, A., O. Axelson, C. Franklin, O.E. Paynter, N. Pearce, D. Stevenson, J.E. Trosko, H. Vaubui, G. Williams, J. Woods, and S.H. Zahm. 1990. Carcinogenic effects of pesticides. Edited by C.F. Wilkinson and S.R. Baker. The Effect of Pesticides on Human Health. Princeton, NJ: Princeton Scientific Pub. Co., Inc., pp. 201–260.

Blair, A., H. Malker, K.P. Cantor, L. Burmeister, and K. Wirklund. 1985. Cancer among farmers: a review. Scand. Jour. Work. Environ. Health **11**:397–407.

Blair, A., and H. Zahm. 1990. Methodologic issues in exposure assessment for case-control studies of cancer and herbicides. Amer. J. of Industrial Med. **18**:285–293.

Blew, J.H. 1990. Breeding Bird Census. 92 Conventional Cash Crop Farm. Jour. Field Ornithology. 61 (Suppl.) **1990**:80–81.

Brock, T., and M. Madigan. 1988. Biology of Microorganisms. London: Prentice Hall.

Brown, L.M., A. Blair, R. Gibson, G.D. Everett, K.P. Cantor, L.M. Schuman, L.F. Burmeister, S.F. Van Lier, and F. Dick. 1990. Pesticide exposures and other agricultural risk factors for leukemia among men in Iowa and Minnesota. Cancer Res. **50**:6585–6591.

Bull, D. 1982. A Growing Problem: Pesticides and the Third World Poor. Oxford: Oxfam.

Carrasco-Tauber, C. 1989. Pesticide Productivity Revisited. Amherst: M.S. Thesis. University of Massachusetts.

CEQ. 1980. The Global 2000 Report to the President. Washington, D.C.: Council on Environmental Quality and the U.S. Department of State.

Chang, W.L. 1965. Comparative study of weed control methods in rice. Jour. Taiwan Agr. Res. **14**:1–13.

Clark, R.B. 1989. Marine Pollution. Oxford: Clarendon Press.

Clark, R.M. 1979. Water supply regionalization: a critical evaluation. Proc. Am. Soc. Civil Eng. **105**:279–294.

Colvin, B.M. 1987. Pesticide uses and animal toxicoses. Vet. and Human Tox. 29 (suppl. 2):15 pp.

Committee, House of Commons Agriculture. 1987. The Effects of Pesticides on Human Health. London: Report and Proceedings of the Committee. 2nd Special Report. Vol. I. Her Majesty's Stationery Office.

Croft, B.A. 1990. Arthropod Biological Control Agents and Pesticides. New York: Wiley.

D'Anieri, P., D.M. Leslie, and M.L. McCormack. 1987. Small mammals in glyphosphate-treated clearcuts in Northern Maine. Canadian Field Nat. **101**:547–550.

DeBach, P. 1964. Biological Control of Insect Pests and Weeds. New York: Rheinhold.

Dennehy, T.J., J.P. Nyrop, R.T. Roush, J.P. Sanderson, and J.G. Scott. 1987. Managing pest resistance to pesticides: A challenge to New York's agriculture. New York's Food and Life Sciences Quarterly **17**:4, 13–17.

Devens, B.H., M.H. Grayson, T. Imamura, and K.E. Rodgers. 1985. O,O,S-trimethyl phosphorothioate effects on immunocompetence. Pestic. Biochem. Physiol. **24**:251–259.

Dobbins, J. 1986. Resources Damage Assessment of the T/V Puerto Rican Oil Spill Incident. Washington, D.C.: James Dobbins Associates, Inc., Report to NOAA, Sanctuary Program Division.

Dogheim, S.M., E.N. Nasr, M.M. Almaz, and M.M. El-Tohamy. 1990. Pesticide residues in milk and fish samples collected from two Egyptian Governorato. Jour. Assoc. Off. Anal. Chem. **73**:19–21.

Eckert, J.W. 1988. Historical development of fungicide resistance in plant pathogens. Edited by C.J. Delp. Fungicide Resistance in North America. St. Paul: APS Press, pp. 1–3.

Ecobichon, D.J., J.E. Davies, J. Doull, M. Ehrich, R. Joy, D. McMillan, R. MacPhail, L.W. Reiter, W. Slikker, and H. Tilson. 1990. Neurotoxic effects of pesticides. Edited by C.F. Wilkinson and S.R. Baker. The Effect of Pesticides on Human Health. Princeton, NJ: Princeton Scientific Pub. Co., Inc., pp. 131–199.

Edwards, C., and J. Lofty. 1977. Biology of Earthworms. London: Chapman and Hall.

Edwards, C.A., and J.R. Lofty. 1982. Nitrogenous fertilizers and earthworm populations in agricultural soils. Soil Biol. Biochem. **14**:515–521.

Elliot, B.R., J.M. Lumb, T.G. Reeves, and T.E. Telford. 1975. Yield losses in weed-free wheat and barley due to post-emergence herbicides. Weed Res. **15**:107–111.

Elliot, J.E., R.J. Norstrom, and J.A. Keith. 1988. Organochlorines and eggshell thinning in Northern Gannets (Sula bassanus) from Eastern Canada 1968–1984. Environ. Poll. **52**:81–102.

EPA. 1977. Minutes of Administrator's Pesticide Policy Advisory Committee. March. Washington, D.C.: U.S. Environmental Protection Agency.

EPA. 1989. Carbofuran: A Special Review Technical Support Document. Washington, D.C.: U.S. Environmental Protection Agency, Office of Pesticides and Toxic Substances.

EPA. 1990a. National Pesticide Survey—Summary. Washington, D.C.: U.S. Environmental Protection Agency.

EPA. 1990b. Fish Kills Caused by Pollution. 1977–1987. Washington, D.C.: Draft Report of U.S. Environmental Protection Agency. Office of Water Regulations and Standards.

FAA. 1988. Census of Civil Aircraft. Washington, D.C.: U.S. Federal Aviation Administration.

FAO. 1986. Production Yearbook. Rome: Food and Agriculture Organization. United Nations. Vol. 40.

FAO. 1988. Integrated Pest Management in Rice in Indonesia. Jakarta: Food and Agriculture Organization. United Nations. May.

FDA. 1989. Food and Drug Administration Pesticide Program Residues in Foods—1988. Jour. Assoc. Off. Anal. Chem. **72**:133A–152A.

FDA. 1990. Food and Drug Administration Pesticide Program Residues in Foods—1989. Jour. Assoc. Off. Anal. Chem. **73**:127A–146A.

Ferro, D.N. 1987. Insect pest outbreaks in agroecosystems. Edited by P. Barbosa and J.C. Schultz. Insect Outbreaks. San Diego: Academic Press, pp. 195–215.

Fiore, M.C., H.A. Anderson, R. Hong, R. Golubjatnikov, J.E. Seiser, D. Nordstrom, L. Hanrahan, and D. Belluck. 1986. Chronic exposure to aldicarb-contaminated groundwater and human immune function. Environ. Res. **41**:633–645.

Fisher, A., L.G. Chestnut, and D.M. Violette. 1989. The value of reducing risks of death: A note on new evidence. Jour. Policy Anal. and Mgt. **8**:88–100.

Flickinger, E.L., K.A. King, W.F. Stout, and M.M. Mohn. 1980. Wildlife hazards from furadan 3G applications to rice in Texas. Jour. Wildlife Mgt. **44**:190–197.

Flickinger, E.L., G. Juenger, T.J. Roffe, M.R. Smith, and R.J. Irwin. 1991. Poisoning of Canada geese in Texas by parathion sprayed for control of Russian wheat aphid. Jour. Wildlife Diseases **27**:265–268.

Foote, R.H., E.C. Schermerhorn, and M.E. Simkin. 1986. Measurement of semen quality, fertility, and reproductive hormones to assess dibromochloropropane (DBCP) effects in live rabbits. Fund. and Appl. Tox. **6**:628–637.

Gaafar, S.M., W.E. Howard, and R. Marsh. 1985. World Animal Science B: Parasites, Pests, and Predators. Amsterdam: Elsevier.

GAO. 1986. Pesticides: EPA's Formidable Task to Assess and Regulate Their Risks. Washington, D.C.: General Accounting Office.

Georghiou, G.P. 1986. The magnitude of the resistance problem. Pesticide Resistance, Strategies and Tactics for Management. Washington, D.C.: National Academy of Sciences, pp. 18–41.

Georghiou, G.P. 1990. Overview of insecticide resistance. Edited by M.B. Green H.M. LeBaron, and W.K. Moberg. Managing Resistance to Agrochemicals: From Fundamental Research to Practical Strategies. Washington, D.C.: Amer. Chem. Soc. pp. 18–41.

Gonzalez, L.M., and F. Hiraldo. 1988. Organochlorines and heavy metal contamination in the eggs of the Spanish Imperial Eagle (Aquila [heliaca] adaberti) and accompanying changes in eggshell morphology and chemistry. Environ. Poll. **51**:241–258.

Grue, C.E., W.J. Fleming, D.G. Busby, and E.F. Hill. 1983. Assessing hazards of organophosphate pesticides to wildlife. Transactions of the North American Wildlife and Natural Resources Conference (48th). Washington, D.C.: Wildlife Management Institute.

Hairston, N.G., F.E. Smith, and L.B. Slobodkin. 1960. Community structure, population control and competition. Amer. Nat. **94**:421–425.

Hall, F.R. 1991. Pesticide application technology and integrated pest management (IPM). Edited by D. Pimentel. Handbook of Pest Management in Agriculture. Boca Raton, FL: CRC Press. pp. 135–167.

Hanner, D. 1984. Herbicide drift prompts state inquiry. Dallas (Texas) Morning News July 25.

Hardy, A.R. 1990. Estimating exposure: the identification of species at risk and routes of exposure. Edited by L. Somerville and C.H. Walker. Pesticide Effects on Terrestrial Wildlife. London: Taylor & Francis, 81–97.

Harper, C.R., and D. Zilberman. 1990. Pesticide regulation: problems in trading off economic benefits against health risks. Edited by D. Zilberman and J.B. Siebert. Economic Perspectives on Pesticide Use in California. Berkeley: University of California. October. pp. 181–208.

Henderson, J. 1968. Legal aspects of crop spraying. Univ. Ill. Agr. Exp. Sta. Circ. 99.

Hoar, S.K., A. Blair, F.F. Holmes, C.D. Boysen, R.J. Robel, R. Hoover, and J.F. Fraumeni. 1986. Agricultural herbicide use and risk of lymphoma and soft-tissue sarcoma. J. Amer. Med. Assoc. **256**:1141–1147.

Hole, F.D. 1981. Effects of animals on soil. Geoderma **25**:75–112.

Holmes, T., E. Nielsen, and L. Lee. 1988. Managing groundwater contamination in rural areas. Rural Development Perspectives. Washington, D.C.: U.S. Dept. of Agr., Econ. Res. Ser. vol. 5 (1). pp. 35–39.

Huffaker, C.B. 1977. Biological Control. New York: Plenum.

Huffaker, C.B., and C.E. Kennett. 1956. Experimental studies on predation: predation and cyclamen mite populations on strawberries in California. Hilgardia **26**:191–222.

Hundley, H.K., T. Cairns, M.A. Luke, and H.T. Masumoto. 1988. Pesticide residue findings by the Luke method in domestic and imported foods and animal feeds for fiscal years 1982–1986. Jour. Assoc. Off. Anal. Chem. **71**:875–877.

ICAITI. 1977. An Environmental and Economic Study of the Consequences of Pesticide Use in Central American Cotton Production. Guatemala City, Guatemala: Final Report, Central American Research Institute for Industry, United Nations Environment Programme.

Ignoffo, C.M., D.L. Hostetter, C. Garcia, and R.E. Pinnelle. 1975. Sensitivity of the entomopathogenic fungus *Nomuraea rileyi* to chemical pesticides used on soybeans. Environ. Ent. **4**:765–768.

Johnson, D.W., L.P. Kish, and G.E. Allen. 1976. Field evaluation of selected pesticides on the natural development of the entomopathogen, *Nomuraea rileyi*, on the velvetbean caterpillar in soybean. Environ. Ent. **5**:964–966.

Keefe, T.J., E.P. Savage, S. Munn, and H.W. Wheeler. 1990. Evaluation of epidemiological factors from two national studies of hospitalized pesticide poisonings, U.S.A. Washington, D.C.: Exposure Assessment Branch, Hazard Evaluation Division, Office of Pesticides and Toxic Substances, U.S. Environmental Protection Agency.

Keeling, J.W., R.W. Lloyd, and J.R. Abernathy. 1989. Rotational crop response to repeated applications of korflurazon. Weed Tech. **3**:122–125.

Kevan, D.K. McE. 1962. Soil Animals. New York: Philosophical Library.

Kizer, K. 1986. California's Fourth of July Food Poisoning Epidemic from Aldicarb-Contaminated Watermelons. Sacramento: State of California Department of Health Services.

Knuth, B.A. 1989. Implementing chemical contamination policies in sport-fisheries: Agency partnerships and constituency influence. Jour. Mgt. Sci. and Policy Anal. **6**:69–81.

Kovach, J., and A.M. Agnello. 1991. Apple pests—pest management system for insects. Edited by D. Pimentel. Handbook of Pest Management in Agriculture. Boca Raton, FL: CRC Press, pp. 107–116.

LaFarge, A.M. 1985. The persistence of resistance. Agrichemical Age **29**:10–12.

Larson, R.J., and R.M. Ventullo. 1983. Biodegradation potential of groundwater bacteria. Edited by D.M. Nielsen. Proceedings of the Third National Symposium on Aquifer Restoration and Groundwater Monitoring, May 25–27. Worthington, Ohio: National Water Well Association, pp. 402–409.

LeBaron, H.M., and J. McFarland. 1990. Herbicide Resistance in Weeds and Crops. Edited by M.B. Green, H.M. LeBaron, and W.K. Moberg. Managing Resistance From Fundamental Research to Practical Strategies. Washington, D.C.: Amer. Chem. Soc., pp. 336–352.

Legislative Counsel's Digest (State of California). 1986. Legislative Assembly Bill No. 2755.

Lijinsky, W. 1989. Statement by D. William Lijinsky before the Committee on Labor and Human Resources. Washington, D.C.: U.S. Senate, June 6, 1989.

Litovitz, T.L., B.F. Schmitz, and K.M. Bailey. 1990. 1989 Annual report of the American Association of Poison Control Centers National Data Collection System. Amer. Jour. Emergency Med. **8**:394–442.

Lotti, M. 1984. The delayed polyneuropathy caused by some organophosphorus esters. Edited by C.L. Galli, L. Manzo and P.S. Spencer. Recent Advances in Nervous Systems Toxicology. Proceedings of a NATO Advanced Study Institute on Toxicology of the Nervous System, Dec. 10–20, 1984, Belgirate, Italy, pp. 247–257.

Luster, M.I., J.A. Blank, and J.H. Dean. 1987. Molecular and cellular basis of chemically induced immunotoxicity. Ann. Rev. Pharmacol. Tox. **27**:23–49.

Mahadevan, V., and K.C. Chandy. 1959. Preliminary studies on the increase in cotton yield due to honeybee pollination. Madras Agr. Jour. **46**:23–26.

Mazariegos, F. 1985. The Use of Pesticides in the Cultivation of Cotton in Central America. Guatemala: United Nations Environment Programme, Industry and Environment. July/August/September.

McEwen, F.L., and G.R. Stephenson. 1979. The Use and Significance of Pesticides in the Environment. New York: John Wiley and Sons, Inc.

McGregor, S.E. 1976. Insect pollination of cultivated crop plants. Washington, D.C.: U.S. Dept. of Agr., Agr. Res. Ser., Agricultural Handbook No. 496.

McGregor, S.E., C. Rhyne, S. Worley, and F.E. Todd. 1955. The role of honeybees in cotton pollination. Agron. Jour. **47**:23–25.

ME & MNR. 1990. Guide to Eating Ontario Sport Fish. Ontario: Ministry of Environment and Ministry of Natural Resources.

Messing, R.H., B.A. Croft, and K. Currans. 1989. Assessing pesticide risk to arthropod natural enemies using expert system technology. Appl. Nat. Resour. Manage., Moscow, Idaho **3**:1–11.

Mineau, P. 1988. Avian mortality in agroecosystems: I. The case against granule insecticides in Canada. Edited by M.P. Greaves, B.D. Smith, and P.W. Greig-Smith. Field Methods for the Study of Environmental Effects of Pesticides. London: British Crop Protection Council (BPCP) Monograph 40, BPCP, Thornton Heath, pp. 3–12.

Mineau, P., and B.T. Collins. 1988. Avian mortality in agro-ecosystems: 2. Methods of detection. In Field Methods for the Study of Environmental Effects of Pesticides. Edited by M.P. Greaves, G. Smith and B.D. Smith. Cambridge, UK: Proceedings of a Symposium Organized by the British Crop Protection Council, held at Churchill College, pp. 13–27.

Mussen, E. 1990. California crop pollination. Gleanings in Bee Culture 118:646–647.

Nanjappa, H.V., and N.M. Hosmani. 1983. Residual effect of herbicides applied in transplanted fingermillet (Eleusive coracana Gaertu.) on the succeeding crops. Indian Jour. of Agron. **28**:42–45.

NAS. 1975. Pest Control: An Assessment of Present and Alternative Technologies. 4 volumes. Washington, D.C.: National Academy of Sciences.

NAS. 1985. Meat and Poultry Inspection. The Scientific Basis of the Nation's Program. Washington, D.C.: National Academy of Sciences.

NAS. 1987. Regulating Pesticides in Food. Washington, D.C.: National Academy of Sciences.

NAS. 1989. Alternative Agriculture. Washington, D.C.: National Academy of Sciences.

NAS. 1991. Malaria Prevention and Control. Washington, D.C.: National Academy of Sciences.

Nielsen, E.G., and L.K. Lee. 1987. The Magnitude and Costs of Groundwater Contamination from Agricultural Chemicals. A National Perspective. Washington, D.C.: U.S. Dept. of Agr., Econ. Res. Ser., Natural Resour. Econ. Div., ERS Staff Report, AGES870318.

NYT. 1988. Shell Loses Suit on Cleanup Cost. New York, NY: New York Times.

OACSJC. 1990. San Joaquin County Agricultural Report 1989. San Joaquin County, CA: San Joaquin County Department of Agriculture.

OACYC. 1990. Yolo County 1989 Agricultural Report. Yolo County, CA: Office of the Agricultural Commissioner, Yolo county.

Oka, I.N. 1991. Success and challenges of the Indonesian national integrated pest management program in the rice-based cropping system. Crop Protection **10**:163–165.

Oka, I.N., and D. Pimentel. 1976. Herbicide (2,4-D) increases insect and pathogen pests on corn. Science **193**:239–240.

Olson, L.J., B.J. Erickson, R.D. Hinsdill, J.A. Wyman, W.P. Porter, R.C. Bidgood, and E.V. Norheim. 1987. Aldicarb immunomodulation in mice: an inverse dose-response to parts per billion in drinking water. Arch. Environ. Contam. Tox. **16**:433–439.

Osteen, C.D., and P.I. Szmedra. 1989. Agricultural Pesticide Use Trends and Policy Issues. Washington, D.C.: U.S. Dept. of Agr., Econ. Res. Ser., Agr. Econ. Report No. 622.

OTA. 1979. Pest Management Strategies. Washington, D.C.: Office of Technology Assessment, Congress of the United States. 2 volumes.

OTA. 1988. Pesticide Residues in Food: Technologies for Detection. Washington, D.C.: Office of Technology Assessment. U.S. Congress.

Pimentel, D. 1971. Ecological Effects of Pesticides on Non-Target Species. Washington, D.C.: U.S. Govt. Printing Office.

Pimentel, D. 1988. Herbivore population feeding pressure on plant host: feedback evolution and host conservation. Oikos **53**:185–238.

Pimentel, D. 1990. Estimated annual world pesticide use. Edited by F. Foundation. Facts and Figures. New York: Ford Foundation.

Pimentel, D., D. Andow, R. Dyson-Hudson, D. Gallahan, S. Jacobson, M. Irish, S. Kroop, A. Moss, I. Schreiner, M. Shepard, T. Thompson, and B. Vinzant. 1980. Environmental and social costs of pesticides: a preliminary assessment. Oikos **34**:127–140.

Pimentel, D., J. Krummel, D. Gallahan, J. Hough, A. Merrill, I. Schreiner, P. Vittum, F. Koziol, E. Back, D. Yen, and S. Fiance. 1978. Benefits and costs of pesticide use in U.S. food production. BioScience **28**:778–784.

Pimentel, D., L. McLaughlin, A. Zepp, B. Lakitan, T. Kraus, P. Kleinman, F. Vancini, W.J. Roach, E. Graap, W.S. Keeton, and G. Selig. 1991. Environmental and economic impacts of reducing U.S. agricultural pesticide use. Edited by D. Pimentel. Handbook on Pest Management in Agriculture. Boca Raton, FL: CRC Press, pp. 679–718.

PN. 1990. Towards a reduction in pesticide use. Pesticide News (March).

Potashnik, G., and I. Yanai-Inbar. 1987. Dibromochloropropane (DBCP): an 8-year reevaluation of testicular function and reproductive performance. Fertility and Sterility **47**:317–323.

Potter, D.A., and S.K. Braman. 1991. Ecology and management of turfgrass insects. Ann. Rev. Entom. **36**:383–406.

Potts, G.R. 1986. The Partridge: Pesticides, Predation and Conservation. London: Collins.

PSAC. 1965. Restoring the Quality of Our Environment. Washington, D.C.: President's Science Advisory Committee. The White House.

Pye, V., and J. Kelley. 1984. The extent of groundwater contamination in the United States. Edited by NAS. Groundwater Contamination. Washington, D.C.: National Academy of Sciences.

Reddy, V.R., D.N. Baker, F.D. Whisler, and R.E. Fye. 1987. Application of GOSSYM to yield decline in cotton, I. Systems analysis of effects of herbicides on growth, development, and yield. Agron. Jour. **79**:42–47.

Reuben, R. 1989. Obstacles to malaria control in India—the human factor. Edited by W.W. Service. Demography and Vector-Borne Diseases. Boca Raton, FL: CRC Press. pp. 143–154.

Risebrough, R.W. 1986. Pesticides and bird populations. Edited by R.F. Johnston. Current Ornithology. New York: Plenum Press. pp. 397–427.

Robinson, W.E., R. Nowogrodzki, and R.A. Morse. 1989. The value of honey bees as pollinators of U.S. crops. Amer. Bee Jour. **129**:477–487.

Rogers, C.B. 1985. Fluometuron carryover and damage to subsequent crops. **45**:8,2375B: Dissertation Abstracts International.

Rovira, A.D., and H.J. McDonald. 1986. Effects of the herbicide chlorsulfuron on Rhizoctonia Bare Patch and Take-all of barley and wheat. Plant Disease **70**:879–882.

Sachs, C., D. Blair, and C. Richter. 1987. Consumer pesticide concerns: a 1965 and 1984 comparison. Jour. Consu. Aff. **21**:96–107.

Salem, M.H., Z. Abo-Elezz, G.A. Abd-Allah, G.A. Hassan, and N. Shakes. 1988. Effect of organophophorus (dimethoate) and pyrethroid (deltamethrin) pesticides on semen characteristics in rabbits. Jour. Environ. Sci. Health **B23**:279–290.

Satchel, J.E. 1967. Lumbricidae. Edited by A. Burges and F. Raw. Soil Biology. New York: Academic Press, pp. 259–322.

Savage, E.P., T.J. Keefe, L.W. Mounce, R.K. Heaton, A. Lewis, and P.J. Burcar. 1988. Chronic neurological seqelae of acute organophosphate pesticide poisoning. Arch. Environ. Health **43**:38–45.

Sebae, A.H. 1977. Incidents of local pesticide hazards and their toxicological interpretation. Alexandria, Egypt: Proceedings of UC/AID University of Alexandria Seminar Workshop in Pesticide Management.

Shaked, I., U.A. Sod-Moriah, J. Kaplanski, G. Potashnik, and O. Buckman. 1988. Reproductive performance of dibromochloropropane-treated female rats. Int. Jour. Fert. **33**:129–133.

Sharp, D.S., B. Eskenazi, R. Harrison, P. Callas, and A.H. Smith. 1986. Delayed health hazards of pesticide exposure. Ann. Rev. Public Health **7**:441–471.

Smith, G.J. 1987. Pesticide Use and Toxicology in Relation to Wildlife: Organophosphorus and Carbamate Compounds. Washington, D.C.: Resource Publication 170, U.S. Dept. of Interior, Fish and Wildlife Service.

Stanier, R., M. Doudoroff, and E. Adelberg. 1970. The Microbial World. London: Prentice Hall.

Stickel, W.H., L.F. Stickel, R.A. Dyrland, and D.L. Hughes. 1984. DDE in birds: Lethal residues and loss rates. Arch. Environ. Contam. and Tox. **13**:1–6.

Stone, W.B., and P.B. Gradoni. 1985. Wildlife mortality related to the use of the pesticide diazinon. Northeastern Environ. Sci. **4**:30–38.

Stringer, A., and C. Lyons. 1974. The effect of benomyl and thiophanate-methyl on earthworm populations in apple orchards. Pesticide Sci. **5**:189–196.

Tabashnik, B.E., and B.A. Croft. 1985. Evolution of pesticide resistance in apple pests and their natural enemies. Entomophaga **30**:37–49.

Taylor, R.B. 1986. State sues three farmers over pesticide use on watermelons. Los Angeles Times, I (CC)-3-4.

Thomas, P.T., and R.V. House. 1989. Pesticide-induced modulation of immune system. Edited by N.N. Ragsdale and R.E. Menzer. Carcinogenicity and Pesticides: Principles, Issues, and Relationships. Washington, D.C.: Amer. Chem. Soc., ACS Symp. Ser. 414, pp. 94–106.

Tinney, R.T. 1981. The oil drilling prohibitions at the Channel Islands and Pt. Reyes-Fallallon Islands National Marine Sanctuaries: some costs and benefits. Washington, D.C.: Report to Center for Environmental Educations.

Trumbel, J.T., W. Carson, H. Nakakihara, and V. Voth. 1988. Impact of pesticides for tomato fruitworm (Lepidoptera:Noctuidae) suppression on photosynthesis, yield, and nontarget arthropods in strawberries. Jour. Econ. Ent. **81**:608–614.

UNEP. 1979. The State of the Environment: Selected topics—1979. Nairobi: United Nations Environment Programme, Governing Council, Seventh Session.

USBC. 1990. Statistical Abstract of the United States. Washington, D.C.: U.S. Bureau of the Census, U.S. Dept. of Commerce.

USDA. 1989a. Agricultural Statistics. Washington, D.C.: U.S. Department of Agriculture.

USDA. 1989b. The Second RCA Appraisal. Soil, Water, and Related Resources on Non-federal Land in the United States. Analysis of Conditions and Trends. Washington, D.C.: U.S. Department of Agriculture.

USFWS. 1988. 1985 Survey of Fishing, Hunting, and Wildlife Associated Recreation. Washington, D.C.: U.S. Fish and Wildlife Service. U.S. Dept. of Interior.

USFWS. 1991. Federal and State Endangered Species Expenditures. Washington, D.C.: U.S. Fish and Wildlife Service.

van den Bosch, R., and P.S. Messenger. 1973. Biological Control. New York: Intext Educational.

van der Leeden, F., F.L. Troise, and D.K. Todd. 1990. The Water Encyclopedia (2nd ed.), Chelsea, MI: Lewis Pub.

Van Ravenswaay, E., and E. Smith. 1986. Food contamination: consumer reactions and producer losses. National Food Review (Spring):14–16.

von Rumker, R., and F. Horay. 1974. Farmers' Pesticide Use Decisions and Attitudes on Alternate Crop Protection Methods. Washington, D.C.: U.S. Environmental Protection Agency.

Walgenbach, F.E. 1979. Economic Damage Assessment of Flora and Fauna Resulting from Unlawful Environmental Degradation. Sacramento, CA: Manuscript, California Department of Fish and Game.

Ware, G.W. 1983. Reducing pesticide application drift-losses. Tucson: University of Arizona, College of Agriculture, Cooperative Extension.

Ware, G.W., W.P. Cahill, B.J. Estesen, W.C. Kronland, and N.A. Buck. 1975. Pesticide drift deposit efficiency from ground sprays on cotton. Jour. Econ. Entomol. **68**:549–550.

Ware, G.W., W.P. Cahill, P.D. Gerhardt, and J.W. Witt. 1970. Pesticide drift IV. On-target deposits from aerial application of insecticides. Journ. Econ. Ent. **63**:1982–1983.

Ware, G.W., B.J. Estesen, W.P. Cahill, P.D. Gerhardt, and K.R. Frost. 1969. Pesticide drift. I. High-clearance vs aerial application of sprays. Jour. Econ. Entomo. **62**:840–843.

White, L. 1967. The historical roots of our ecological crisis. Science **155**:1203–1207.

White, D.H., C.A. Mitchell, L.D. Wynn, E.L. Flickinger, and E.J. Kolbe. 1982. Organophosphate insecticide poisoning of Canada geese in the Texas Panhandle. Jour. Field Ornithology **53**:22–27.

WHO. 1990 March 28. World Health Organization Press Release. Tropical Diseases News **31**:3.

WHO/UNEP. 1989. Public Health Impact of Pesticides Used in Agriculture. Geneva: World Health Organization/United Nations Environment Programme.

Whorton, D., R.M. Krauss, S. Marshall, and T.H. Milby. 1977. Infertility in male pesticide workers. The Lancet Dec. 17, **1977**:1259–1261.

Wigle, D.T., R.M. Samenciw, K. Wilkins, D. Riedel, L. Ritter, H.I. Morrison, and Y. Mao. 1990. Mortality study of Canadian male farm operators: non-Hodgkin's lymphoma mortality and agricultural practices in Saskatchewan. J. Nat. Canc. Instit. **82**:575–582.

Wilkinson, C.F. 1990. Introduction and overview. Edited by C.F. Wilkinson and S.R. Baker. The Effect of Pesticides on Human Health. Princeton, NJ: Princeton Scientific Pub. Co., Inc., pp. 1–33.

Zahm, S.H., D.D. Weisenburger, P.A. Babbitt, R.C. Saal, J.B. Vaught, K.P. Cantor, and A. Blair. 1990. A case-control study of non-Hodgkin's lymphoma and the herbicide 2,4-dichlorophenoxyacetic acid (2,4-D) in Eastern Nebraska. Epidemiology **1**:349–356.

4

The Relationship Between "Cosmetic Standards" for Foods and Pesticide Use

David Pimentel, Colleen Kirby, and Anoop Shroff

Introduction

The American marketplace features nearly perfect fruits and vegetables. Gone are the apples with an occasional blemish, a slightly russetted orange, or fresh spinach with a leaf miner. Less apparent but present in fresh and processed fruits and vegetables are a few small insects and mites. This increase in the "cosmetic standards" of fruits and vegetables has resulted from the development of new pesticide technologies and the efforts of the Food and Drug Administration (FDA) and U.S. Department of Agriculture (USDA) to limit the levels of insects and mites in fruits and vegetables, and as a consequence of new standards established by wholesalers, processors, and retailers. Consumer preferences have probably influenced these changes.

The Food and Drug Administration sets defect action levels (DAL) for insects and mites allowed in fruits and vegetables and in products made from these foods. During the past 40 years, as the FDA has been lowering these tolerance levels, more pesticides have been used to insure that crop produce meet the more stringent defect levels (FDA, 1972a,b, 1974, 1989a; *Federal Register*, 1973). In addition, wholesalers, processors, and retailers have been increasing their "cosmetic standards" for various reasons, including perceived consumer demand. The results have been higher economic costs for pest control, widespread environmental and human health problems caused by pesticides, as well as higher contamination levels of insecticides and miticides in fruits and vegetables (Steinman, 1990). Clearly, the economic, public health, and environmental values behind these changes need to be reexamined.

In this study, the following factors are examined: the legal tolerance levels of nonharmful insects and mites allowed in foods; the health and nutritional aspects of consuming these insects and mites; related trends in pesticide use and crop loss; and fossil energy costs of producing pesticides. In addition, the environmental and health hazards associated with increased pesticide use are compared with the benefits of having fewer insects and mites in foods.

Governmental Regulation of Insects and Mites Found in Food

Because American consumers are strongly disposed to purchase produce which is not damaged by pests or does not show the presence of insects and mites in or on their produce, the FDA established *defect action levels* (DAL) to keep insects and mites in foods to a minimum (FDA, 1974). In addition to the visual prejudice against insects, there is the well-placed concern that some insects, like houseflies and cockroaches, may transmit disease organisms.

The dominant consideration in establishing the defect action levels DALs (FDA, 1972b) was to reduce insect and mite infestation to a reasonable level, based on the existing state of insect and mite control technology (provided that the insects and mites are not easily seen). As detailed in *Food Purity Perspectives* (Anon., 1974), FDA standards for small insects and mites in fruits, vegetables, and products made from them were established because the presence of insects and mites indicated that crops had insufficient insect and mite control, were improperly washed, were unsatisfactorily inspected, and/or contained small insects and mites harmful to human health.

FDA and USDA inspectors check a small sample of food lots during processing and before transport to market. If any lot is found to have an insect infestation above the DAL, the lot is seized and destroyed. During 1950, one of the peak years for quantities of food seized, for example, only about 0.2% of the total crop of spinach and broccoli was seized (FDA, 1944–66). At that time, neither food processors nor the FDA issued reports as to the actual level of insects and mites found in or on fresh or processed fruits and vegetables. The defect action levels for insects and mites present in broccoli, spinach, and other crops for 1949 and 1950 were listed only in restricted FDA administrative guidelines. The established DALs were not published by FDA until 1972, but have been available to the public since then (FDA, 1972a, 1989a).

Even under the DAL regulatory guidelines, a few insects and mites remain in or on produce. For instance, the DAL for apple butter is an "average of 5 whole insects or equivalents . . . per 100 grams" . . . "not counting mites, aphids, thrips, or scale insects" (FDA, 1989a).

The DAL for canned sweet corn is similar. It states that if "2 or more 3 mm or longer larvae, cast skins, larval or cast skin fragments of corn ear worm or corn borer and the aggregate length of such larvae, cast skins, larval or cast skin fragments exceeds 12 mm in 24 pounds" (11 kg) then the DAL is exceeded (FDA, 1989a). For shelled peanuts, the DALs are an average of 5% insect infested, while an "average of 30 insect fragments per 100 grams" is permitted in peanut butter (FDA, 1989a).

Tomatoes commonly are infested with insects, especially by fruit flies (*Drosophila*). Thus, some insects are permitted in the produce. For processed tomato paste and other sauces the DALs are an "average of 30 or more fly eggs per 100 grams; or 15 or more fly eggs and 1 or more maggots per 100 grams; or 2 or more maggots per 100 grams" (FDA, 1989a). Likewise, for processed spinach, the DAL is an "average of 50 or more aphids and/or thrips and/or mites per 100 grams" or "leaf miners of any size average 8 or more per 100 grams or leaf miners 3 mm or longer average 4 or more per 100 grams" (FDA, 1989a).

Indeed, thrips, aphids, and mites, all minute in size, are practically impossible to eliminate from most vegetables as well as fruits. Consider raspberries and blackberries, which consist of clusters of many individual fruits from which it is impossible entirely to exclude these tiny organisms. Recognizing that it is impossible to intensely spray and/or clean these berries without destroying the fruit, the DAL for such berries permits up to 4 larvae [insect] per 500 grams (not counting mites, aphids, thrips, or scale insects) (FDA, 1989a).

The DALs for other fruits and vegetables are similar to those listed above but are tailored to reflect the pests of particular crops. It is obvious that although the numbers of insects and their parts are severely limited, some will remain, generally unseen, and will be eaten.

Changes in the DALs and Pesticide Use

The DALs have become more rigorous over time according to a statement by FDA administrators published in the *Federal Register* (1973). The reduced DALs for broccoli and spinach, which were especially well documented, illustrate this. Between 1938 and 1973 the DALs for aphids, thrips, and/or mites in broccoli were 80 per 100 grams (R. Angelotti, Associate Director, Compliance, Bureau of Foods, FDA, PC [personal communication], January 19, 1976). Then in 1974 the level was reduced to 60 aphids/thrips/mites per 100 grams and continues in effect (FDA, 1974, 1989a; USDA, 1983).

During the 1930s the FDA's "confidential figure" for spinach was 110 aphids allowed per 100 grams (FDA, 1972b). This guideline was based on "information on market sample findings." Successful aphid control was achieved in "fresh spinach by immersion in a dilute pyrethrum [insecticide] solution to loosen the insects from the leaves, followed by a detergent wash" (FDA, 1972b). In the early 1940s pressure for stricter standards from FDA's district laboratories resulted in a reduction in the DALs to a level of 60 aphids allowed per 100 grams of spinach (FDA, 1972b). This DAL remained in effect until 1974, when it was further reduced to 50 aphids per 100 grams (FDA, 1974), or to less than half the 1930s guideline for aphids in spinach.

Over time the DALs for leaf miners in spinach have also been reduced. During the 1930s based on what was termed, "a guide to repulsiveness," 40 leaf miners were allowed per 100 grams of spinach (USDA, 1969; FDA, 1972b). The FDA reported that "numerous seizures for leaf miners in spinach were effected as early as 1938 based on findings which 'appear sufficiently repulsive to warrant consideration of regulatory action' " (FDA, 1972b; and the same was true for the USDA [memo of R. Angellotti, FDA, 1972]). After the severe "leaf miner outbreak in California in 1956," a lower level of 9 leaf miners per 100 grams of spinach was adopted (FDA, 1972a,b). The DAL remained at this level until 1974, when it was reduced further to eight leaf miners per 100 grams (FDA, 1974), a level five-fold less than that allowed in the 1930s. The eight-leaf-miners-allowed-per-100-grams level continues today for spinach (USDA, 1983; FDA, 1989a).

In an effort to meet the FDA and USDA DAL regulations and their increasing stringency, farmers have used increasing amounts of pesticides on their crops and also instituted other pest control measures. The FDA (1972b) reported that the 1956 pest outbreak in the spinach crop "stimulated research by the University of California (Davis) Department of Entomology to develop more effective field programs to control leaf miner damage in spinach. Control programs have apparently been effective since we [FDA] have had little or no regulatory action on this problem in recent years [through 1972]." Altered FDA DALs appear to have influenced insect control procedures and the amount of insecticide used in spinach production, and probably used on other crops as well. W. H. Lange, Jr. (University of California [Davis], PC, 1976) reported that the reduction of the leaf miner problem since 1956 was made possible because of several interrelated programs: a three-fold increase in the use of insecticides on spinach, from 1–2 to 3–6 treatments per season; a new bait-spray program for control of adult leaf miners; planting in spring and late fall, instead of in fall, when leaf miners are most severe; and growing fewer crops that act as alternate hosts for leaf miners in the spinach production area.

Is it realistic to aim for ever-more-stringent DALs, until no pests are allowed? The possibility of reducing the presence of apple maggots in apple sauce to zero has been discussed in New York State. But if this were accomplished, the amount of insecticide used in apple orchards would increase substantially and would undermine the current integrated pest management (IPM) program now operating in New York State (Tette and Koplinka-Loehr, 1989). This currently used IPM program is generally successful in controlling major apple pests, while keeping pesticide applications to a relatively low but effective level. Only with greatly augmented insecticide use could insect-free applesauce be produced.

If a *zero* tolerance for insects and mites in fresh and processed foods were established, many foods, like raspberries and strawberries, would be totally eliminated from the market because it is impossible to produce these products without any insects and mites present. Furthermore, as mentioned, the absolute elimination of insects and mites from other fruits and vegetables would require enormous amounts of insecticides and miticides. This would result in a "very real danger" of exposing the public and the environment to hazardous levels of pesticides (FDA, 1974). Even with high levels of pesticides, it is probably impossible to reach the goal of no insects and mites in fruits and vegetables, and as discussed here, this may be not only an unattainable goal but an unwarranted one as well.

Cosmetic Appearance

In addition to restricting the numbers of pests found in and on produce, the minor surface blemishes found on fruits and vegetables caused by pests, are a part of "cosmetic standards." The growing emphasis given to the "cosmetic appearance" of fruits and vegetables is alleged to reflect consumer preference (van den Bosch et al., 1975). Since 1945, food processors, wholesalers, and retailers, following the lead of the FDA and USDA, have placed increasing importance on improving the "cosmetic appearance" of fruits and vegetables. As mentioned, the achievement of producing almost "perfect" produce has been possible because of the increased availability and use of insecticides and miticides. As a result apples, oranges, tomatoes, cabbage, and other fruits and vegetables found in U.S. supermarkets today have little or no insect damage on their surfaces.

Clearly, cosmetic appearance of produce is one of the primary factors used by consumers in assessing the overall quality of the produce they buy. Certainly visually perfect produce is appealing. Unfortunately consumers are not provided with more substantive measures of quality such as nutritional values or pesticide residue levels. In considering produce to purchase, the consumer is left to make selections based solely on surface cosmetic appearance and, or course, price of the produce (EPA, 1990).

In general, the public has not been aware of the connection between cosmetic appearance and increased pesticide use. However, some recent evidence suggests that when consumers are made aware of the trade-offs, they will purchase produce that is not cosmetically perfect because it has less or no pesticide residues (Lynch, 1991).

Evidence suggests that distributors and wholesalers desire to propagate the idea that consumers will not tolerate any cosmetic damage on their produce. With fresh produce, contracts between growers and buyers (i.e., distributors and wholesalers) permit buyers to make subjective evaluations of produce based on cosmetic appearance. This enables buyers to reject produce when supply is excessive. Growers agree to such contracts because of the market power of buyers. That is, many small growers face monopolistic buyers who are dominant in the marketplace (EPA, 1990). Given such contractual agreements, growers feel assured of sales, and to achieve this, they have a strong incentive to produce cosmetically perfect produce and thus resort to heavy pesticide use (EPA, 1990).

Marketing order arrangements, currently present in the produce industry, also play a role in increasing pesticide use by growers. Although the original intent of marketing orders was to improve price stability and grower profitability, marketing orders have had the unintentional result of raising the cosmetic standards of produce. The establishment of voluntary grading standards by federal marketing orders has resulted in grading of produce (e.g., USDA Extra Fancy) that, over time, has evolved into mandatory industry requirements. Distributors and wholesalers supplying retail supermarket chains will only purchase the highest grade of produce (i.e., cosmetically perfect produce) from growers, especially during times of abundant supply. In this way marketing orders raise the cosmetic standards of produce and exclude cosmetically less-perfect but nutritious produce from entering the fresh market (EPA, 1990).

Moreover, retail supermarket chains, claiming to satisfy consumer preference for perfect cosmetic appearance, demand fresh produce with a specified maximum level of cosmetic damage. To meet the demands of the supermarket chains, distributors and wholesalers require growers to meet an even higher cosmetic standard. Growers, in turn, must set even higher standards to have a margin of safety. The result is that growers must apply more pesticide to achieve these marketplace demands (EPA, 1990).

The presence of surface blemishes on fruits and vegetables generally does not affect their nutritional content, storage life, or even their flavor (van den Bosch et al., 1975). For example, citrus rust mites cause "russetting" or "bronzing" on Florida oranges. Unless the mite population is extremely high (Allen, 1979; McCoy et al., 1988), the internal quality of the oranges, determined by the content of sugars and other nutrients, is virtually unaffected by the russetting (McCoy and Albrigo, 1975; Krummel and Hough, 1980). About 80% of the citrus acreage in Florida (Krummel and Hough, 1980) is sprayed for rust mites, usually about three times during the season, at a cost to the grower of about $200/ha. The rust mites cause little or no reduction in the yield of oranges, unless the mites become highly abundant (Lye et al., 1990).

One has to question why Florida oranges are excessively treated for rust mites when 95% of these oranges are ground up for juice. Apparently the reason orange growers continue to treat their oranges for rust mites is because juice processors require a russet-free external appearance (NAS, 1980; Ziegler and Wolfe, 1975). Further, it appears that processors use the presence of rust mite injury to downgrade the price of oranges purchased from growers when the supply of oranges is abundant. In this way the processors use the presence of rust mite injury to their economic advantage. This seems to be a common strategy among processors who process other fruits and vegetables (EPA, 1990).

In contrast to oranges, fresh grapefruit with russetting is classified as "golden" and sells for a higher price in the market than unblemished grapefruit that is classified as "bright" (Krummel and Hough, 1980). For example, it is reported that in the Chicago and Boston markets, "golden" grapefruit sold for more than $2/box higher than "bright" grapefruit. The reason for the price differential is that russetted fruits are reported to be sweeter than "bright" fruits. Some evidence supports this idea—mite russetting is reported to allow some moisture to escape the grapefruit (Krummel and Hough, 1980; McCoy et al., 1988). When this occurs the sugars and solids in the russetted grapefruit and oranges are concentrated. This suggests that education of the public is possible in order to obtain new, sound cosmetic standards.

Another example of how stringent cosmetic standards have influenced pesticide use concerns the control of citrus thrips on California oranges that "scar" the skin of the fruit. Scarred fruit receives a lower grade from wholesalers/distributors, and therefore, sells at a lower price in the marketplace (Flaherty et al., 1973; Tanigoshi et al., 1985). As with the mite blemishes, thrip blemishes do not affect the nutritional or eating quality of oranges as measured by percentage moisture and ratio of soluble solids to acid (van den Bosch et al., 1975). Nonetheless, citrus thrips are considered one of the most serious pests of oranges because of the scarring problem, and as a result large quantities of insecticides are applied for thrips control. Currently, control of thrips and other pests in California orange groves is estimated to average about $600/ha per year (Teague et al., 1988).

Similarly, on tomatoes grown for processing, about two-thirds of all insecticide applied is to control the tomato fruitworm (EPA, 1990), which is "essentially a cosmetic pest" because it damages only the tomato skin (van den Bosch et al., 1975; Walgenbach et al., 1989). Most processors allow no more than from 0.5% to 2% fruitworm damage to the surface of tomatoes by weight, while many accept only perfect fruit (van den Bosch et al., 1975; Zalom et al., 1986; Metcalf, 1986; Feenstra, 1988). Yet 90% of the processed tomatoes are peeled and then used for paste, sauce, catsup, juice, and puree—products with no skins (van den Bosch et al., 1975).

To date, consumers, processors, and regulators have not clearly understood that the nutritional quality of surface-scarred or -blemished fruit and vegetables is not inferior to the "perfect fruit or vegetable" (van den Bosch et al., 1975), except under conditions of excessive pest damage, when nutritional quality may be affected (McCoy et al., 1988; Gorham, 1981). They also seem to be unconcerned about the hazards of ingesting pesticide residues and/or that there is a direct correlation between the "perfect" produce and pesticide residues. Equipped with an understanding of this connection between blemish-free produce and pesticide-contaminated produce, perhaps more people would accept slightly blemished produce, which is less likely to contain insecticide and miticide residues than perfect produce. There is some evidence that suggests that changes are in progress. Recently several state farm bills and the federal farm bill defined what food can be certified as "organically" grown (Gates, 1990).

Consuming Insects—Health Effects

According to the FDA (1974), the DALs for insects and mites in produce were established to prevent any "hazard to health." This goal appears to have been met, because in recent reports no mention is made of health hazards related to presence of insects and mites found in foods (USDA, 1983; FDA, 1989a,c). The only exception to this would be for insects like houseflies and cockroaches, which could invade foods stored prior to processing (Gorham, 1989; 1991). Indeed, all herbivorous insects and mites that are found in and on harvested fruits and vegetables are harmless to humans (Pimentel et al., 1977; Phelps et al., 1975; Taylor, 1975; Defoliart, 1989; Gorham, 1991). Further evidence of their safety is demonstrated in countries throughout the world where insects are a part of the normal diet and contribute important nutrients to peoples' daily nutrition (Bodenheimer, 1951; Gorham, 1976; Dufour, 1987; Posey, 1987; Brickey and Gorham, 1989).

In many places, pest insects of crops also are important foods for humans. Defoliart (1989) suggests that harvesting insect pests for food could be part of pest management programs and thereby reduce the need for insecticides. Although some insects, like cockroaches, that invade produce during processing, may present a health hazard (Gorham, 1989, 1991), it is the harmless herbivores (pests of crops) that are the target of increased insecticide use designed to produce "perfect" fruits and vegetables. If the choice is between "perfect" produce with increased insecticide residues or less-than-"perfect" produce with the presence of a few insects and mites, then indeed it might be safer to tolerate a few insects and mites (Pimentel et al., 1977; Gorham, 1991).

The EPA regulates pesticide use. The USDA and FDA regulate the levels of insects and mites in processed foods (USDA, 1983; FDA, 1989a). Because the FDA regulations identify the type of insect/mite contaminant in foods it allows, it should be able to develop specific regulations for houseflies and cockroaches while enabling harmless insect and mite residues to be regulated less stringently (Gorham, 1991). There appears to be a lack of consideration concerning the trade-offs of pesticide use and insect and mite levels in foods both within and between the federal agencies.

Defoliart (1975), Taylor (1975), Finke et al. (1987, 1989), Nakagaki et al. (1987), and Gorham (1989) have assembled data on the nutritional values of several insects, which compare favorably with those of shrimp, lobster, and crawfish. The latter are also arthropods but are often considered food delicacies. Digestible protein content of the insects ranges from 40% to 65% (Defoliart, 1975; Taylor, 1975; Kok et al., 1988) compared to 75% to 84% for shrimp, lobster, and crawfish (USDA, 1986) and 30% to 75% for trimmed beef, lamb, pork, chicken, and fish (USDA, 1986).

Given the conclusion that most insects found on produce are probably not any more of a health hazard than beef or chicken, consumers must decide whether they are willing to tolerate the presence of a few insects rather than "perfect" produce that has required the use of high levels of pesticides.

Pesticide Usage

An estimated 434 million kg of pesticides are used in the United States annually (Fig. 4.1). These pesticides consist of 69% herbicides, 19% insecticides, and 12% fungicides (Pimentel et al., 1991). Of this, agriculture uses about 320 million kg of pesticides with about 3 kg applied per ha on 100 million ha of farm land (Pimentel and Levitan, 1986). The remaining pesticides are used by the public, industries, and government.

The application of pesticides for pest control is not evenly distributed among crops. For example, 93% of all row-crop hectarage, like corn, cotton, and soybeans, is treated with some type of pesticide (Pimentel and Levitan, 1986). In contrast, less than 10% of forage-crop hectarage is treated.

About 62 million kg of insecticides are applied to 5% of the U.S. agricultural land (Pimentel and Levitan, 1986). Vegetable crops that have from 85% to 100% of their acreage treated include potatoes, tomatoes, sweet corn, onions, and sweet potatoes (Pimentel et al., 1991). The heaviest

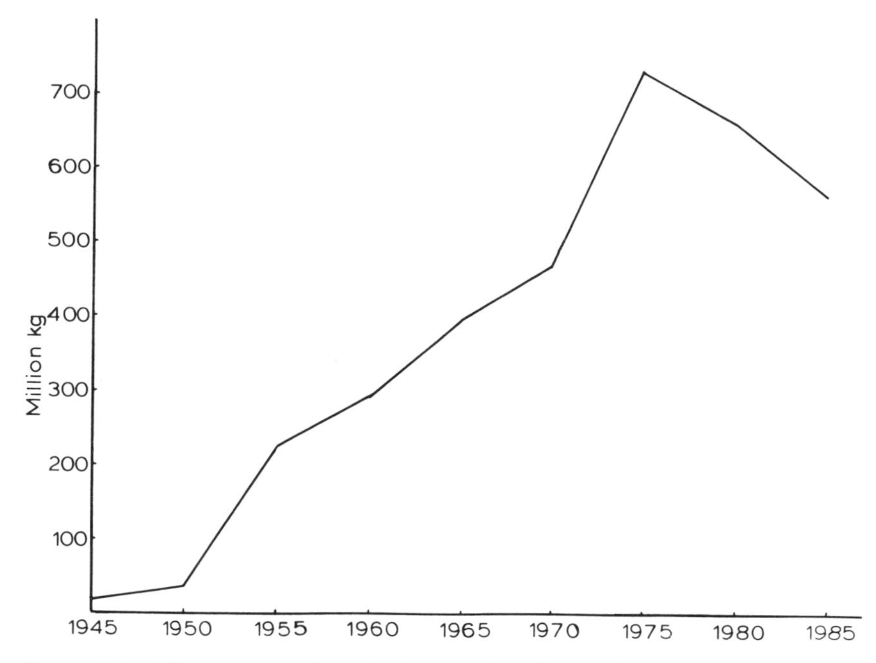

Figure 4.1. The amount of synthetic pesticides (insecticides, herbicides, and fungicides) produced in the United States (Arrington, 1956; USBC, 1970, 1990). About 90% is sold in the United States. The decline in total amount produced since 1975 is in large part due to the 10- to 100-fold increased toxicity and effectiveness of the newer pesticides.

insecticide-treated fruit crops include apples, cherries, peaches, pears, and grapefruit.

Fungicides are primarily applied to fruit and vegetable crops. Vegetable crops having 85% to 97% of their acreage fungicide-treated include lettuce, potatoes, tomatoes, and onions (Pimentel et al., 1991). The fruit crops having 85% or more of their acreage treated include apples, cherries, peaches, pears, grapes, oranges, and grapefruit (Pimentel et al., 1991).

Various experts estimate that from 60% to 80% of the pesticide applied to oranges and 40% to 60% of the pesticide applied to tomatoes to be processed are used only to improve cosmetic standards (CALPIRG, 1991). Overall, on fruits and vegetables we estimate that from 10% to 20% of insecticides and fungicides used are applied to comply to strict "cosmetic standards" now in force.

Pesticides and Crop Losses

Synthetic pesticide use in the United States has increased about 33-fold since 1945 (Fig. 4.1). Thc amounts of insecticides, herbicides, and fungicides used have changed with time due not only to changes in agricultural practices but also because cosmetic standards have been raised. Concurrently, the toxicity of new pesticides to pests and their biological effectiveness have increased at least 10-fold. For example, in 1945 DDT was applied at a rate of about 2 kg/ha; at present, similar effective insect control is achieved with pyrethroids and aldicarb applied at 0.1 kg/ha and 0.05 kg/ha, respectively.

In spite of the current use of pesticides plus some nonchemical controls, an estimated 37% of all crop production is lost annually to all pests (13% to insects and mites, 12% to plant pathogens, and 12% to weeds) (Pimentel et al., 1991). The share of crop yields lost to insects has nearly doubled (7% to 13%) during the last 40 years despite a more than 10-fold increase in the amount and toxicity of synthetic insecticides used (Arrington, 1956; USBC, 1970, 1990).

This dramatic rise in crop losses despite increased insecticide use can be partially explained by some of the major changes that have taken place in agricultural practices. These include the reduction in crop rotations; the increase in monocultures and reduced crop diversity (Pimentel, 1961; Pimentel et al., 1977); reduction in tillage with more crop residues left on the land; the planting of some crop varieties that are highly susceptible to insect and mite pests; the increased use of aircraft for pesticide application; the reduction in field sanitation (Pimentel, 1986); pesticides causing some destruction of natural enemies, thereby creating the need for additional pesticide treatments (van den Bosch and Messenger, 1973); the increase in the number of pests becoming resistant to pesticides (Georghiou, 1986); the culture of crops in climatic regions where insects are serious pests; the lowering of FDA tolerances for insects and mites in foods plus the enforcement of more stringent "cosmetic standards" by fruit and vegetable processors and retailers (Pimentel et al., 1977); and the use of pesticides that have altered the physiology of crop plants, making them more susceptible to insect attack (Oka and Pimentel, 1976).

The best example of how change in agricultural practices has led to greater crop losses, despite increased pesticide use, is illustrated with the culture of field corn. In 1945, most field corn was grown in rotation after soybeans, wheat, oats, and other noncorn host crops (USDA, 1954). During the early 1940s, little or no insecticide was applied to corn, and losses to insects averaged only 3.5% (USDA, 1954). Since then, a major portion of the corn has been cultured without rotation, insecticide use on corn has grown more than 1,000-fold, and losses due to insects have increased to 12%, or nearly fourfold over 1940 levels (Schwartz and Klassen, 1981). With no rotation, the corn rootworm population continued to increase on the stand of corn.

Health Hazards

In the United States, increased public health and environmental hazards have been associated with the increased use of insecticides and other pesticides on fruits and vegetables (NAS, 1987, 1989). About 67,000 humans are poisoned by pesticides annually (Litovitz et al., 1990). Yearly about 27 fatalities (Blondell, 1987; Litovitz et al., 1990) and less than 10,000 cases of cancer are associated with pesticide use (Pimentel et al., 1991). This sharply contrasts with no known case of human poisoning or death from ingesting insects and mites in or on foods.

Annual studies are conducted by the FDA to determine the kinds and amounts of pesticide residues in typical daily diets (FDA, 1990). The FDA reports that about 35% of the foods eaten contain detectable pesticide residues (FDA, 1990). From about 1% to 3% of the foods contain pesticide residues above the legal tolerance level (Hundley et al., 1988; FDA 1990).

However, a major concern about the "acceptable tolerance levels" remains because significant gaps exist in the data concerning tumor production in animals for a majority of the pesticides that are currently registered and used in agriculture (GAO, 1986). Thus, the absolute safety of the currently accepted levels of pesticide residues that occur in our foods has not been proven.

Under existing regulations foods are seized for exceeding FDA regulatory tolerances for pesticide residues as well as for exceeding the DALs for insects and mites. From 45% to 60% of this food is sold and consumed before it can be recalled (GAO, 1986; Mott and Snyder, 1987). Note, the OTA (1988) reports that of 743 pesticides and their breakdown products that can be found in foods, the best analytical chemical methods are capable of detecting less than one-third of these. The five major methods used by the FDA only detect 290 pesticides and their breakdown products or 40% of all that can be found in foods.

Economic Costs and Returns

Each year U.S. farmers use an estimated 320 million kg of pesticides for all crops at an approximate cost of $4.1 billion (Pimentel et al., 1991). This investment in pesticides saves farmers less than $16 billion in crops or about 10% of their total crop yield (Pimentel et al., 1978). This savings, however, does not take into account the indirect or environmental and social costs associated with pesticide use, which may have a total annual cost of nearly $4 billion (Pimentel et al., 1991).

The direct benefits of pesticides are about $4 per dollar invested in pesticides, as indicated above. A much higher return, however, could be realized through the implementation of nonchemical alternatives for pest control. For example, crop rotations, biological control, and breeding for host-plant resistance would return on the average about $30 per dollar invested in pest control (Pimentel, 1986).

Several recent studies suggest that it is technologically feasible to reduce pesticide use in the United States 35% to 50% without reducing yields (PSAC, 1965; OTA, 1979; NAS, 1989; Palladino, 1989; Pimentel et al., 1991). Such a policy in pest control was begun in 1985 in Denmark where an action plan was developed to reduce the use of pesticides by 50% before 1997 (B. Mogensen, National Environmental Research Institute, Denmark, PC, 1989). Subsequently, in 1988 Sweden approved a program to reduce pesticide use by 50% within 5 years (NBA, 1988). The Netherlands has also developed a program to reduce pesticide use by 50% within the next 10 years (A. Pronk, Wageningen University, Netherlands, PC, 1990). Similarly the Province of Ontario, Canada, in 1987 developed a program to reduce pesticide use by 50% during the following 15-year period (G. Surgeoner, University of Guelph, Canada, PC, 1990).

An assessment of the impact of a program to reduce pesticide use in the United States by 50% suggested that it would cause no reduction in crop yields and for some crops it would increase yields (Pimentel et al., 1991). The estimated coincident increase in food costs to the consumer would be only 0.6%. This marketplace cost increase did not take into account the positive environmental and social benefits that would accrue if pesticide use were reduced. If the environmental and social benefits are considered, the 0.6% increase in consumer costs would be more than offset by the environmental and social benefits associated with reduced pesticide use.

Ethical and Moral Issues

Recently, an FDA (1989b) survey found that 97% of the public preferred food without pesticides. Other reports also indicate that the public is becoming truly concerned about pesticide use and residues in their foods (Lecos, 1984; Steele, 1990). In another survey, from 50% to 66% of the people polled indicated that they would be willing to pay higher prices for pesticide-free food (Ott, 1990; Anon., 1991). Few doubt the desire of the public for foods untreated by pesticides or treated with minimum amounts of pesticides, but the unanswered question is, Will they purchase foods that have a few blemishes? This dilemma illustrates the different values held by individual people that make up the population.

But another related message from the public concerning pesticides is clearer than it has been in recent decades. The public has less confidence in government and less confidence that food is safe because of current levels of pesticides. Sachs et al. (1987) documented in a recent survey that from 1965 to 1985 the public became increasingly concerned about the safety of the food they purchased. In 1965, about 98% of those surveyed were confident that pesticide regulations were sound and were being effectively implemented (Sachs et al., 1987). In 1985, however, less than 46% felt that their food was safe and that pesticide regulations were being effectively implemented.

Several major incidents associated with pesticides may have contributed to the public's changing views. For instance, in the "watermelon poisonings" that took place in California in 1985, when farmers illegally treated watermelons with aldicarb, the more than 1,000 human poisonings clearly shook public confidence in government regulations (Taylor, 1986). Equally important was the more recent "Alar incident" (Hathaway, 1988, 1989). Although Alar is a plant growth regulator, it is regulated by EPA as a pesticide. Alar was used not only to keep fruit on the tree until harvest time, but also for cosmetic purposes to enhance the redness of apples. Many years prior to its final removal from use, questions had been raised concerning its safety by New York State Health officials and the EPA, which had surveyed several scientists concerning Alar's safety (D. Pimentel, unpublished, 1987). The results of the EPA survey were never published. However, alarm over the use of Alar continued to grow, particularly because apple juice and applesauce are consumed in large quantities by infants and young children. Public alarm escalated, but the EPA did not act. It was not until 1990 that Uniroyal Company, the producer of Alar, decided to withdraw Alar from the market. In hindsight, government action to restrict Alar should have been taken when enough data had accumulated to suggest that Alar was suspected to be a carcinogen. Because of the delay, farmers lost millions of dollars once the danger of Alar was aired in the press, and the public boycotted apples and apple products whether they were treated with Alar or not.

Without question the inept handling of Alar further eroded public confidence in the government's ability to regulate pesticide use. Furthermore, the public now appears to have the opinion that chemical industries determine pesticide policies in the United States. Surely the EPA, FDA, USDA, and other agencies of the federal government and their state counterparts have the obligation to represent and understand public concerns about pesticide use. Yet in fairness, it must be pointed out that individuals appear to have differing values concerning blemish-free produce vs. pesticide-free produce. These opinions send conflicting messages to government regulators.

All concerned, including farmers and chemical companies, should be heard and their viewpoints should receive attention. However, the general impression given by government agencies is that their primary concern is fear of lawsuits from chemical companies because of their regulatory decisions and that inadequate attention is given to public safety and health concerns.

The following example illustrates the problem. Insects, such as apple maggots, are a concern of processors because they fear consumer lawsuits or negative publicity. This is the view of processors but not that of the FDA (1989a). To achieve zero tolerance of apple maggots in applesauce requires the use of enormous amounts of insecticides. This would cause higher insecticide-residue contamination of fruit sold in the market and the environment, and also high economic pest control costs. In addition, substantially increasing insecticide use on apples would undermine various IPM programs that have been established in apple-growing states like New York. In regulating pesticides, government agencies should carefully consider the views and concerns of chemical companies, processors, farmers, and consumers. Equally vital is the consideration of all the environmental and public health aspects that are adversely affected by heavy pesticide use.

Hopefully state and federal governments will work together to develop educational programs that better inform the public about the relationships that exist between blemish-free fruits and vegetables and heavy pesticide use. Another consideration is the high economic costs associated with heavy pesticide use. For example, both farmers and consumers understand the fact that the overall quality of russetted oranges is not decreased. Further, of particular interest to farmers is the fact that russetted orange yields are not reduced and pesticide treatment costs could be reduced by $200/ha. With this situation both the farmer and consumer would benefit from less pesticide and lower production costs.

Government policies need to be carefully monitored to avoid the possibility of inadvertently encouraging pesticide use. For example, past price support policies have encouraged the use of pesticides in cotton, corn, and numerous other crops because various high-pesticide-use technologies were inadvertently legislated (NAS, 1989).

Conclusions

From the 1930s to 1976, the FDA and USDA gradually reduced the defect action levels (DALs) for insects and mites found in foods, even though there was no proven health hazard associated with the presence of small herbivorous arthropods in foods. Since 1976, both the FDA and USDA have maintained the established DALs. This is encouraging. However, food processors, wholesalers, and retailers seem to be placing even greater emphasis on blemish-free, perfect produce. Not only has this pressure caused substantial crop losses because large portions of food crops are now being classified as unsuitable for commercial sale, but it has also contributed to heavy pesticide usage by farmers who feel compelled to spray to reduce the incidence of insects and mites in foods to meet these "cosmetic appearance" standards (Lichtenberg and Zilberman, 1986).

The estimated 10–20% additional insecticide and/or miticide used on fruits and vegetables to meet the new "cosmetic appearance" standards (Van den Bosch et al., 1975; Pimentel et al., 1977) has caused substantial increases in pesticide use. This has resulted in a greater portion of the foods being contaminated with pesticide residues. Concurrently, the number of human pesticide poisonings and illnesses have increased, and there also has been further contamination of the environment. In addition, more fossil energy was used for spraying and producing pesticides and food costs for the consumer increased.

Further investigation of the widespread impact that "cosmetic appearance" standards are having on pesticide usage is needed in order to understand the possible trade-offs between insect and mite damage of food and quantities of pesticide residues in food. Although the presently enforced DALs are stringent, overall the FDA appears to be realistic in their regulations. However, for some crops the DALs could be safely raised. In addition, food processors and wholesalers need to reassess their market policies concerning "perfect" produce, especially as related to how the produce is to be used, e.g., whole tomatoes skinned and made into sauce. The retailers also must be realistic in assessing the trade-offs between the "perfect" produce, "pesticide" contamination, and the relative prices of the foods. Such changes certainly involve consumers who need to understand the pesticide consequences of buying only "perfect" produce.

In this chapter many of the factors related to maintaining stringent "cosmetic standards" for produce were analyzed. All sectors of society are involved in the relationships that have developed in response to government regulations. To find safe and equitable solutions will require knowledge and compromise. We hope this analysis will be helpful to all concerned as they endeavor to make wise, safe, and fair choices for the betterment of agriculture, the environment, public health, and society as a whole.

Acknowledgments

We thank the following people for reading an earlier draft of this manuscript, for their many helpful suggestions, and, in some cases, for providing additional information: S.E. Bomer, USDA; H. Lehman and G.A. Surgeoner, University of Guelph; O. Pettersson, Swedish University of Agricultural Sciences; J.R. Gorman, U.S. Food and Drug Administration; and E.R. Figueroa, Cornell University.

References

Allen, J.C. 1979. The effect of citrus rust mite damage on citrus tree growth. Jour. Econ. Entomol. **72**:195–201.

Anon. 1974. Amounts and kinds of filth in foods and the parallel methods for assessing filth and insanitation. Food Purity Perspectives **3**:19–20 (August).

Anon. 1991. Appearance of produce versus pesticide use. Chemecology 20 (**4**):11.

Arrington, L.G. 1956. World Survey of Pest Control Products. Washington, D.C.: U.S. Government Printing Office.

Blondell, J. 1987. Accidental pesticide related deaths in the United States, 1980 to 1985. Washington, D.C.: Report of U.S. Environmental Protection Agency. Office of Pesticides and Toxic Substances. Health Effects Div.

Bodenheimer, F.S. 1951. Insects as Human Food. The Hague: Dr. W. Junk.

Brickey, P.M., and J.R. Gorham. 1989. Preliminary comments on federal regulations pertaining to insects as food. Food Insects Newsletter **2**:1, 7.

CALPIRG. 1991. Who Chooses Your Food. Los Angeles, CA: CALPIRG.

Defoliart, G.R. 1975. Insects as a source of protein. Bull. Entomol. Soc. Amer. **21**:161–163.

Defoliart, G.R. 1989. The human use of insects as food and as animal feed. Bull. Entomol. Soc. Am. **35**:22–35.

Dufour, D.L. 1987. Insects as food: a case study from the northwest Amazon. Am. Anthropologist **89**:383–397.

EPA. 1990. An Overview of Food Cosmetic Standards & Agricultural Pesticide Use. Washington, D.C.: Office of Policy, Planning, & Evaluation, U.S. Environmental Protection Agency.

FDA. 1944–66. Notices of Judgement under the Federal Food, Drug, and Cosmetic Act. Washington, D.C.: Issued monthly by U.S. Dept. of Health Education and Welfare.

FDA. 1972a. Current Levels for Natural or Unavoidable Defects in Food for Human Use that Present no Health Hazard. Rockville, MD: Office of the Assistant Commissioner for Public Affairs. Food and Drug Administration, 31 October.

FDA. 1972b. Revision of Defect Action Levels for Spinach. Washington, D.C.: In-house memorandum, 14 Dec. U.S. Dept. of Health Education and Welfare.

FDA. 1974. Current Levels for Natural or Unavoidable Defects in Food for Human Use that Present no Health Hazard. Rockville, MD: Dept. of Health Education and Welfare. U.S. Public Health Service (fifth revision).

FDA. 1989a. Defect Action Levels. Washington, D.C.: Food and Drug Administration. U.S. Dept. Health, Education, and Welfare.

FDA. 1989b. Food and Drug Administration Pesticide Program Residues in Foods—1988. Jour. Assoc. Off. Anal. Chem. **72**:133A–152A.

FDA. 1989c. The Food Defect Action Levels. Washington, D.C.: Food and Drug Administration.

FDA. 1990. Food and Drug Administration Pesticide Program Residues in Foods—1989. Jour. Assoc. Off. Anal. Chem. **73**:127A–146A.

Federal Register. 1973. Human foods; current good manufacturing practice (sanitation) in manufacture, processing, packaging, or holding. Federal Register. **38** (3: Part 1): 854–855.

Feenstra, G.A. 1988. Who Chooses Your Food: A Study of the Effects of Cosmetic Standards on the Quality of Produce. Los Angeles, CA: CALPIRG.

Flaherty, D.L., J.E. Pehrson, and C.E. Kennett. 1973. Citrus pest management studies in Tolare County. Calif. Agric. **27**:3–7.

GAO. 1986. Pesticides: EPA's Formidable Task to Assess and Regulate Their Risks. Washington, D.C.: U.S. General Accounting Office.

Gates, J.P. 1990. Organic certification. Spec. Ref. Briefs Natl. Agric. Libr., U.S., Beltsville, MD: The Library, Jan. 1990. issue 90-04, 9 pp.

Georghiou, G.P. 1986. The magnitude of the resistance problem. Pesticide Resistance, Strategies and Tactics for Management. Washington, D.C.: National Academy of Sciences.

Gorham, J.R. 1976. Insects as food. Bull. Soc. Vector Ecol. **3**:11–16.

Gorham, J.R. 1981. Principles of food analysis for filth-decomposition and foreign matter. In Principles of Food Analysis. Edited by J.R. Gorham. Washington, D.C.: Food and Drug Administration Tech. Bull. 1, U.S. Dept. of Health and Human Services. Publ. #(FDA) 80–2128, pp. 63-124.

Gorham, J.R. 1989. Foodborne filth and human disease. Jour. Food Protection **52**:674–677.

Gorham, J.R. 1991. Filth and extraneous matter in food. In Encyclopedia of Food Science and Technology. Edited by Y.H. Hui. New York: Wiley, pp. 847–868.

Hathaway, J.S. 1988. Agriculture and public health: why we aren't protected from pesticides in food. New Engl. Fruit Meet. Proc. Annu. Meet. Mass. Fruit Grow. Assoc.

Hathaway, J.S. 1989. An environmentalist's perspective on the magnitude of the health risk from pesticide residues in foods. Food Drug Cosmet. Law Jour. **44**:659–670.

Hundley, H.K., T. Cairns, M.A. Luke, and H.T. Masumoto. 1988. Pesticide residue findings by the Luke method in domestic and imported foods and animal feeds for fiscal years 1982–1986. Jour. Assoc. Off. Anal. Chem. **71**:875–877.

Kok, R., K. Lomaliza, and U.S. Shivhare. 1988. The design and performance of an insect farm-chemical reactor for human food production. Can. Agric. Eng. **30**:307–318.

Krummel, J., and Hough, J. 1980. Pesticides and controversies: benefits versus costs. Edited by D. Pimentel and J. H. Perkins. Pest Control: Cultural and Environmental Aspects. Boulder, Colorado: Westview Press, pp. 159–180.

Lecos, C. 1984. Pesticides and food public worry No. 1. FDA Consumer **18**:12–15.

Lichtenberg, E., and D. Zilberman. 1986. Problems with pesticide regulation: health and environment versus food and fiber. Edited by T.T. Phipps and P.R. Crosson. Agriculture and the Environment: An Overview. Washington, D.C.: Resources for the Future, pp. 123–145.

Litovitz, T.L., B.F. Schmitz, and K.M. Bailey. 1990. 1989 Annual report of the American Association of Poison Control Centers National Data Collection System. Amer. Jour. Emergency Med. **8**:394–442.

Lye, B.H., C.W. McCoy, and J. Fojtik. 1990. Effect of copper on the residual efficacy of acaricides and population dynamics of citrus mite, Acari, Eriophyidae. Fla. Entomol. **73**:230–237.

Lynch, L. 1991. Consumers choose lower pesticide use over picture-perfect produce. Food Review January–March: 9–11.

McCoy, C.W., and L.G. Albrigo. 1975. Feeding injury to the orange caused by the citrus rust mite, Phylocoptruta oleivora (Prostigmata: Eriophyidae). Ann. Entomol. Soc. Am. **68**:289–297.

McCoy, C.W., L.G. Albrigo, and J.C. Allen. 1988. The biology of citrus rust mites and its effects on fruit quality. The Citrus Industry, September: 44–54.

Metcalf, R. 1986. The ecology of insecticides and the chemical control of insects. In Ecological Theory and Integrated Control of Insects. Edited by M. Kogan. New York: Wiley, pp. 251–297.

Mott, L., and K. Snyder. 1987. Pesticide Alert: A Guide to Pesticides in Fruit and Vegetables. San Francisco: Sierra Club Books.

Nakagaki, B.J., M.L. Sunde, and G.R. Defoliart. 1987. Protein quality of the house cricket, **Acheta domesticus**, when fed to broiler chicks. Poult. Sci. **66**:1367–1371.

NAS. 1987. Regulating Pesticides in Food. Washington, D.C.: National Academy of Sciences.

NAS. 1989. Alternative Agriculture. Washington, D.C: National Academy of Sciences.

NAS. 1980. Regulating Pesticides. Washington, D.C.: National Academy of Sciences.

NBA. 1988. Action Programme to Reduce the Risks to Health and the Environment in the use of Pesticides in Agriculture. Stockholm, Sweden: The National Board of Agriculture.

Oka, I.N., and D. Pimentel. 1976. Herbicide (2,4-D) increases insect and pathogen pests on corn. Science **193**:239–240.

OTA. 1979. Pest Management Strategies. Washington, D.C.: Office of Technology Assessment, Congress of the United States. 2 volumes.

OTA. 1988. Pesticide Residues in Food: Technologies for Detection. Washington, D.C.: Office of Technology Assessment. U.S. Congress.

Ott, S.L. 1990. Supermarket shopper's pesticide concerns and willingness to purchase certified pesticide residue-free fresh produce. Agribusiness **6** (6):593–602.

Palladino, P.S.A. 1989. Entomology and Ecology: The Ecology of Entomology. The "Insecticide Crisis" and the Entomological Research in the United States in the 1960s and 1970s: Political, Institutional, and Conceptual Dimensions. St. Paul, MN: Ph.D. Thesis. University of Minnesota.

Phelps, R.J., J.K. Struthers, and S.J.L. Moyo. 1975. Investigations into the nutritive value of Macrotermes falciger (Isoptera: Termitidae). Zool. Afr. **10**:123–132.

Pimentel, D. 1961. Species diversity and insect population outbreaks. Ann. Entomol. Soc. Amer. **54**:76–86.

Pimentel, D. 1986. Agroecology and economics. In Ecological Theory and Integrated Pest Management Practice. Edited by M. Kogan. New York: John Wiley and Sons, pp. 299–319.

Pimentel, D., J. Krummel, D. Gallahan, J. Hough, A. Merrill, I. Schreiner, P. Vittum, F. Koziol, E. Back, D. Yen, and S. Fiance. 1978. Benefits and costs of pesticide use in U.S. food production. BioScience **28**:778–784.

Pimentel, D., and L. Levitan. 1986. Pesticides: amounts applied and amounts reaching pests. BioScience **36**:86–91.

Pimentel, D., L. McLaughlin, A. Zepp, B. Lakitan, T. Kraus, P. Kleinman, F. Vancini, W.J. Roach, E. Graap, W.S. Keeton, and G. Selig. 1991. Environmental and economic impacts of reducing U.S. agricultural pesticide use. In Handbook of Pest Management in Agriculture. Edited by D. Pimentel. Boca Raton, FL: CRC Press, pp. 679–718.

Pimentel, D., E. Terhune, W. Dritschilo, D. Gallahan, N. Kinner, D. Nafus, R. Peterson, N. Zareh, J. Misiti, and O. Haber-Schaim. 1977. Pesticides, insects in foods, and cosmetic standards. BioScience **27**:178–185.

Posey, D.A. 1987. Ethoentomological survey of Brazilian indians. Entomol. Gen. **12**:191–202.

PSAC. 1965. Restoring the Quality of Our Environment. Washington, D.C.: President's Science Advisory Committee. The White House.

Sachs, C., D. Blair, and C. Richter. 1987. Consumer pesticide concerns: a 1965 and 1984 comparison. Jour. Consu. Aff. **21**: 96–107.

Schwartz, P.H., and W. Klassen. 1981. Estimate of losses caused by insects and mites to agricultural crops. In Handbook of Pest Management in Agriculture. Edited by D. Pimentel. Boca Raton, FL: CRC Press, pp. 15–77.

Steele, J.H. 1990. Pesticides and food safety. Stockholm: Proceedings, World Association of Veterinary Food Hygienists 10th (Jubilee) International Symposium.

Steinman, D. 1990. Diet for a Poisoned Planet. New York: Harmony Book.

Tanigoshi, L.K., J. Fargerlund, and J.Y. Nishio-Wong. 1985. Biological control of citrus thrips, Scirtothrips citri (Thysanoptera: Thripidae), in southern California citrus groves. Environ. Entomol. **14**:733–741.

Taylor, R.B. 1986. State sues three farmers over pesticide use on watermelons. Los Angeles Times, I (CC)-3–4.

Taylor, R.L. 1975. Butterflies in My Stomach. Santa Barbara, CA: Woodbridge Press.

Teague, P.W., G.S. Smith, D. Swietlik, and J.V. French. 1988. Survey of citrus producers in the Rio Grande Valley: results and analysis. Jour. Rio Grande Val. Hortic. Soc. **41**:97–109.

Tette, J.P., and C. Koplinka-Loehr. 1989. New York State Integrated Pest Management Program: 1988 Annual Report. Geneva, NY: IPM House, New York State Agric. Exp. Sta.

USBC. 1990. Statistical Abstract of the United States. Washington, D.C.: U.S. Bureau of the Census, U.S. Dept. of Commerce.

USBC. 1970. Statistical Abstracts of the United States. Washington, D.C.: Bureau of the Census, U.S. Dept. of Commerce.

USDA. 1969. Consumer Marketing Service. Washington, D.C.: U.S. Department of Agriculture, Fruit and Vegetable Division, Processed Products Standardization and Inspection Branch.

USDA. 1983. Inspection procedures for foreign material. Washington, D.C.: U.S. Department of Agriculture, Agr. Marketing Serv., Fruit and Vegetable Division, Processed Products Branch.

USDA. 1954. Losses in Agriculture. Washington, D.C.: U.S. Dept. of Agric., Agric. Res. Serv.

USDA. 1986. Nutritive Value of Foods. Washington, D.C.: U.S. Dept. of Agriculture, Human Nutrition Inf. Serv., Home and Garden Bull. No. 72.

van den Bosch, R., M. Brown, C. McGowan, A. Miller, M. Moran, D. Petzer, and J. Swartz. 1975. Investigation of the Effects of Food Standards on Pesticide Use. Washington, D.C.: U.S. Environmental Protection Agency. Draft Report.

van den Bosch, R., and P.S. Messenger. 1973. Biological Control. New York: Intext Educational.

Walgenbach, J.F., P.B. Shoemaker, and K.A. Sorensen. 1989. Timing pesticide applications for control of Heliothis zea Boddie Lepidoptera Noctuidae Alternaaria solani Ell. and G. Martin Sor. and Phytophthora infestans Mont. de Bary on tomatoes in western North Carolina USA. Jour. Agric Entomol. **6**:159–168.

Zalom, F.G., L.T. Wilson, and M.P. Hoffman. 1986. Impact of feeding by tomato fruitworm, Heliothis zea (Boddie) (Lepidoptera: Noctuidae), and the beet armyworm, Spodoptera exigua (Hubner) (Lepidoptera: Noctuidae), on processing tomato fruit quality. Jour. Econ. Entomol. **79**:822–826.

Ziegler, L.W., and H.S. Wolfe. 1975. Citrus Growing in Florida. Gainesville, FL: The University Presses of Florida.

5

Risk of Pesticide-Related Health Effects: An Epidemiologic Approach

Stanley Schuman

Introduction

Members of the public are concerned about developing cancer or other serious illnesses such as neurological disorders as a result of exposure to pesticides. Such public concerns may lead, through political processes, to decisions to restrict pesticide usage. Sometimes decisions are made which would not be made were the ignorance of the members of the public replaced by scientific knowledge. Sometimes government regulators are induced to make judgments about the potential danger of specific pesticides in the absence of scientific knowledge supporting those specific claims (Doll, 1991). In the past, decisions regarding prevention of disease have sometimes been made which, while not based on rigorous science, have met with some degree of success. Included among these are decisions referred to in the Bible regarding sanitary encampments and diet. Also included are the adoption in the 19th century of sanitary measures in London (Dubos, 1959). Nonetheless, it is reasonable to expect that decisions of this sort, based on intuition, will not lead to consequences which are best overall. In this paper I describe briefly an epidemiologic approach to determining whether pesticides are causing illness or injury. This approach is contrasted with other less-rigorous approaches and various studies are cited which illustrate these approaches. I conclude that epidemiological studies should be undertaken in an effort both to understand the etiology of agricultural illnesses and to determine such steps as are warranted to prevent such illnesses.

The Epidemiologic Approach

The epidemiologic approach is essentially a rigorous scientific approach. It consists of several distinct phases. The symptoms of the alleged health effect must be precisely specified. Criteria of illness need to be formulated well enough so that accurate counts of cases can be made. It is not sufficient to have loose definitions or to count cases of injury based on reports of unqualified individuals, as was the case in a report concerning injury due to improperly applied aldicarb to melons in California in 1985 (Anon, 1986). Loose criteria for recognizing alleged health effects lead to exaggerated claims of the number of injuries. Providing the alleged health effect is precisely specified the epidemiologist can carefully collect and analyze data. This should include the following: (1) measurement of exposure, (2) follow-up examinations, and (3) laboratory confirmations. Prior to affirming that some herbicide is the cause of a particular injury, meticulous care should be taken to determine that injured people indeed have ben exposed to the pesticide. Scientific data must be based on careful observations. It is not sufficient that exposure be inferred on the basis of hearsay or to extrapolate exposure from casually collected samples. Claims of exposure to pesticide should be verified by careful observations of qualified observers and not simply be based on judgments given by people who believe that they were exposed. Cases of injury should be confirmed by diagnoses by properly qualified personnel and by laboratory analyses. A rigorous diagnosis of an allergic skin reaction, for example, is one involving an appropriate skin test capable of detecting the suspected antigen. Finally, the conclusion that a particular injury is due to pesticide exposure is merely an opinion or conjecture unless the biological mechanism through which the substance causes injury is known. Further, the injury should be confirmed by appropriate laboratory analyses, which should be included in the case definition. Many examples of properly conducted epidemiological studies are available in the literature (Clifford and Niew, 1989; Bethea et al., 1988; Schuman and Dobson, 1985; Une et al., 1987).

The demand for rigor characteristic of a scientific approach to determining causes places logical restrictions on conclusions that can be scientifically obtained. It is unlikely that such criteria will be able to be met where the incidence of the exposure to a chemical is too low, providing few cases. It will be impossible to reach conclusions when the symptoms of alleged injury are too vaguely defined, when it is not possible to have properly qualified personnel examine suspected cases, when requisite laboratory facilities are not utilized, when there is insufficient opportunity to make careful determinations of exposure, or when there is insufficient understanding of causal mechanisms. One source of data on alleged pesticide poisonings, data obtained from poison control centers, should be handled with caution. In spite of the difficulties inherent in application of scientific methods to diagnosis of pesticide-related health effects, much is known of such effects and more can be learned.

Pesticide-Related Health Effects (PRHEs)

There are six major categories of human health effects of overexposure to pesticides, beginning with the simplest, acute toxicity, and ending with the more complex and least-understood categories of hypersensitivity and psychological or presumed immunologic effects. The categories include the following: (1) *Acute toxicity*, refers to acute exposure with direct damage to known organ systems or metabolic consequences, resulting in a range of symptoms from discomfort to hospitalization to death. (2) *Subacute, delayed toxicity* refers to delayed renal, hepatic, hematologic, or neurologic effects requiring weeks or months after the acute exposure to be detected clinically; an example is the puzzling neuropathy which occurs weeks to months after overexposure to organophosphates in a small minority of patients. Another example is delayed contact chemical dermatitis which takes 24–48 hours for the cellular immune system to react. (3) *Chronic, cumulative toxicity* is most difficult to establish for any chemical due to the long latency period between exposure(s) and detectable health effects. An example is accumulation of organochlorine residues in body fat or breast milk, which are easily measured but which are not specific for delayed health effects. A notable exception is a rare but well-documented chronic effect (chloracne), a specific skin manifestation (clogged sebaceous pores, etc.) of excessive exposure to dioxins. Another example of a chronic, cumulative effect would be carcinogenesis, but with the exception of arsenates and vinyl chloride (a once widely used propellant) there are no widely accepted human pesticide carcinogens (Council on Scientific Affairs, AMA, 1988). (4) *Reproductive effects* include sterility and birth defects. Examples include acute depression of spermatozoa in workers overexposed to DBCP (dibromochloropropane), a fumigant, which has caused fertility problems that fortunately have been reversed after workers have been removed from the contaminated environment. No birth defect syndrome has yet been detected or documented despite decades of exposure and surveillance for effects, hypothesized from animal experimentation using heavy loading models. (5) *Hypersensitivity* to low-dose exposure to one or more pesticides is difficult to diagnose, as are most allergies including chemical, food, plant, mold, and insect intolerances. Irritant (high-dose) effects need to be separated from dose-independent effects of micro-amounts of antigen (true allergy), which may be cell mediated or serum mediated in the human host. Skin, eye, and lungs are favorite targets and cardiovascular anaphylactic shock is a common result. (6) *Psychological or psychiatric conditioning* to time/place exposures to any chemical is well documented and difficult to diagnose and treat.

A seventh category is emerging from a gray zone of anecdotal medical histories, a theory of *multiple chemical sensitivity* which so far fails to meet sound medical criteria of a new syndrome. The theory developed decades ago by Dr. Theron Randolph presumed an "overload" of exposure to man-made chemicals which overwhelms the resistance of the host through presumed immunologic and/or neuroendocrinologic stress and adaptation syndromes. This theory has yet to develop beyond anecdotal histories, cures, testimonials, support groups, and medicolegal and governmental responses to public opinion. Occupational medicine specialists are confronted with workplace exposures with multiple presumed interacting or cumulative effects with unproven biologic mechanisms. Psychiatrists oppose loose criteria for this "20th century disease syndrome" on the grounds that patients can waste valuable time and effort with misguided treatments when treatable anxiety, depression, agoraphobia, and hypochondriasis are overlooked or denied.

Each of these categories (with the exception of the seventh) has a detailed body of medical literature based on human data: volunteer studies, occupational exposures, studies of employees under working conditions, emergency room experiences, and a growing toxicologic data base derived from human and animal systems including tissue culture and microbiologic assays. Often the data for a single pesticide hinges on a sizable amount of industrial and governmental animal testing with relatively few data on humans. The lack of documented adverse health effects data on a specific pesticide may indicate (after many years of widespread use and high-volume production) a relatively broad margin of safety for the pesticide. It may indicate a lack of interest in looking for adverse health effects, although this is less likely to be true for the United States in recent years. Some groups are trying to establish a system of reporting pesticide poisoning cases related to agriculture to a health agency in the same fashion mandatory reports of communicable disease are made. California is a model, but many limitations are built into the system which the surveillance team acknowledges. Industries are conducting follow-up studies of employees who were formulators of pesticides under conditions of heavy exposure. While the numbers of such workers are small, and precise levels of industrial time-weighted exposure are usually absent, broad categories of risk can be used to estimate delayed or chronic presumed health effects (Delzell et al., 1990).

Sources of Pesticide Health Data

Occupational epidemiology has been the most revealing source for estimating dose-response effects (Levine, 1991; Bell et al., 1990). The combination of toxicology, measurement of exposure, clinical examinations, and epidemiological follow-up of pesticide workers including mixers, load-ers, agricultural pilots, and others has provided a useful data base for most of the widely used chemical categories (organochlorines, organophosphates, carbamates and phenoxy compounds, pyrethroids, etc.). Statistically the number of workers who are at risk for assessing rare or delayed events such as reproductive or neoplastic PRHE (with statistical confidence) is small. Also the exposure of pesticide workers is rarely unifactorial (single chemical) or quantitative beyond years-of-exposure or type of crop worked. Other sources of data on PRHE are not as useful scientifically. Environmental exposure assessment, differential diagnosis, and laboratory confirmation are all essential to provide valid information.

Some sources which are cited actually fall short of acceptable scientific validity. This deserves careful attention.

Poison Control Center Data

Currently members of the American Association of Poison Control Centers (n = 70) routinely receive telephone calls for assistance with presumed or actual exposure to man-made or natural toxic substances. In the report for 1989, 1,581,540 "exposures" (telephone contacts) were documented (Litovitz et al., 1990).

This annual report published in the *American Journal of Emergency Medicine* provides a mass of information including 23 tables and 95 clinical case histories. Each exposure represents a phone call for help for a situation which may be trivial, moderate, or severe. Reassurance or first-aid advice from pharmacists-on-duty is a key function of the centers, as is their ability to network with other health providers. Reassurance is reflected in the report showing that 66% of the 1,581,540 calls were either "asymptomatic" or "symptomatic, unrelated to exposure" (Litovitz et al.: Table 10). In Table 12 of Litovitz et al. (1990), the medical outcomes of the calls include 26% of "no effect" and 23% of "unknown presumably non-toxic effect," totaling 782,364 human poison-exposure cases. The spectrum of "illness" ranges from 590 deaths (compared to 545 in 1988) to home treatment with ipecac (50,707 cases) and is cited in Table 15.

What do these fatality numbers mean in regard to acute pesticide poisoning? In 1989, 14–18 deaths are listed depending on which table you use (Table 16 or Table 21). Only two deaths were related to "accidental environmental exposure"—specifically, "methyl bromide inhalation." This gas is usually used as a fumigant for treating soil and commodities. In California, several "accidental" deaths occur from year to year, usually from breaking-and-entering posted, treated property (Mehler et al., 1990). The remainder of the 1989 pesticide fatalities were all *suicides*, with one or more toxic substances ingested by the victim. Incidentally, the poison center tally of fatalities will be somewhat less than a tally based on death certificates, which is compiled more slowly than this relatively current annual report to the medical journal.

To the nonepidemiologist searching for a useful number of acute pesticide poisonings in the United States in 1989, Table 20 of the report provides a tempting number: 60,045 exposures representing 3.8% of all phone calls received. What does the number mean? Perhaps it is no more significant in terms of human morbidity than the number of calls received for plant vegetation exposures: 100,704, representing 6.4% of all phone calls received. In the words of the authors: "the reader is cautioned to interpret this [number] as frequency of involvement in calls to poison centers with no correlation to severity of toxicity. Indeed, several of the plants pose little if any ingestion hazard." Should we equate "plant" calls with "pesticide" calls? No, but as the authors warn, we cannot equate telephone calls for assistance with estimates of documented PRHEs. Three-fourths of the 1,581,540 exposures were handled at the phone-caller's home; this suggests a low level of case severity and high level of public or parental concern and professional service (Table 11).

Anyone who requires useful estimates of PRHE in the United States must use data from poison control centers with great caution and recognize the limits of confidence in the annual numbers, as urged by the authors who routinely collect the information (Pimentel, 1991). For example, for the year 1989, Litovitz and colleagues list, in Table 17, 442 cases of nonmedicinal fungicides, 1,743 cases of herbicides, 12,147 cases of insecticides, and 4,186 cases of rodenticides as being seen or treated in a health care facility. This adds up to a total of 18,518 cases of acute pesticide exposure. This is in contrast to the total of 60,045 exposures to pesticides including rodenticides listed in Table 20 of the same report; the difference is between "exposures" and "visits to health care facility," a ratio of 60,045/18,518 = 3.24/1.0.

Perhaps some believe that *a nonspecific approach to PRHE* ("a clean-broom" approach to the misuse of pesticides) is required to eliminate nonspecified, assumed, presumed, and hypothetical PRHEs. This chapter will not defend misuse of pesticides. The ethics of pesticide use must be applied to the best available data on risk assessment and risk management including cost benefit to society. Objectivity must prevail over subjectivity. We all agree that much is known, some is unknown, and some will be learned in the future about pesticides, PRHEs, and integrated pest management in a hungry world which depends on nutrition not only for survival but for its quality of life. Our technology can provide biological mechanisms for our sanitary measures in the 20th century which were not available in the 17th century. Also, we should be able to monitor health effects in a prospective quantitative way.

The Art and Science of Epidemiology

Epidemiologists define cases and noncases of illness (infectious or non-infectious) by using biological and mathematical expressions of rates to arrive at logical risk estimates. "Nonepidemiology" is used in this chapter as a short term for examples of poor science: using faulty assumptions about the mechanisms of pesticide toxicology, extrapolating numbers with wide ranges of error, and failing to integrate biologic mechanisms with statistical associations.

For example, compare the number of annual cases hospitalized (for at least 1 day) by year due to agricultural pesticide exposure in two agricultural states (California and South Carolina) by using the same international Classification of Diseases and hospital record codes for pesticide poisoning (Figure 5.1). It is interesting to note that the total number of hospitalized pesticide poisonings are higher in California than South Carolina but the rate per 10,000 farm workers is more than sixfold lower in California than in South Carolina.

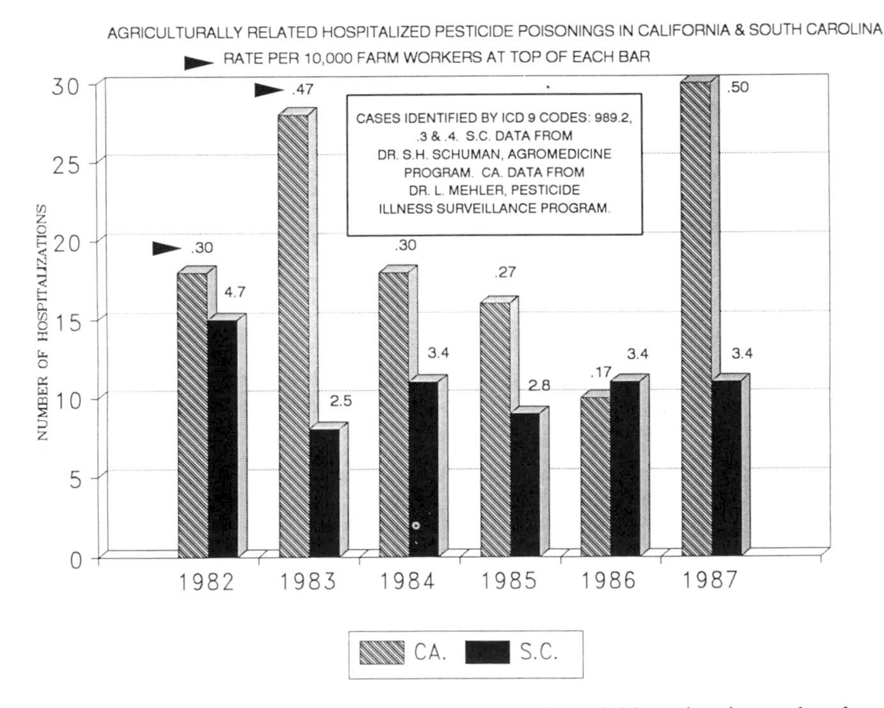

Figure 5.1. Comparison of hospitalized cases of pesticide poisonings related to agricultural exposure in California and South Carolina, 1982–1987, as a number and as a rate per estimated 10,000 farm workers at risk.

Consider how few alleged pesticide-related dermatoses are actually diagnosed and differentiated clinically from other medical causes. For example, a person who is sensitive to a fungicide (such as anilazine or benomyl, for example) is defined by an *outcome* (contact dermatitis) and by an objective immunologic skin test with 99.5% chemically pure antigen used at a 1:1,000 dilution applied to unbroken skin and read at 48 hours. In one outbreak of tomato field workers, 30–40% experienced the rash and most were skin test positive. A noncase is a person with similar work exposure but without a history of contact dermatitis and negative skin test (Schuman and Dobson, 1985).

Most likely, a cancer epidemiologist will not expect his/her information to be extrapolated in a univariate way. For example, 100 cancer experts could be asked to estimate what percent of U.S. cancer cases might be due to (1) passive smoking, (2) diesel fumes from traffic, (3) medical x-rays, (4) electromagnetic power lines, (5) food dyes, (6) video display terminals, (7) background radon gas in homes, (8) feline leukemia virus, (9) UV light, (10) environmental asbestos, etc. If each of 100 experts estimated "1%" etiology, then logically one could prevent 100% of cancer deaths by eliminating or controlling 100 × 1% of 100 etiologic agents.

This logic exposes the fallacy of the univariate approach of regulatory science to control population exposure to man-made chemicals. Heredity, ageing, diet, exercise, immunity, viruses, radiation, and lifestyle (smoking and alcohol) plus access to medical care account a great deal for more than 1% and they operate in an interactive, multivariate, multistage manner. The percentages (frequently promoted for specific etiologic agents) imply a specificity used by governmental regulators, not by critical-thinking scientists (Doll, 1991).

Another example of epidemiology is a farmer mortality study of occupationally coded death certificates in two states, South Carolina (Une et al., 1987) and North Carolina. Using the method of proportionate mortality (proportional mortality ratios—PMRs) within 5-year age intervals and examining the different causes of death, one discovers that cancer deaths (of all types) are *less* in male farmers than in nonfarmers who died between age 34 and 84, for both whites and blacks; SC (W/B = .83/.95); NC (W/B = .86/.88); three of the four PMRs are statistically significant. Furthermore, the hematopoietic tumors of concern with herbicide exposure in Sweden and the Midwest (leukemia, non-Hodgkin's lymphoma, malignant myeloma, Hodgkin's disease and soft-tissue sarcoma) are notably *not* elevated in white or black farmers in these two states. These data are based on the years 1983–1987, derived from death certificates of 20,526 farmers and 132,206 nonfarmers (Schuman, 1991). These data suggest that the heredity/environmental factors for cancer among farmers in the Southeast are different from the etiologic factors in the Midwest. In fact, the leukemia and non-Hodgkin's lymphoma concern with groundwater nitrites and herbicides in the Midwest cannot be generalized with confidence to other regions of the country without additional evidence.

Preliminary PMR data from Georgia comparing 18,848 farmer deaths and 292,958 nonfarmer deaths for the years 1984–1989 also reveal less total malignancy for farmers than nonfarmers; whites = .92, significantly lower; and blacks = .96, not significant. No excess of non-Hodgkin's lymphoma, Hodgkin's disease, multiple myeloma, or leukemia is observed in either race. Observed are excess skin cancer in white farmers (+66%), external cause of death (trauma), +36%, and less cancer of the colon/rectum, −24% for whites and −35% for blacks (Dever, 1991).

Bacillus thuringiensis var. Kurstaki (B.t.-K.) is a microbial pesticide used in agricultural and forestry operations and by home gardeners. Microbial pesticides are bacteria or viruses that are toxic or infectious to targeted pests. They are generally safe and considered to be of low risk to animals. B.t.-K. has been in use for over 30 years, and despite its wide use, it has never been assessed for its potential to cause infection in exposed human populations. A recent epidemiologic study examined the potential in a large population over a 2-year period. The authors were not able to document any case of human infection due to B.t.-K. However, B.t.-K. was suspected in three cases of immunocompromised patients. The authors suggest that all biological agents used in pest control be evaluated for their use around seriously compromised hosts (Green et al., 1990).

Epidemiology and the Public and the Media

It is hard to find a better example of the pitfalls of imprecise definition of PRHE than the episode of aldicarb misuse on watermelons in California during the summer of 1985. Under the pressure of consumer fears of involuntary exposure to aldicarb in illegally treated melons, "working" definitions were developed by health and agriculture officials of California. These definitions were expanded beyond physician-diagnosed, atropine-treated, hospitalized cases with lab-positive residues in melon samples. The California Department of Health Services defined a case of watermelon-related illness as anyone with "cholinergic symptoms" (gastrointestinal, peripheral autonomic, skeletal muscle, or central nervous system) self-reported or outpatient-reported which occurred within 2 hours of consuming a melon or within 12 hours of consuming a melon causing more than one person to become sick (cluster cases).

Consider the fact that the food alert was broadcast on a July 4th weekend with massive watermelon consumption statewide. Consider the media coverage and the fact that 550 fields were tested and "70,000" melons were sample tested for misapplication. Consider that all large growers' melons had to be "safety stickered"; that 95 emergency rooms were notified; that 6,849 markets, 6,081 restaurants, and 245 street vendors were contacted; and that 24,000 physicians and other health professionals were alerted. Considering the statewide alert, and the "working" definition of "watermelon-related illness" in the absence of adequate, skilled, coordinated aldicarb-measurement laboratory support, is it surprising that as many as 1,350 cases were reported to the state health department? There were 493 probable cases before July 10 and 197 probable cases on or after July 10, 1985, the date of an official ban (Fig. 5.2). What is surprising is the fact *that, in retrospect, only ten of 250 melon tests (4%) were positive*. The "toll" of 17 hospitalizations, 6 deaths, and 2 stillbirths cited by the CDC and later quoted in textbooks were not lab-confirmed as aldicarb poisonings. Only three hospitalized cases were linked to positive melons.

The discrepancy between lab confirmations and presumed clinical and cluster poisonings in California should bother anyone concerned with scientific criteria. Nonetheless, the episode was characterized by the CDC as "the largest recorded North American outbreak of foodborne pesticide illness" (Anon, 1986). The previous Nebraska outbreak of hydroponically grown cucumbers accounted for five cases within 1 hour of ingestion (Goes et al., 1980). Evidently, as few as a dozen confirmed cases could count as the largest outbreak.

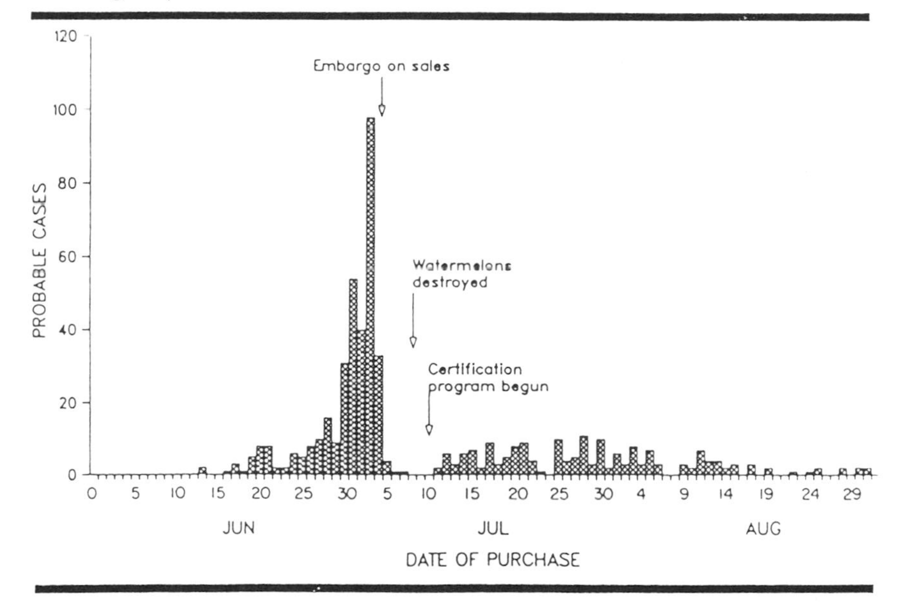

Figure 5.2. CDC time-plot of "probable" watermelon-related illness, California June–August, 1985.

Let us consider a parallel "overcount" of cases during the same summer of 1985 in Oregon. M.A. Green and colleagues reported 264 possible cases, 61 of whom were "definite," 25 of whom could be linked to 10 positive melons out of 16 melons tested; but *none* of 61 cases were hospitalized and none of the 61 "definite" cases were treated with atropine. Regardless of the media and the potential harm, the actual cases of melon poisoning were mild and uncomplicated (Green et al., 1987).

One would hesitate to use this kind of circumstantial diagnosis (clinical or epidemiologic) as a model. In retrospect, who can criticize the stressed health and agricultural authorities for "prudent" safeguarding of the consumer-at-risk statewide? Nevertheless, the misdeeds of a few criminal growers created watermelon panic, not only in California but nationwide, ruining the watermelon market in South Carolina, where choice, locally grown, safe melons could not sell for 75 cents!

What is the proper approach to a massive food health threat? Surely one cannot justify exaggeration of the number of cases, or claim that pesticide residues cause human illness at levels below the threshold of laboratory analysis.

Extrapolation of doses and cases and stretching criteria of illness add up to circumstantial, not scientifically credible, evidence. To be specific, in Figure 5.2, cases of watermelon-related illness were still being counted beyond the ban of July 5–10, 1985, with 197 "probable" cases being counted up to September 1st statewide, and up to September 15 in L.A. County (Weiss et al., 1986).

Finally, L. Goldman and colleagues go to extraordinary lengths to estimate and extrapolate presumed doses of cucumber and slices of watermelon to estimated body weight (by age and sex) to compute "new" LOEL (lowest observed effect level) and NOEL (no observed effect level) toxic doses based on the above shaky health department case definitions (Goldman et al., 1990). Would this exercise in theoretical pharmacology pass the criteria of a serious toxicology journal board of editors?

The aldicarb-tainted watermelon story illustrates the unnecessarily obtuse definitions for PRHEs used by some who only confuse other scientists. The fact is that infectious disease epidemiologists tend to use *precise* criteria, agreed upon, repeatable, lab-confirmable, useful for effective prevention and control. The exceptions (such as Lyme disease, Rocky Mountain spotted fever, and chronic fatigue syndrome) make the point because they currently lack the availability of lab confirmation for the primary-care physician to verify his cases. Pesticide chemists have powerful analytic tools; physicians and epidemiologists have to do a better job of diagnosis to advance our knowledge of actual PRHEs. These tools need to be practical for use at the work site, health clinic, and emergency room.

Florists and Pesticides

A growing percentage of floral designers have been afflicted with severe and recurring attacks of contact dermatitis of the hands (but not the face) in Maryland, Pennsylvania, and South Carolina since 1985. This was linked to exposure to a newly introduced flower: alstroemeria, which comes from The Netherlands and other countries.

One theory suggested dermatitis-causing pesticides used in poorly regulated countries. Clinical-epidemiologic investigation failed to find *any* pesticide link. The cause was hypersensitization to butyrolactone, which is naturally present in the stem, sap, leaf, and flower. Twelve percent of the florists in South Carolina were affected. This was confirmed by skin test results and preventable by withdrawal from exposure to the flower (Bethea et al., 1988). This information serves to protect a significant number of floral designers from disabling skin disease and delay in diagnosis.

Health Problems Associated With Pesticides

Since "pesticide" is defined by its use, not by its formula, it is difficult to move from one category of pesticides to another in a logical way to prevent "lumping." For example, rotenone and pyrethrums are plant-derived as is nicotine from tobacco plants—all except nicotine have agricultural uses as well as toxic side effects. There are over 10,000 registered pesticide formulations in South Carolina.

Consider, for simplicity, *a single* man-made chemical, the herbicide, 2,4,5-T as manufactured in the 1960's and early 1970's with trace amounts of the contaminant and highly toxic chemical, dioxin:

In Alsea, Oregon a study was made of chlorphenoxy herbicides and their toxicity in three preselected areas of (A) high exposure, (B) low exposure, and (C) also low exposure to forests treated with the herbicide. No excess in rates of reproductive damage was found when (A) and (B) were compared, but a small statistical difference was found between (A) and (C). Contrary to epidemiologic practice, where confounding variables, such as access to medical care, etc., are weighed appropriately, the (A–C) comparison was selectively used to persuade the court to ban 2,4,5-TP use as an emergency measure. The EPA report was not accepted for a peer-review publication. Nevertheless it worked in a court of law in 1978 to suspend the use of the herbicide (Schuman, 1986a).

In contrast to the inconclusive Alsea study, parathion handlers in a pesticide formulation plant were unnecessarily exposed to a single pair of overalls (contaminated by spill) when residues were not removed by the usual laundering cycle. As a result, not one but three employees were hospitalized sequentially for skin absorption of the pesticide. Even after three wash cycles, the uniform fabric still contained 7% parathion. (The original product had 76% parathion.) A specific health effect (cholinesterase depression) was documented and effectively treated (2-PAM and atropine), and the mode of exposure was defined and corrected by industrial management (one-use disposable uniforms instead of the previous uniform-disposal system) (Clifford and Niew, 1989).

World Pesticide Poisonings

The World Health Organization estimated there are 3.0 million pesticide-related cases, of which 2.0 million are suicidal (WHO, 1989). These numbers are used as an example of PRHEs and preventable morbidity. If 2.0 million persons want to commit suicide per year and pesticides are somehow eliminated, what choice will they have but to use old-fashioned hanging, drowning, plant poisons, firearms, etc.? This is not to suggest that pesticides and pharmaceuticals should be available over the counter. However, anyone who lumps accidental pesticide poisoning with intentional poisoning is confusing environmental medicine with psychiatry or is using numbers inappropriately.

The approach in assessing human pesticide poisonings should be conducted with care. For example, a western Washington (Woods et al., 1987) study of agricultural and forestry workers for PRHEs (non-Hodgkin's lymphoma, soft-tissue sarcoma, and Hodgkin's disease) estimated work-related exposure to herbicides, including 2,4,5-T. Meticulous estimates of each worker-category of exposure (by person-days) were calculated *before* the illness rates were computed. This represents a model of difficult, but logical, weighting of exposure. The results stand in contrast to a widely quoted Kansas study (Hoar et al., 1986) in which exposure was estimated by telephone interviews with cancer cases and surviving relatives months to years after the exposure. It is not surprising that the Kansas study and Washington study reached different conclusions as to the risk of the three tumors: in Washington 0 out of 3 and in Kansas 1 out of 3 tumors could be liked statistically and retrospectively to herbicide use. No wonder scientists who are not epidemiologists have difficulty evaluating the differing results and interpretations (CAST, 1987; Colton, 1986).

Best Estimates of PRHEs

Two approaches to PRHEs in two states result in remarkably similar estimates of hospitalized risk of agricultural exposure to pesticides.

Physician reports of PRHEs to the California Worker Health and Safety staff and to the California Department of Health Services are required before services are reimbursed. This provides a valuable data set for confirmation and analyses. Cases are categorized by severity (hospitalized or not), by organ system (skin, eyes, systemic, etc.), by individual pesticide (a list of over 100 compounds including sulfur dust, etc.), and by occupational or nonoccupational category (fumigator, flagger, pilot, picker, etc.). Data from the California Pesticide Illness Surveillance Program for the years 1982–1988 indicate an average of about 1,000 agricultural pesticide illnesses each year. These cases include those categorized as "definite," "probable," or "possibly" related to pesticide exposure. During a typical year (1988) only 130 out of 594 definite occupational exposures (22%) were medically identified as "systemic" poisoning. The rest were identified as 64% eye cases, 11% skin cases, and 3% both eye and skin. How many persons who use California morbidity data are aware of these severity-and-specificity percentages?

Also in South Carolina, since 1972 the data set has been used to document inpatient (hospital records) and estimate outpatient rates of pesticide poisonings. South Carolina and most other states do not have the complex outpatient reporting system of California (Schuman et al., 1986, 1989).

Our method in South Carolina is characterized by (1) statewide hospital record access, (2) standardized criteria, and (3) detailed audit of hospital records. Limitations include occasional uncoded death certificates or miscoded hospital discharges (Fig. 5.1).

In addition to the above data from California and South Carolina, research from Florida concurs with a relatively low rate of worker-reported, confirmed poisonings. In fact in 1985–1986, California reported 220 cases per 100,000 farmworkers while a special study in Florida revealed 219 cases per 100,000 (Griffin et al., 1985).

Current available data suggest three trends:

1. Declining rate of agriculturally related pesticide poisoning cases hospitalized in spite of greater third-party payments for hospitalization. Progress in labeling, less-toxic formulations, better training, certification, reenforcement by Extension, and health education are evident. Much remains to be accomplished.
2. Misuse and accidents in the use of pesticides account for a great majority of hospitalized cases. There is a disturbing increase in misuse of pesticides by mentally disturbed people (homicide, suicide).
3. There is a need to protect children from accidental ingestion of pesticides.

Data from the three states of California, Florida, and South Carolina need to be expanded to include other states with similar definitions of PRHEs and criteria (Mehler, 1991). At least 30 of the 50 U.S. states have computerized hospital discharge data sets for monitoring pesticide poisonings.

Pesticides in Foods

Space does not permit a detailed discussion of the issues of food safety and pesticide residues consumed in the diet. Readers are referred to a cross section of opinions, position statements, and some facts on pesticide-food residues collected by the EPA (Heritage, 1990). A forum on risk assessment by the American Chemical Association also covers a cross section of opinion and fact (Heylin et al., 1991).

Time will tell whether the "alert" raised by the Natural Resources Defense Council over Alar reflects logical pediatric concerns or the illogical application of computer models of rodent carcinogenesis developed by the EPA and others. Caught in the decision dilemma are government officials, pediatricians, toxicologists, growers, food processors, and a confused, fearful public (Rhodes, 1990). Pediatricians and family doctors are in a special bind because their training does not prepare them as yet for dealing with the complex issues involving PRHEs, toxicology, pharmacokinetics, nutrition, and food processing.

Diet *will* emerge, as predicted by Sir Richard Doll, Bruce Ames, Saxon Graham, Ernst Wynder, Walter Willett, Takeshi Hirayama, and Takashi Sugimura, as one way to prevent some of the major cancers. Diet, obesity, and lack of exercise are estimated to account for at least 35% of cancer beyond the lifestyle proportion due to abuse of alcohol and tobacco (33%).

Nature's own daily dietary intake of carcinogens and anticarcinogens in milligram amounts will provide increasingly practical clues to cancer prevention and to the role of diet and food preparation (Knudsen, 1986). The roles of trace amounts of pesticide residues at the ppm and ppb levels in food will continue to be debated. Technology now can measure parts per billion, trillion, and quadrillion. In the absence of a trustworthy animal, tissue culture, or human biomarker model of biologic evidence to provide objective evaluation, how can we interpret such tiny amounts pharmacologically in the human body? There are in fact no experts in the potential field of generalizing human health effects at dosages of parts per million or billion.

Ethics

One principle that I recommend is that of epidemiologic evidence (Dubos, 1959). This sort of evidence demands logic and biologic plausibility, and an acknowledgment of the inherent complexity involved in defining PRHEs—a complexity which resists oversimplification. In other words, claims for PRHE should be supported by strong scientific evidence—not hearsay, not anecdotal, not presumptive evidence. This position is in contrast to that of the seekers of a risk-free environment, which presumes that any or all man-made chemicals are toxic until proven otherwise, especially if they can be measured at parts per billion or quadrillion with modern instrumentation.

Sound epidemiologic or clinical evidence of a true PRHE is (1) not that hard to obtain, (2) can be proven and replicated by other observers, (3) and can be applied rigorously to prevent and control PRHEs. One example is lead arsenate: once widely used in vineyards, it is a proven heavy-metal toxin *and* arsenical carcinogen. There is no question: this constituted a human health hazard.

One statesman-epidemiologist, Ernst Wynder, put the need for sound evidence this way:

> We will be cautious that the law does not run ahead of scientific evidence. The Delaney Clause and the unfounded carcinogen scares from Alar[R] to fluoride all require our attention. We must do better in assessing and managing low-level risks, like carcinogen exposure, to ensure that the public is not unduly frightened, concerned, and ultimately desensitized to *actual* risks. Moreover, surveillance by scientists can prevent industries from needlessly being damaged by recommendations that have no established scientific base. . . . Quantitative aspects of the exposure and risk equation are important, yet are often ignored, a key element in view of the power of analytical chemistry to measure minute amounts of environmental chemicals.

Sound guidelines for PRHEs are proposed below:

a. *Oncogenes and oncogene suppressors* are emerging as the key to understanding the complex etiology of carcinogenesis.

b. *No single agent* can account for the failure of the human host to defend itself immunologically at the cell level against the cancerous process. Cancer is a multifactorial process.

c. *It is unlikely* that the rodent-loading model of birth defects and carcinogenesis will last beyond the present decade as a reliable or useful index of human risk assessment (Jollow, 1991).

d. Epidemiologic studies of cases and populations will provide valuable, though quantitatively limited, *surveillance* information and guidelines. Follow-up studies of heavily exposed industrial populations will continue to provide objective epidemiologic data on PRHEs.

e. *Species difference* between mice, hamsters, guinea pigs, strains of rats, salmonella (Ames test), and man are unlikely to be overcome for pesticides, since this has never been achieved for pharmaceuticals. (Consider thalidomide and arsenic, for which there are no animal models comparable to human disease.)

f. *Integrated pest management* (IPM) is the logical extension of scientific agriculture, preventive medicine, economic welfare, stewardship of the land, and respect for the environment. IPM provides the formula for sustainable agriculture and feeding a hungry world or a hungry neighborhood (Brittain, 1989).

g. *Central and peripheral nervous system disorders* will receive increased clinical and epidemiologic study (Rosenstock et al., 1991; Schuman and Wagner, 1991). Exposure to pesticides will be a part, but only a part, of greater research on toxic and immunologic factors. Physicians need to have better training to recognize early stages of PRHEs, including dermatology, toxicology, neurology, and pharmacokinetics.

Conclusion

Good ethics requires good judgment; good judgment requires good science. Poor science does not serve judgment, ethics, or the public interest. Epidemiology provides a tool for accurately assessing chemical risk in biologic systems concerned with human health. When this tool is ignored or used imperfectly, society is at risk of errors in judgment, in policy, and in economics which may have far-reaching adverse consequences.

If and when epidemiologic science provides credible support for presumed PRHEs such as multiple chemical sensitivity, low-dose cumulative pesticide neurotoxicity, low-dose carcinogenesis from pesticide residues in food or water sources, or for reproductive damage from low-dose pesticide exposure, we will agree on the same principle: to prevent or control unnecessary adverse health effects.

This may not prevent general public fear of accidental injury as a result of exposure to pesticides. Risk of accidental injury should not be attributed merely to the pesticide itself. The risk, if any, arises from the manner of handling, transporting, applying, etc., the pesticide. The incidence of pesticide-related injuries and deaths can often be reduced by developing safer ways of sorting and transporting the pesticide, by developing more precise ways of applying the pesticide, and by taking steps to insure that applicators are properly trained in use of the pesticide and that they are protected by suitable clothing or other gear from accidental spills, etc.

It should also be noted that, in making decisions regarding the acceptability of using a particular pesticide, it is not sufficient to consider only the risk of accidental pesticide poisonings. To live is to be exposed to risk of injury or death. While there is risk of injury or death from use of a pesticide there may be equal or greater risk, in some cases, of injury or death from failure to use a pesticide. In light of this consideration it appears unwise to follow the recommendations of those who advocate a nonspecific approach to eliminating pesticide-induced injuries or diseases. Such an approach would consist of eliminating pesticide use. Rather, it is recommended here that pesticides be used judiciously in light of scientific knowledge of specific causal relations between pesticides and injury and disease.

References

American College of Physicians: Clinical ecology. Ann Intern Med 1989, **111**:168–178.

Anon: Aldicarb food poisoning from contaminated melons—California, MMWR 1986, 35:254–258, cited in JAMA 1986, **256**:175–176.

Bell CA, Storet NA, Bender TR, et al.: Fatal occupational injuries in the U.S.; 1980–1985. JAMA 1990, **263**:3047–3050.

Bethea LK, Schuman SH, Smith-Phillips SE, et al.: South Carolina florists dermatitis: case report and survey results. J S C Medical Assn 1988, **84**:446–448.

Brittain JA: LISA, Haven't we met some place before? Lecture at VI Annual Convention of the SC Horticulture Society, November 1989, Clemson, SC 29634.

Bullman TA, Kang HK, Watanaba KK: Proportionate mortality among US Army Vietnam veterans who served in Military Region I. Am J Epidemiol 1990, **132**:670–674.

CAST (Council for Agricultural Science and Technology): Perspectives on the safety of 2,4-D. *Comments from CAST*, 1987-3, Ames, Iowa 50010.

Clifford NJ, Niew AS: Organophosphate poisoning from wearing a laundered uniform previously contaminated with parathion. JAMA 1989, **262**:3035–3036.

Colton T: Herbicide exposure and cancer. (Editorial) JAMA 1986, **256**:1176–1178.

Council on Scientific Affairs, American Medical Association: Cancer risk of pesticides in agricultural workers. JAMA 1988, **260**:959–966.

Delzell E, Cole P, Sathiakumar, et al.: A retrospective follow-up study of mortality among workers at Ciba-Geigy Corporation McIntosh plant. UAB School of Public Health, Dept of Epidemiology, July 27, 1990.

Dever GEA: Preliminary Communication, April 91, Mercer University School of Medicine, Dept. of Family and Community Medicine, Macon, GA 31207.

Doll R: The lessons of life: keynote address, *Nutrition and Cancer*, Proc. International Conference on Research in Human Cancer, Cancer Research (Suppl.) **52**:2024–2029, 1992.

Dubos R: *Mirage of Health: Utopias, Progress and Biologic Change*, Harper & Bros., NY, 1959.

Goes EA, Savage EP, Gibbons G, et al: Suspected food borne carbamate intoxication associated with hydroponic cucumbers. Am J Epidemol 1980, **III**:254–260.

Goldman LR, Beller M, Jackson RJ: Aldicarb food poisonings in California, 1985–1988; toxicity estimates for humans, Arch Environ Health 1990, **45**:141–147.

Green MA, Heimann MA, Wehy HM, et al.: Public health implications of the microbial pesticide bacillus thuringiensis: an epidemiologic study, Oregon, 1985–6. AJPH 1990, **80**:848–852.

Green MA, Heimann MA, Wehy HM, et al.: An outbreak of watermelon-borne pesticide toxicity. Am J Public Health 1987, **77**:1431–1434.

Griffith J, Duncan RC, Konefel J: Pesticide poisonings reported by Florida citrus field workers. J Env Sci Health 1985, **B20**:701–727.

Henderson TW: Toxic tort litigation: medical and scientific principles in causation. Am J Epidemiol 1990, **132**:569–578.

Heritage J, Editor: Pesticides and food safety, EPA Journal 16: No. 3, 1-54, 1990, EPA, GPO, Washington, D.C. 20402.

Heylin M, Editor: Risk assessment of pesticides, CEN Forum, Chemical & Engineering News 1991, **69**:27–52. American Chemical Society, Washington, DC 20036.

Hoar S, Blair A, Holmes FF, et al.: Agricultural herbicide use and risk of lymphoma and soft tissue sarcoma. JAMA 1986, **256**:1141–1147.

Jollow D: Personal communication, Department of Pharmacology, MUSC, Charleston, SC 29425, 1991.

Knudsen I, Editor: *Genetic Toxicology of the Diet*, Alan R. Liss, Inc., New York, NY, 1986.

Kurtz PH, Esser TE: A variant of mass (epidemic) psychogenic illness in the agricultural work setting. J Occup Med 1989, **31**:331–334.

Levine R: Recognized and possible effects of pesticides in humans, Ch. 7, Vol. 1, pp 275–360, in Hayes WJ, Laws ER: *Handbook of Pesticide Toxicology*, Academic Press, San Diego, CA, 1991.

Litovitz TB, Schmitz BF, Bailey KM: 1989 Annual Report of the American Association of Poison Control Centers National Data Collection System, Am J Emerg Med 1990, **8**:394–442.

Loevinsohn ME: Insecticide use and increased mortality in rural central Luzon, Philippines, Lancet 1987, **1**:1359–1362.

Mehler L, Edmiston S, Richmond D, et al.: Summary of illnesses and injuries reported by California physicians as potentially related to pesticides, 1988. HS-15-41, April 30, 1990, California Dept of Food and Agriculture, Worker Health and Safety Branch, Sacramento, CA 94271.

Mehler L: May 1991, personal communication, Department of Food and Agriculture, Sacramento, CA 94271.

Pimentel D: Pesticide use (letter) Science 1991, **222**:358–359.

Rhodes ME: Food safety in the 1990's: Gnats, camels, and Noah, pp 7–13 in Stallings SF, Stewart MP: *Proceedings of An Agromedicine Model: Dissemination Symposium*, Oct 31–Nov 1, 1990, Clemson University, Clemson, SC.

Rosenstock L, Keifer M, Daniell WE, et al.: Chronic central nervous system effects of acute organophosphate intoxication. Lancet 1991, **338**:223–227.

Schuman SH, Wagner SL: Pesticide intoxication and chronic CNS effects (letter). Lancet 1991, **338**:948.

Schuman SH, Dobson RL: An outbreak of contact dermatitis in farmworkers. J Am Acad Derm 1985, **13**:220–223.

Schuman SH: The "Mirage of Health" epidemiologic aspects of risk assessment emphasizing birth defects. The Agrarian—Proceedings of a Clemson University Symposium "Perspectives on Risk," February 1986a, pp 14–35.

Schuman SH, Caldwell ST, Whitlock NH, Brittain JA: Etiology of hospitalized pesticide poisonings in South Carolina, 1979–1982. JSC Med Assn 1986, **82**:73–77.

Schuman SH: An apparent cluster of aplastic anemia: Credible science? Letter to the Editor. Arch Intern Med 1986b, **146**:809.

Schuman SH: Where is the birth defects epidemic? Agricultural Aviation 1986c, **13**:20–22.

Schuman SH, McEvoy M: Chronic neurotoxicity of carbaryl. Letter to the Editor. Am J of Med 1986, **81**:1124–1125.

Schuman SH: *Practice-Based Epidemiology*, Gordon & Breach, New York, 1986d.

Schuman SH: Unpublished data, 1991.

Schuman SH, Whitlock NH, Caldwell ST, Horton PM: Update on hospitalized pesticide poisonings in South Carolina, 1983–1987. JSC Med Assoc **85**(2):62–66, 1989.

Schuman SH: Practice-based agromedicine: The need for client-centered research. Am J Ind Med 1990, **18**:405–408.

Une H, Schuman SH, Whitlock NH, Caldwell ST: Agricultural lifestyle: A mortality study among male farmers in South Carolina, 1983–1984. Southern Medical Journal 1987, **80**:1137–1140.

Weiss BP, Fong-Huie, Strasburg MA, et al.: Outbreak of illness associated with pesticide contaminated watermelon, Los Angeles, CA. Presentation APHA, 114th Annual Meeting, Las Vegas Nevada, Sept 29, 1986; LA County Dept Health Services, LA, CA 90012.

WHO/UNEP Working Group: Public health impact of pesticides used in agriculture: Abstract: V, Table 7:1, 1987 conference, WHO Geneva, 1989.

Woods JS, Polissar L, Severson RK, et al.: Soft tissue sarcoma and non-Hodgkin's lymphoma in relation to phenoxy herbicide and chlorinated phenol exposure in western Washington. JNCI 1987, **78**:899–910.

6

Pesticides and Natural Toxicants in Foods*

Thomas W. Culliney, David Pimentel, and Marcia H. Pimentel

Humans living in the 20th century not only are exposed to the natural chemicals that are basic components of the foods they eat, but to a wide variety of synthetic chemicals that find their way into foods, water supplies, and the atmosphere. Included are residues from fertilizers and pesticides and antibiotic residues in foods as well as pollutants from automobiles and industries. Additional chemical additives are incorporated into processed foods to serve such functions as nutrient enrichment; improvement of color, texture, and flavor; as well as enhancement of shelf life.

Given the diverse, fragmentary, and often contradictory information disseminated by scientists, government officials, and the news media, public apprehension about chemicals is understandable. Such concerns were highlighted in the results of a recent consumer survey of public views about food supply safety. Warren et al. (1990) reported that 51% of the people surveyed felt there was no good reason to put additives in foods; 64% were concerned that supermarket meat contained chemical residues, while 74% felt that fresh produce contained toxic chemicals. This survey also highlighted the fact that individuals do not understand the concepts of risks and benefits. Equally important, it is clear that the public is distrustful of the efforts of government regulatory agencies to insure a safe food supply. Furthermore, according to Warren et al. (1990), "fear of uncontrolled (involuntary) exposure may be an important factor contributing to the respondents' concerns about the use of pesticides and antibiotics."

Widespread press reports have also stirred the public controversy over pesticide contamination, as illustrated by the recent Natural Resources Defense Council (NRDC) report on the cancer risks associated with the plant growth regulator daminozide (Alar) (Sewall and Whyatt, 1989). Further, public concerns are being translated into political action. For example, in California two existing laws (Safe Drinking Water and Toxic Enforcement Act of 1986 [Proposition 65] and the Federal Insecticide, Fungicide,

*Reprinted from Agriculture, Ecosystems & Environment, 1992 (Elsevier), with permission.

and Rodenticide Act as amended in 1988 [FIFRA 1988]) have been implemented and enforced. In addition, a recent ballot initiative, the Environmental Protection Act of 1990, had it been passed by voters, would have curtailed the use of many pesticides now used in California agriculture (Stimmann and Ferguson, 1990; Pimentel, 1990).

In addition to the risks associated with pesticides, some investigators claim that the health risks from natural chemicals in foods are greater than those from pesticide residues in foods (Ames et al., 1987; Ames and Gold, 1989). Although they have not considered as proven that human exposure to natural toxicants is a major health concern to humans, they maintain that specific constituents of commonly eaten vegetables, like cabbage and broccoli, are toxic, and even more toxic than chemical pesticides in foods (Ames and Gold, 1989; Ames, 1989).

These opinions not only alarm and confuse the public as to what food choices they safely can make, but do not follow the advice on food selection given the U.S. public by nutritional authorities. Indeed, the importance to human health of a nutritious food supply was reemphasized by the 1988 surgeon general's report on nutrition and health and the 1989 report of the National Academy of Sciences (NAS, 1989a), *Diet and Health: Implications for Reducing Chronic Disease Risk.* Both reports presented data concerning the relationships between diet and such chronic diseases as heart disease, stroke, cancer, and diabetes. Among their nutritional recommendations were that the public consume more complete carbohydrates and fruits and vegetables, especially those high in carotene (e.g., cabbage, broccoli, and other brassicas). Also, these reports reconfirmed the dietary guidelines (HHS, 1985) and the nutritional advice given by the National Cancer Institute (NCI, 1988a).

The aim of this paper is to examine available scientific data concerning the known public health risks of naturally occurring toxicants in foods and also the health risks associated with current pesticide use in the United States. In addition, the most recent recommendations made to Americans concerning their diets and food choices made by recognized nutritional authorities are presented.

Diet and Human Health

All plant and animal foods consumed by humans are composed of various kinds and amounts of chemicals. Included are carbohydrates, fats, and proteins as well as the minerals and vitamins needed by humans and other animals to grow, reproduce, and remain healthy. Plants, which need chemicals (nutrients) to grow and flourish, also contain certain chemicals that help protect them against pest attack (Levin, 1976; Pimentel, 1988).

Some chemicals found in plant foods, if eaten in large dosages, are known to cause human illness and even death. Over time, probably by trial and error, humans have learned to avoid the particular plants that have proven unsafe to eat. Plant foods, however, for both vegetarians and nonvegetarians, have made and continue to make up a large and nutritionally vital part of the human diet.

Historically, nutrition research focused on the disease states that were caused by diets deficient in various specific nutrients, many of which were provided by plant foods. Vitamin C (ascorbic acid, found in citrus fruits and vegetables), which prevents scurvy, and vitamin B (thiamine, found in whole grains), which prevents beriberi, are classical examples of the important role of plant-based nutrients in human nutrition. More recently, nutrition research has focused on the associations among foods and other dietary components and the risk of chronic diseases prevalent in the human population.

Numerous scientific studies, on both animals and humans, have established that diet affects the risk of several chronic diseases, e.g., atherosclerotic cardiovascular diseases and hypertension as well as some cancers (HHS, 1988; NAS, 1989a). The specific food-related factors identified as associated with those health problems are high fat intake, saturated fats, cholesterol, alcohol, sodium, and diets unbalanced as to calories and individual energy needs. Categories of foods that promote health include the complex carbohydrate-containing grains and legumes and other vegetables and fruits, especially those high in vitamins A and C (HHS, 1988; NAS, 1989a).

As mentioned, the safety of a large variety of commonly eaten fruits and vegetables was questioned by Ames and Gold (1989). The fruits they mentioned included apples, bananas, cantaloupes, grapefruit, orange juice, honeydew melon, peaches, pineapples, and raspberries. The vegetables included members of the *Brassica* or cole group (broccoli, brussels sprouts, cabbage, cauliflower, collards, turnips, and kale), celery, and some herbs and spices. Note, the cole and citrus foods are outstanding sources of vitamins A (beta-carotene) and C (USDA, 1986) and the foods advocated by the nutritional authorities mentioned previously.

In contrast to the views expressed by Ames and Gold (1989), many recent scientific studies identify the positive role that vitamin A– and C–rich fruits and vegetables play in reducing the incidence of human cancer (HHS, 1988; NCI, 1988b; NAS, 1989a; Birt, 1989). Vitamin A, especially in plant form, beta carotene, has been found to be protective against cancers of the larynx, lung, stomach, bladder, and prostate (HHS, 1988; NAS, 1989a; Tufts, 1989). Additional studies illustrate the protective effect of cole vegetables against colon and rectal cancers (Lee et al., 1989). However, further studies are needed to clarify whether other components, like indoles or unmeasured carotenoids, also present in the vitamin A–rich foods, especially the brassicas, are contributing to their protective effect against cancer (NAS, 1989a).

Vitamin C, present in many commonly eaten fruits, especially citrus as well as a wide variety of vegetables, is being evaluated for its possible beneficial effects relative to the incidence of cancer. Some human studies have demonstrated a protective association between vitamin C–rich foods and cancer of the esophagus and stomach (HHS, 1988). In biochemical studies, vitamin C prevents the reactions of nitrates with amines and/or amides to form nitrosamines, known carcinogens (NAS, 1982; HHS, 1988). Tumors in test animals were reduced when they were given vitamin C (NAS, 1989a). The beneficial effect of vitamin C has not been quantified, and there is no evidence that amounts of vitamin C above the Recommended Dietary Allowance are beneficial (NAS, 1989b).

Current dietary recommendations made to the public emphasize increasing both the frequency of eating and serving size of green and yellow vegetables as well as citrus fruits (AICR, 1984; NCI, 1988a,b; HHS, 1988; NAS, 1989a). Note, these groups of foods, which contain both vitamins A and C as well as various other nutritionally important components, are the ones cited by Ames et al. (1987) as having health risks greater than those of pesticides. To date, no nutritional authority has suggested that Americans are at any health risk from eating the fruits and vegetables that Ames and Gold (1989) indict as potentially more hazardous than pesticide residues because they contain natural toxicants.

Natural Toxicants in Foods

As mentioned, all human food is a complex mixture of chemicals including carbohydrates, amino acids, fats, oils, vitamins, and other chemicals, some of which may be toxic if consumed in large quantities (Strong, 1974). Plant foods contain some chemicals that are known to be toxic to animals, including humans. Some of these chemicals evolved in plants to protect them from insect, plant pathogen, and other organism attack (Pimentel, 1988; Ames, 1989). In general, the adverse effects of toxic chemicals in plants are related to interference with nutrient availability, metabolic processes, detoxification mechanisms, and allergic reactions in particular animals and humans. A small number of the chemicals, like hydrazines in a few mushrooms, are highly carcinogenic.

Reviews by the National Academy of Sciences (1973), Strong (1974), and Liener (1986) summarize available data about the possible health risks of food toxicants. Many toxicants are found in staple foods of the human diet, like grains and legumes. Some of these are discussed below.

Protease inhibitors are widely distributed throughout the plant kingdom, particularly in the Leguminosae and to a lesser extent in cereal grains and tubers. These substances inhibit the digestive enzymes trypsin and chymotrypsin (Bender, 1987). For example, raw soybeans contain a protein that inactivates trypsin and results in a characteristic enlargement of the pancreas and an increase in its secretory activity. It is this latter effect, mediated by trypsin inhibition, that depresses growth. Clearly, soybeans and other related legumes should be properly cooked and processed before being eaten.

Lectin proteins (phytohemagglutinins) are present in varying amounts in legumes and cereals and in very small amounts in tomatoes, raw vegetables, fruits, and nuts. The lectin ricin, which is extremely toxic and can be fatal to humans, was used as an insecticide at one time. When untreated lectins are eaten, they agglutinate red blood cells and bind to the epithelial cells of the intestinal tract, impairing nutrient absorption. Fortunately, heat destroys the toxicity of lectins.

Lathyrogens, found in legumes like chick peas and vetch, are derivatives of amino acids that act as metabolic antagonists of glutamic acid, a neurotransmitter in the brain (NAS, 1973). When foods containing these chemicals are eaten in large amounts by either humans or other animals, they cause a crippling paralysis of the lower limbs and may result in death. Lathyrism is primarily a problem occurring in some areas of India.

Goitrogens (glucosinolates), which inhibit the uptake of iodine by the thyroid, are present in many commonly consumed plant foods. They are estimated to contribute approximately 4% to the worldwide incidence of goiter in humans (Liener, 1986). Cabbage, cauliflower, brussels sprouts, broccoli, kale, kohlrabi, turnips, radish, mustard, rutabaga, and oil seed meals from rape and turnip all possess some goitrogenic activity (Coon, 1975). Effects of thyroid inhibition are not counteracted by the consumption of dietary iodine. The nature and extent of toxicity of glucosinolates are still the subject of debate. Although there are few, if any, acute illnesses in humans, chronic and subchronic effects remain a possibility (Heaney and Fenwick, 1987).

Additional foods with the potential for antithyroid activity include plants in the genus *Allium* (onion group); other vegetables, such as chard, spinach, lettuce, celery, green pepper, beets, carrots, and radishes; legumes, such as soybeans, peas, lentils, beans, and peanuts; nuts, such as filberts and walnuts; fruits, such as pears, peaches, apricots, strawberries, and raisins; and animal products, such as milk, clams, oysters, and liver (Coon, 1975). However, it has not been proven that a diet of these foods would be goitrogenic unless these foods comprise an excessively high proportion of the diet and if a substantial amount of these foods were eaten raw or not well cooked. Goitrogens in foods are largely destroyed by thorough cooking, but it must be acknowledged that many of the above-listed foods are eaten uncooked (Coon, 1975).

Cyanogenic glycosides occur in many food plants like cassava, lima beans, and seeds of some fruits, like peaches. Because of their cyanide content, ingestion of large amounts of cassava and, to a lesser extent, lima beans, can be fatal if they are eaten raw or are not prepared correctly (Strong 1974). Cassava toxicity is much reduced by peeling, washing in running water to remove the cyanogen, and then cooking and/or fermenting to inactivate the enzymes and to volatilize the cyanide. In regions like Africa where cassava is a staple food, care is taken in its preparation for human consumption.

For some individuals, eating field or broad bean, *Vicia faba*, precipitates the condition known as favism, characterized by anemia caused by the hemolysis of red blood cells (NAS, 1973). Genetic factors in humans as well as the concentration of toxic factors in the beans account for the different individual responses to the bean (Coon, 1975).

When potatoes, which contain two major glycoalkaloid fractions, alpha-solanine and alpha-chaconine, are exposed to sunlight, a significant increase in the alkaloid content results (NAS, 1973). Solanine is a cholinesterase inhibitor and can cause neurologic and gastrointestinal symptoms (Oser, 1978), with death being caused by depression of the activity of the central nervous system. According to Litovitz et al. (1989), 1,844 humans are known to have ingested solanine during 1988. Most cases (1,503) involved children under the age of 6 and were accidental in nature. Only 352 cases required treatment, and there were no deaths.

Estrogenic compounds are present in many plants like soybeans and palm kernels as well as various forages, yet there are no data substantiating human illness or cancer activity of plant estrogens (NAS, 1982).

Chemical substances like hydrazines found in the mushrooms *Aqauius besperus* (commonly eaten mushroom) and *Gyromitra esculenta* (false mold) have proven to be carcinogenic in animal studies (Oser, 1978; NAS, 1982). The carcinogenicity of these compounds to humans has not been established due to the lack of epidemiologic studies. However, the ingestion of *G. esculenta* has caused human poisonings and death (NAS, 1982, 1989a).

Oxalates are widely distributed throughout the plant kingdom. Appreciable amounts are found in spinach, rhubarb, beet leaves, tea, and cocoa. Most of the other commonly eaten fruits and vegetables contain small amounts of oxalates. The presence of large amounts of oxalates is thought to diminish the absorption of dietary calcium, e.g., calcium in spinach (Coon, 1975). Of the 17,159 human exposures to oxalates during 1988 almost all were attributed to accidental exposure for children under the age of 6 years, only 725 required attention in a health facility, and none was considered serious (Litovitz et al., 1989). There is some evidence to suggest that the acute toxicity of rhubarb leaves and other oxalate-containing plants may be caused by toxicants other than oxalates (NAS, 1973; Coon, 1975).

Bracken fern, which contains quercetin, is consumed by humans throughout the world and especially in Japan, where it is used both as a gruel and as a green. Hirayama (1979) reported a higher incidence of human esophageal carcinoma with increased consumption of fresh fern eaten as a gruel. Rats fed bracken fern or a quercetin-containing diet showed a significantly higher incidence of intestinal and bladder tumors than rats fed a standard diet (Pamukcu et al., 1980). Cooking reduces, but does not eliminate, the toxicity of quercetin (NAS, 1982).

The most potent natural toxicants responsible for human suffering and death are the mycotoxins, which are not strictly plant compounds but toxic metabolites produced by fungi infesting foodstuffs, especially cereals and nuts, which have been stored under conditions of elevated temperature and high humidity (NAS, 1982, 1989a). Among the ailments caused by these mycotoxins, the most notable, historically, is ergotism, or "St. Anthony's fire," which afflicted people centuries ago. This was caused by ergot alkaloids produced by *Claviceps purpurea* growing on cereal grains (NAS, 1973).

At present, the most widely studied mycotoxins are the aflatoxins produced by species of the common mold *Aspergillus* (e.g., *A. flavus*). Aflatoxin B has been identified as a potent liver carcinogen in experimental animals, such as the monkey, rat, ferret, and duckling, from which the only reliable data have come (NAS, 1982, 1989a; Neal, 1987). Ingestion of highly aflatoxin-contaminated foods has been correlated with the incidence of liver cancer in certain regions of Asia and Africa (NAS, 1989a). In the United States, where residue levels of aflatoxin are regulated by the FDA, it is not considered a human cancer risk (NAS, 1982). Although several other mycotoxins have been found to be mutagenic in bacterial systems and carcinogenic in laboratory animals, their role as human carcinogens has not been established.

In addition to the contaminants of microbial origin, other potentially dangerous components in plant foods can originate from the uptake of such chemicals as nitrate from soil and drinking water (Coon, 1975). Nitrate is not considered a human carcinogen, but nitrosamines which are formed from nitrates and nitrites, like those used in curing meats, are carcinogenic in animals (NAS, 1989a). Other hazardous chemicals like lead, iodine, mercury, zinc, arsenic, copper, and selenium are found in varying quantities in foods, and if consumed in large amounts, can cause health problems or death.

The extent of the risk to human health associated with naturally occurring toxicants remains a scientifically contentious matter (Watson, 1987). Debate on this subject has been clouded by the absence of a systematic approach to defining and, particularly, quantifying human hazards. Although many data have been assembled on the chemical properties and biological sources of most of these compounds, their risks to public health have not been established. In particular, there is almost a complete lack of data on the effects on human populations of long-term ingestion of natural toxicants in foods. Above all, it is important to reemphasize that there is, at present, no firm evidence to demonstrate a link between long-term ingestion of natural toxicants in commonly eaten foods and any type of chronic human illness (Watson, 1987).

This view is upheld in the NAS report (1989a), which states that although "naturally occurring compounds found to be carcinogenic in animals have been found in small amounts in the average U.S. diet . . . there is no evidence thus far that any of these substances individually makes a major contribution to cancer risk in the United States."

Pesticide Exposure and Potential Risk to Humans

Human exposure to a wide variety of pesticides (insecticides, herbicides, and fungicides) is extensive, and comes not only from sprayed residues left on food crops, but also from residues found in drinking water and air. Yearly, about 1 billion pounds of pesticides are used in the United States, with approximately three-quarters of this applied to agricultural crops (Pimentel and Levitan, 1986; USDA, 1989). Fruit and vegetable crops receive most of these pesticides.

Pesticide Poisonings and Fatalities

In the United States, pesticide poisonings reported by the American Association of Poison Control Centers (AAPCC) total 67,000 each year (Litovitz et al., 1990). J. Blondell (EPA, personal communication [PC], 1990) has indicated that because of demographic gaps, this figure represents only 73% of the total. The number of accidental fatalities is about 27 per year (J. Blondell, EPA, PC, 1990).

Most human pesticide poisonings reported by the American Association of Poison Control Centers during 1988 were attributed to insecticides (Litovitz et al., 1989). Note that 15,000 cases of pesticide poisonings were treated in health care facilities and 20 deaths were reported (Table 6.1). In contrast, following the ingestion of plant materials, including hallucinogens, stimulants, plant material exposures to cyanogenic glycosides, gastric irritants, oxalates, solanines, etc., only 6,700 cases required treatment in health care facilities and only one death occurred (Table 6.1).

Table 6.1. Demographic profile of exposure cases by category of substances and products (data for 1988, after Litovitz et al., 1989).

Category	Number of exposures	Number treated in health care facilities	Deaths
Fungicides	1,496	378	1
Herbicides	1,657	420	2
Diquat/paraquat	4,549	1,435	5
Insecticides	41,499	10,451	12
Rodenticides	10,626	3,387	2
Plants	93,975	6,660	1?
TOTAL	153,802	22,731	23

About 40% of the foods purchased by U.S. consumers have detectable levels of pesticide residues (FDA, 1989). As expected, the contamination rate is higher for fruits than for any other commodity (Sewell and Whyatt, 1989). Some of these residues remain on some foods even after processing. The highest levels of exposures to pesticides, however, occur to pesticide applicators, to farm workers, and to people who live adjacent to heavily treated agricultural land (WHO/UNEP, 1989).

Induction of Cancers

Approximately 65% of all pesticides used in the United States are handled by farmers (Pimentel et al., 1991). Epidemiological studies of the health of farm workers have constituted an indirect, although consistent, line of inquiry into the carcinogenicity of pesticides. Although evidence is often circumstantial, studies generally have indicated increases in certain cancers in agricultural workers. In reviewing the evidence, Sharp et al. (1986) and Blair et al. (1985) reported significantly higher cancer incidence among farmers in the United States and Europe than among non–farm workers. They found that Wisconsin farmers from counties using large amounts of herbicides and insecticides showed higher risks of non-Hodgkin's lymphoma and multiple myeloma than other groups of farmers. Also, Iowa farmers in counties using large amounts of herbicides and insecticides were at significantly higher risk for multiple myeloma as well. Evidence from Sweden and New Zealand indicated increased risk of soft-tissue sarcomas among farmers and foresters exposed to herbicides (Blair et al., 1985). Further, they suggest that high nitrate intake from food (naturally occurring and from nitrogen fertilizers) and water polluted with pesticide runoff may be associated with stomach cancer. Nitrates may interact with common herbicides, such as atrazine, to produce nitrosamine, a known mutagen. An increased risk of leukemia was positively correlated with insecticide use in studies from Nebraska and Wisconsin and with herbicide use in Iowa (Blair et al., 1985). Their data also suggested an association between pesticide exposure and brain cancer. In Maryland, contact with insecticides by farm children resulted in more brain tumors than among control children (Blair et al., 1985). Further, an Italian study showed a sample of glioma patients to be more likely to have been employed in agricultural occupations after 1960 (when modern agricultural chemicals became commonplace in Italian agriculture) and to have worked for more than 10 years (Blair et al., 1985). Higher levels of chlorinated hydrocarbon compounds in adipose tissues were reported from glioblastoma patients than from noncancer controls, further supporting the association between pesticides and brain cancer (Blair et al., 1985).

Other recent studies have provided additional evidence linking pesticide exposure of farm workers to cancer incidence. In a population-based, case-control study of residents of Iowa and Minnesota, Brown et al. (1990) found a small, but significant, risk for all leukemias (odds ratio = 1.2) among persons who lived or worked on a farm as an adult. A significant correlation existed between risk for development of acute nonlymphocytic leukemia and use of any fungicide and between risk for chronic lymphocytic leukemia and use of any herbicide, any insecticide, and any animal insecticide. Leukemia risk was significantly elevated among farmers reporting use of natural product and organophosphate (OP) insecticides on animals. Within the OP class, a significant leukemia risk approximately twice the norm was associated with the use of halogenated aliphatic (dichlorvos, trichlorfon), halogenated aromatic (chlorpyrifos, coumaphos, crufomate, ronnel, tetrachlorvinphos), and nonhalogenated aromatic OPs (azinphos methyl, crotoxyphos, dioxathion, famphur, fensulfothion, methyl parathion, parathion, phosmet). Significant, elevated risks for chronic lymphocytic leukemia were associated with use of carbamates on crops and animals and with use of OPs on animals. Specific pesticides associated with increased risks included carbaryl (sevin), malathion, crotoxyphos, DDT, dichlorvos, famphur, methoxychlor, nicotine, and pyrethrins.

A study comparing male siblings of similar occupation found that patients with primary lung cancer had been exposed more frequently to herbicides, grains, and diesel fumes than siblings (McDuffie et al., 1988). Results suggested that farmers who develop lung cancer may have had more extensive exposures to agricultural chemicals than their farming, non-cancer-affected brothers.

In a study of Swedish grain millers, Alavanja et al. (1987) found significant excesses of non-Hodgkin's lymphoma associated with flour-milling, which is the sector of the milling industry thought to be the most frequent user of insecticides and fumigants, including aluminum phosphide (phosphine), carbon disulfide, carbon tetrachloride, ethylene dibromide, ethylene dichloride, malathion, and methyl bromide. Results suggested that the cancer patterns in the study population were consistent with the effects of pesticide exposure. However, the confounding influences of other important risk factors for cancer could not be discounted.

Morris et al. (1986) reported an almost threefold increase in risk for multiple myeloma in subjects exposed to pesticides as compared to controls. Because of the small number of subjects reporting exposure to various specific compounds, no firm conclusions could be drawn concerning the particular class or classes of pesticides responsible for the overall increase in cancer risk.

See et al. (1990) investigated genotoxicity in the urine of orchardists occupationally exposed to pesticides. Genotoxic activity (as indicated by clastogenic activity in Chinese hamster ovary cells) of urine specimens collected from workers during the spraying period was significantly elevated. Results suggested the potential mutagenic and carcinogenic hazard posed by occupational pesticide exposure. In a survey including halogenated hydrocarbons, organophosphates, carbamates, and other classes of pesticides, Börzsönyi et al. (1984) found 29 pesticides to be definite or suspected genotoxic carcinogens.

Regarding the incidence of cancer, the International Agency for Research on Cancer found "sufficient" evidence of carcinogenicity for 18 pesticides and "limited" evidence of carcinogenicity for an additional 16 pesticides based on animal studies (Lijinsky, 1989; WHO/UNEP, 1989). With humans the evidence concerning cancer is mixed. For example, a recent study in Saskatchewan indicated no significant difference in non-Hodgkin's lymphoma mortality between farmers and nonfarmers, whereas others have reported some human cancer (WHO/UNEP, 1989). A realistic estimate of the number of U.S. cases of cancer in humans due to pesticides is given by D. Schottenfeld (University of Michigan, PC, 1991), who estimated that less than 1% of the nation's cancer cases are caused by exposure to pesticides. Considering that there are approximately 1 million cancer cases/year (USBC, 1990), Schottenfeld's assessment suggests less than 10,000 cases of cancer due to pesticides per year.

Experimental data do confirm that several organochlorine pesticides cause cancer in mice and other animal species (NAS, 1982). Also, a differential susceptibility between the sexes may in some cases be operative. For example, in a recent study, in which treatments of chlordecone (Kepone) were administered to rats at concentrations to produce rat liver levels of the insecticide comparable to those measured in liver biopsies of male chemical workers previously heavily exposed, Sirica et al. (1989) found an obvious sex-related difference in the capacity of chlordecone to promote liver tumorigenesis. Unlike males, which exhibited only precancerous focal and nodular lesions in the liver, a significant proportion of females had grossly visible tumors that proved to be well-differentiated hepatocellular carcinomas. Results suggested that the higher liver tumor incidence in female rats might arise from a more extensive differential effect of the chlordecone promotion treatment on the peculiar hormonal milieu in this sex, particularly as it might affect estrogenic activity. The organophosphate pesticides, except for parathion, have not been found to be carcinogenic in laboratory animals. However, the NAS report (1982), warns that both organochlorine and organophosphate pesticides "have the potential to modify the activity of microsomal enzymes and to engage synergistic interactions."

Effects on Reproduction

Many other acute and chronic maladies are beginning to be associated with pesticide use. For example, the recently banned pesticide dibromochloropropane (DBCP) caused testicular dysfunction in animal studies (Foote et al., 1986; Sharp et al., 1986; Shaked et al., 1988) and was linked with infertility among human workers exposed to DBCP (Whorton et al., 1977; Potashnik and Yanai-Inbar, 1987). Also, a large body of evidence has been accumulated over recent years from animal studies suggesting pesticides can produce immune dysfunction (Devens et al., 1985; Olson et al., 1987; Luster et al., 1987; Thomas and House, 1989).

According to Thomas (1981), organophosphates generally do not pose the same, potentially adverse effects on female reproductive systems as the organochlorines. Parathion was shown to alter hepatic sex steroid metabolism and to induce some chromosomal damage to cells of the seminiferous tubules of laboratory animals.

Pesticides are implicated in fetal malformations, resorptions, and growth retardation in a variety of laboratory species (Thomas, 1981). Although the herbicide Agent Orange has been implicated, available evidence is inconsistent and inconclusive. Similarly, an association between exposure to organochlorine pesticides like DDT and spontaneous abortion and premature delivery is only suspect (Sharp et al., 1986). In general, fungicides do not appear to be highly embryotoxic.

However suggestive the link is between pesticide exposure and potential reproductive consequences in animals, caution should be exercised in projecting this relationship to humans. DBCP is the only pesticide conclusively found to have an adverse effect on human reproduction (Sharp et al., 1986).

However, results of large-scale statistical studies do offer suggestive evidence supporting a link between pesticide exposure and adverse reproductive effects in humans. For example, White et al. (1988) found evidence for an association between maternal exposure to agricultural chemicals and three major reproductive anomalies combined (neural tube defects, facial clefts, and bilateral renal agenesis), as well as spina bifida without hydrocephalus. Stronger evidence was found for an association between stillbirths and such exposure during the second trimester, and an exposure-response gradient was indicated. Risks were assessed based on an agricultural chemical exposure opportunity (ACEO) index, which took into account seven pesticide categories: fenitrothion formulations, aminocarb formulations, other insecticides, herbicides with some phenoxy component, herbicides with only phenoxy, chlorinated herbicides, and nonchlorinated herbicides.

Much of the lack of consistency in reproductive studies stems from problems of sample selection, methodology, and data analysis and interpretation (Rosenberg et al., 1987). According to Sharp et al. (1986), either the effects are weak at the exposure levels encountered by humans and thus difficult to detect epidemiologically, or there are no effects, and the apparent findings result from multiple comparisons, each study examining a variety of outcomes, some of which are by chance statistically significant. Often the necessarily large samples required for statistical testing of the rare reproductive effects are not met (Rosenberg et al., 1987). Then too, animal models are conducted under well-planned and controlled experimental conditions and thus may not bear any relationship to the doses and exposure duration to a toxin present under existing exposure conditions in the environment (Thomas, 1981).

Effects on the Immune and Nervous Systems

A large body of evidence has accumulated over recent years from laboratory animal studies suggesting that many pesticides and pesticide-related compounds, as well as impurities carried in by-products, can produce immune dysfunction (Devens et al., 1985; Olson et al., 1987; Luster et al., 1987; Thomas and House, 1989). The studies have involved acute and subchronic exposure regimes exposing animals to both high and low doses of test agents. These results, in turn, have prompted the study of immune function in human subjects inadvertently exposed to some pesticides. For example, in a study of women who had chronically ingested low levels of aldicarb-contaminated groundwater (mean = 16.6 ppb), Fiore et al. (1986) reported the subjects showed a significantly altered immune response, but did not have any adverse health problems.

Numerous animal studies have shown that the administration of toxic polyhalogenated aromatic hydrocarbons (particularly tetrachlorodibenzodioxin [TCDD]) to laboratory animals causes lymphoid atrophy, immunosuppression, and alterations to host resistance to challenge with infectious agents or transplantable tumor cells. Thymic atrophy, immunosuppression, and bone-marrow alterations are dominant symptoms of TCDD toxicity that occur in almost all animal species examined (Luster et al., 1987).

Results of animal studies also suggest that the immune system is more susceptible to immunomodulation by some pesticides (e.g., chlordane) during prenatal development than it is in the adult (Porter et al., 1984; Thomas and House, 1989).

Some studies suggest that pesticides may also be responsible for the development of hypersensitivity (allergy) as well as autoimmunity (Hamilton et al., 1978; Thomas and House, 1989).

Recent studies have also suggested a possible link between pesticide exposure and incidence of neurological disease. For example, in a comparison of 150 patients with Parkinson's disease with an equal number of age- and sex-matched controls, Koller et al. (1990) found a significant association between drinking water from rural wells and increased risk for the disease. The patients were more likely to have drunk well water and to have lived in a rural environment than the controls. Since many agricultural chemicals are leached from soil into groundwater, where concentrations may increase due to low rate of turnover of the water, results of the study suggested a potential role for pesticidal contamination in the induction of Parkinsonism.

Dietary Exposure to Pesticides

Residue analyses have shown that the main source of pesticide exposure for the general public is through food (Barthel, 1983). For example, it has been estimated that >90% of the DDT accumulated in human tissue came from food (Barthel, 1983). The EPA considers a substantial fraction (at least 62%) of all pesticides to be carcinogenic or potentially carcinogenic based on animal studies (GAO, 1986). On the basis of quantities applied, approximately 60% of all herbicides, 90% of fungicides, and about 30% of all insecticides fall into these categories (NAS, 1987). The EPA's classification system for carcinogens in humans is detailed in its recent report (EPA, 1987). The list of pesticides for which evidence is sufficient to indicate an association with carcinogenesis includes Amitrol, aramite, arsenic, chlordecone, BHC, Mirex, and toxaphene; those showing limited evidence are chlordane, 2,4-D, DDT, dieldrin, and lindane (Barthel, 1983; Börzsönyi et al., 1984).

For the 28 oncogenic pesticides included in the NAS (1987) assessment of risks, about 80% of the oncogenic risk is associated with raw foods and only 20% with processed foods.

Attempts have been made to estimate the potential cancer risk from pesticidal residues in foods. A list of the 10 major pesticides that the NAS (1987) identifies as posing risks includes alachlor, oxadiazon (herbicides);

chlordimeform (insecticide); and Zineb, Captafol, Captan, Maneb, Mancozeb, Folpet, and metirm (fungicides). For meat, milk, and other dairy and poultry products, herbicidal residues pose the highest risk, whereas the risk from fungicides are estimated to be negligible. However, when residues in all foods are considered, nearly 60% of the estimated carcinogenic risk is from fungicides, 27% from herbicides, and 13% from insecticides (NAS, 1987).

Approximately 20% of the current cancer risk is associated with consumption of processed foods; fungicides account for about 75% of this risk (NAS, 1987).

A single herbicide, linuron, accounts for >98% of the estimated dietary cancer risks caused by herbicides (NAS, 1987). Two insecticides, chlordimeform and permethrin, contribute >95% of the estimated dietary risk from insecticides (NAS, 1987).

Fifteen foods are associated with about 80% of all estimated dietary cancer risks from pesticide residues (NAS, 1987). The major foods include tomatoes, beef, potatoes, oranges, and lettuce. Herbicide residues in beef, potatoes, and pork account for >20% of the estimated dietary risk (mainly from linuron) (NAS, 1987). The risks from insecticides were distributed relatively evenly among the following 10 products: lettuce, chicken, beef, cottonseed, milk, tomatoes, pork, peaches, spinach, and cabbage. Fungicide residues constitute the highest risk (NAS, 1987). The estimated dietary carcinogenic risk from processed tomato products may be as high as 15% of the total carcinogenic risk attributed to all pesticide residues because of the high use of fungicides on tomatoes.

Some dietary components, such as fish, may pose a substantial cancer risk through bioaccumulation of pesticidal residues. For example, the FDA action level for DDT in fish is 5.0 mg/kg (ppm) and for dieldrin it is 0.3 ppm. Although concentrations of DDT and dieldrin in edible tissue of Great Lakes sport fish were below action levels in 1986, Foran et al. (1989) determined that cancer risk projections (based on guidelines of the U.S. EPA and Office of Science and Technology [OSTP] and the International Agency for Research on Cancer [IARC]) associated with tissue concentrations of 1.0 ppm DDT or 0.1 ppm dieldrin would range as high as 5.7×10^{-4} for DDT and 3.7×10^{-3} for dieldrin (or approximately four excess occurrences of cancer during the lifetimes of 1,000 exposed individuals living to age 70) at a consumption rate of 96 g/day (approximately three 227-g [one-half pound] portions/week) (the U.S. national average fish consumption is 6.5 g/day or less than 1 one-half-pound portion/month). At a fish consumption rate of 32 g/day (within the 10–50 g/day range that surveys suggest for an average consumption rate of sport fish by anglers and their families in the Greak Lakes basin), risk projections associated with these tissue concentrations would range as high as 2.0×10^{-4} for DDT and 1.5

$\times 10^{-3}$ for dieldrin. At the highest likely consumption rate (96 g/day) of fish contaminated with the two insecticides at their respective action levels, cancer risk projections would range as high as 2.8×10^{-3} for DDT and 1.05×10^{-2} for dieldrin. Based on these projections, Foran et al. (1989) suggested that current consumption advisories for Great Lakes sport fish are inadequate and that consumers may face significant cancer risks when tissue concentrations of DDT or dieldrin in fish are at or near the FDA action level; considerable risk might be incurred at one-third to one-fifth the action level if fish is consumed at least weekly.

Considering data from all sources, overall risk from pesticidal residues in food would appear to be relatively low. The NAS (1987) study, based on EPA methodology, concluded that the estimated additive cancer risk for the American public averages no greater than 1×10^{-3}.

The EPA method of estimating pesticide residues in food employs a defined set of assumptions, such as chemical application by growers at maximal recommended rates and at recommended frequencies, which may bias figures higher than actual levels (NAS, 1987). Some assays have found lower levels of residues in food than expected and such findings have implications for estimating cancer risks. For example, Hankin (1988) reported that when 283 samples of 44 different types of produce from Connecticut and elsewhere in the United States were tested for pesticide residues, traces of pesticides were found in 43% of the samples, but none contained residues above the allowable tolerances. The FDA (1989) reported that about 40% of the foods checked had detectable pesticide residues and that 3% exceeded the tolerance level. Thiodan was the pesticide most commonly detected.

Some uncertainty exists concerning the magnitude of dietary pesticide exposure of the general population and the significance this exposure has for public health. The cancer risks of several pesticides, as estimated by Archibald and Winter (1989), are lower than those of the NAS (1987) report. The major difference between the NAS and Archibald/Winter studies was that NAS assumed that the residues in the foods were at the accepted tolerance level, whereas Archibald and Winter used FDA data on actual residue levels (Anon., 1988). Whether the NAS or the Archibald/Winter study is closer to being correct in predicting cancer risk to the public needs to be studied.

Any assessment of cancer risk from exposure to carcinogens must take into consideration the role of dosage. Here, however, estimation of hazard posed by carcinogens has been hampered by a lack of data. Few dose-response studies over a large range of doses have been carried out, but in those that have, there have been significant tumor responses even at the lowest doses used (Lijinsky, 1989). These studies (e.g., Pitot et al., 1987; Lijinsky et al., 1988) do not suggest a threshold for carcinogens. Rather, a linear relationship of increasing risk with increasing exposure has been generally seen. Recent work suggests that the carcinogenic potential of pesticides may follow the same pattern. For example, Schröter et al. (1987) found that levels of the a, b, and g isomers of hexachlorocyclohexane (HCH) in rat tissue rose in an approximately linear fashion with increasing dose. At doses of 2–3 ppm or more, each of the isomers led to distinct increases of both the number of preneoplastic foci in the liver and their total area after initiation by N-nitrosomorpholine. Results suggested that HCH, at doses of 2 ppm/day and more, enhances phenotypic divergence between normal and altered hepatocytes, effects characteristic of liver tumor promoters. g-HCH (lindane) appeared to be severalfold more potent in stimulating foci expansion than the other isomers. Schröter et al. (1987:87) concluded that

> Since human milk contains relatively high levels of HCH human sucklings represent another subpopulation exposed to higher than average doses of HCH. Furthermore, quantitative risk estimates should take into account the presence of several other persistent organochlorine compounds in human foodstuffs; in human milk this contamination is 10-fold higher than HCH levels. The possibility of additive effects of many of these organochlorine contaminants should be considered in risk estimates.

This study and others suggest that it may not be possible to conduct valid experiments that allow establishment of a safe threshold for exposure of people to any carcinogen identified through animal experiments (Lijinsky, 1989). For this reason, U.S. regulatory agencies employ linear, nonthreshold models (Perera et al., 1991).

The scientific data concerning the overall safety of pesticides are not reassuring, and the health risks are sufficient to require the continued testing of pesticides and also to control applications to make certain public health is protected (Barthel, 1983; NAS, 1987). When the safety of a pesticide is suspected, ideally it should be removed from use until additional tests are conducted to confirm its safety. This strategy would alleviate public anxiety and avoid situations like the recent Alar/apple scare, which diminished the credibility of both government and industry scientists.

Conclusion

About 35% of the foods purchased by U.S. consumers have detectable levels of pesticide residues (FDA, 1990). Of this from 1% to 3% of the foods have pesticide residue levels above the legal tolerance level (Hundley et al., 1988; FDA, 1990). These residue levels may well be higher because the U.S. analytical methods now employed detect only about one-third of the more than 600 pesticides in use (OTA, 1988). Certainly the contamination rate is higher for fruits and vegetables because these foods receive the highest dosages of pesticides. Therefore, there are many reasons why 97% of the public is genuinely concerned about pesticide residues in their food (FDA, 1989).

Individuals differ in their perception of risk and the degree of risk they are willing to accept. With food selection, most individuals have the option of making personal choices. However, the acknowledged health benefits of the foods recommended by nutritional authorities are such that consumers should not be frightened into eliminating them from their diets because of the implied threat of danger from naturally occurring toxicants.

In drawing conclusions from risk analyses of dietary exposure to toxicants, important caveats should be noted. Short-term screens (e.g., the "Ames test"), whether for genetic damage or increased cell proliferation, are far from 100% predictive of carcinogenicity and, thus, not a replacement for long-term bioassays (Cohen and Ellwein, 1991). Also, no matter how suggestive epidemiological or experimental studies may be, they cannot provide unequivocal proof that a certain diet will increase the risk of cancer. No study has directly demonstrated that implementing dietary changes in a given individual inhibited the onset of cancer or kept an established cancer from spreading. As Cohen (1987:48) has remarked: "Definitive data could be obtained from experiments involving people. . . . [However,] the logistics would be formidable and the cost astronomical, and the outcome might not be known for from 10 to 20 years."

Furthermore, unlike animal experimentation, humans cannot be kept physically isolated for long periods of time and fed diets containing possible toxic substances. Nor can heredity or environmental factors be controlled. Data from laboratory animal tests and epidemiological studies with humans must serve as guides for assessing the safety of the food supply. Ultimately, it is extremely difficult, in the absence of further information, to predict the sensitivity of humans to the tumor-promoting, mitogenic, or cytotoxic potential of a given compound. Thus, risk extrapolation under conditions in which individuals are exposed to multiple factors (the situation in the real world), and in heterogeneous populations, is much more complicated than envisioned by some authors (e.g., Ames and Gold, 1990; Weinstein, 1991).

Risk from pesticide residues as well as naturally occurring toxicants in foods depends on the dosage of the chemical, time of exposure, and susceptibility of the individual human. These data along with sound experimental investigations of the particular pesticide and natural toxicant are essential to estimate potential risks to humans of various toxic chemical exposure in human foods.

Plant foods do contain many chemicals, some of which (e.g., hydrazines and mycotoxins) are highly toxic to animals and humans. While these compounds may play important roles in influencing the incidence of certain types of human cancer, the exact proportion of cancers that are due to "natural" vs. synthetic carcinogens is not known (Perera et al., 1991). Again, referring to most other natural toxicants, the National Academy of Sciences (NAS, 1989a) reports there "is no evidence that any of these substances individually makes a major contribution to cancer risk in the United States." Further, as Davis (1987) notes, "humans have been eating complex foods for longer than they have been exposed to synthetic, organic carcinogens." However, there is evidence to suggest that synthetic chemicals present in food may increase cancer risk over that which may be posed by the presence of natural toxins alone. For example, laboratory rodent diets also contain many of the same naturally occurring toxins present in the human diet. Nevertheless, several compounds, such as aflatoxin, TCDD, and DBCP, when added to the diet of mice and rats, significantly increase tumor incidence, even when present at very low levels. This suggests that, in several cases, the risk of tumorigenesis from certain synthetic food contaminants is increased in the animal over any risk presented by the background level of "natural pesticides" (Weinstein, 1991). Lacking contrary evidence, there is no reason to assume a difference in humans.

With pesticides, most individuals feel they have little or no choice but must depend on the integrity of scientists and government agencies to ban dangerous pesticides and regulate the dosages and application procedures of those permitted. The last decade has witnessed a growing awareness in the public sector about the chemicals they are exposed to in their foods, air, and water. These are perceived as added risks, ones over which the individual has no control. Some of this concern is being translated into action, individually to buy pesticide-free food and collectively—as the "Big Green" initiative in California—through the ballot box, to eliminate the most toxic pesticides.

The causes of chronic illnesses, including cancers, are extremely complex. In their lifetime, individuals, each differing in genetic makeup and susceptibility, are exposed to a wide variety of carcinogens. Some chemicals by themselves are safe, but may well act as synergists or promoters, in concert with other chemicals, to cause illness. Future research as to how human health is affected by increasing exposure to all chemicals is of prime importance. As Perera et al. (1991:904) have suggested:

> Risks from both natural and synthetic carcinogens are of concern. The appropriate policy for natural carcinogens is to test suspect constituents and to advise and educate the public about dietary factors that may be either hazardous or protective. Indeed, the American Cancer Society, the National Cancer Institute, and other organizations are already doing this. This policy for synthetic carcinogens is testing and regulation of those that pose significant risks, with use of the most cost-effective meas-

> ures to reduce human exposure. This, in fact, is also the current policy of U.S. regulatory agencies. Ignoring the potential health hazards of synthetic carcinogens is antithetical to current preventive public health policies in the United States and many other countries.

The public is confused and is ill served by the conflicting information and risk assessments it receives (e.g., Ames vs. NAS). As Lijinsky (1989:569) has stated:

> In view of the small amount of information about the mechanisms by which chemicals give rise to cancer (and the uncertainty about the relevance of that information), it is unwise to permit officials or experts to calculate tolerable or 'safe' exposures for humans to carcinogens. All of us are fallible even when armed with sound information. Reliable information about carcinogens is limited almost to whether or not the substance is one.

No wonder the public is skeptical of what it reads and hears, and is becoming more wary about exposure to chemicals, like pesticides, that cannot be avoided.

Acknowledgments

We thank J. Blondell (U.S. Environmental Protection Agency), B. Robinson (Connecticut Department of Environmental Protection), J. Thomson (Connecticut Poison Control Center, University of Connecticut), and D. Brown (Connecticut Health Department) for providing information.

References

AICR. 1984. Planning meals that lower cancer risk: a reference guide. Washington, D.C.: Amer. Inst. for Cancer Res. 88 pp.

Alavanja, M.C.R., H. Malker, and R.B. Hayes. 1987. Occupational cancer risk associated with the storage and bulk handling of agricultural foodstuff. J. Toxicol. Environ. Health **22**:247–254.

Ames, B.N. 1989. Pesticide residues and cancer causation, pp. 223–237. *In* N.N. Ragsdale and R.E. Menzer (eds.). Carcinogenicity and pesticides: principles, issues, and relationships (ACS Symp. Ser. 414). Washington, D.C.: Amer. Chem. Soc. 246 pp.

Ames, B.N. and L.S. Gold. 1989. Pesticides, risks and applesauce. Science **244**:755–757.

Ames, B.N. and L.S. Gold. 1990. Too many rodent carcinogens: mitogenesis increases mutagenesis. Science **249**:970–971.

Ames, B.N., R. Magaw, and L.S. Gold. 1987. Ranking possible carcinogenic hazards. Science **236**:271–280.

Anonymous. 1988. Food and Drug Administration pesticide program: residues in foods—1987. J. Assoc. Off. Anal. Chem. **71**:156A–158A.

Archibald, S.O. and C.K. Winter. 1989. Pesticide residues and cancer risks. Calif. Agric. **43**(6):6–9.

Barthel, E. 1983. Pesticides and cancer risk. Z. Erkrank. Atm.-Org. **161**:257–265.

Bender, A.E. 1987. Effects on nutritional balance: antinutrients, pp. 110–124. *In* D.H. Watson (ed.). Natural toxicants in food: progress and prospects. Chichester, England: Ellis Horwood Ltd. 254 pp.

Blair, A., H. Malker, K.P. Cantor, L. Burmeister, and K. Wiklund. 1985. Cancer among farmers: a review. Scand. J. Work Environ. Health **11**:397–407.

Börzsönyi, M., G. Török, A. Pintér, and A. Surján. 1984. Agriculturally-related carcinogenic risk. IARC Scient. Publ. **56**:465–486.

Brown, L.M., A. Blair, R. Gibson, G.D. Everett, K.P. Cantor, L.M. Schuman, L.F. Burmeister, S.F. Van Lier, and F. Dick. 1990. Pesticide exposures and other agricultural risk factors for leukemia among men in Iowa and Minnesota. Cancer Res. **50**:6585–6591.

Cohen, L.A. 1987. Diet and cancer. Sci. Am. **257**(5): 42–48.

Cohen, S.M. and L.B. Ellwein. 1991. Carcinogenesis mechanisms: the debate continues. Letters. Science **252**:902–903.

Coon, J.M. 1975. Natural toxicants in foods. J. Am. Dietet. Assoc. **67**:213–218.

Davis, D.L. 1987. Paleolithic diet, evolution, and carcinogens. Science **238**:1633.

Devens, B.H., M.H. Grayson, T. Imamura, and K.E. Rodgers. 1985. O,O,S-trimethyl phosphorothioate effects on immunocompetence. Pestic. Biochem. Physiol. **24**:251–259.

EPA. 1987. Pesticide poisoning summary. Unpublished report. U.S. Environmental Protection Agency. Office of Pesticides and Toxic Substances. Health Effects Div., Washington, D.C.

FDA. 1989. Food and drug administration pesticide program. Residues in foods. 1988. J. Assoc. Off. Anal. Chem. **72**:133A–152A.

FDA. 1990. Food and Drug Administration Pesticide Program Residues in Foods—1989. J. Assoc. Off. Anal. Chem. **73**:127A–146A.

Fiore, M.C., H.A. Anderson, R. Hong, R. Golubjatnikov, J.E. Seiser, D. Nordstrom, L. Hanrahan, and D. Belluck. 1986. Chronic exposure to aldicarb-contaminated groundwater and human immune function. Environ. Res. **41**:633–645.

Foote, R.H., E.C. Schermerhorn, and M.E. Simkin. 1986. Measurement of semen quality, fertility, and reproductive hormones to assess dibromochloropropane (DBCP) effects in live rabbits. Fund. and Appl. Tox. **6**:628–637.

GAO. 1986. Pesticides: EPA's formidable task to assess and regulate their risks. GAO/Rced-86-125. Washington, D.C.: Govt. Accounting Off. April 1986.

Hamilton, H.E., D.P. Morgan, and A. Simmons. 1978. A pesticide (dieldrin)-induced immunohemolytic anemia. Environ. Res. **17**:155–164.

Hankin, L. 1988. Pesticide residues in produce sold in Connecticut. Conn. Agric. Exp. Stn. Bull. 863. 7 pp.

Heaney, R.K. and G.R. Fenwick. 1987. Identifying toxins and their effects: glucosinolates, pp. 76–109. *In* D.H. Watson (ed.). Natural toxicants in food: progress and prospects. Chichester, England: Ellis Horwood Ltd. 254 pp.

HHS. 1985. Nutrition and your health: dietary guidelines for Americans, 2nd ed. Washington, D.C.: Dep. of Health and Human Services.

HHS. 1988. The Surgeon General's report on nutrition and health. HHS Publ. 88-50210. Washington, D.C.: Dep. of Health and Human Services. 246 pp.

Hirayama, T. 1979. Diet and cancer. Nutr. Cancer **1**:67–81.

Hundley, H.K., T. Cairns, M.A. Luke, and H.T. Masumoto. 1988. Pesticide residue findings by the Luke method in domestic and imported foods and animal feeds for the fiscal years 1982–1986. J. Assoc. Off. Anal. Chem. **71**:875–877.

Koller, W., B. Vetere-Overfield, C. Gray, C. Alexander, T. Chin, J. Dolezal, R. Hassanein, and C. Tanner. 1990. Environmental risk factors in Parkinson's disease. Neurology **40**:1218–1221.

Lee, H.P., I. Gourley, S.W. Duffey, J. Esteve, J. Lee, and N.E. Day. 1989. Colorectal cancer and diet in an Asian population—a case control study among Singapore Chinese. Int. J. Cancer **43**:1007–1016.

Levin, D.A. 1976. Chemical defenses of plants to pathogens and herbivores. Ann. Rev. Ecol. Syst. **7**:121–159.

Liener, I.E. 1986. The nutritional significance of naturally occuring toxins in plant foodstuffs, pp. 72–94. *In* J.B. Harris (ed.). Natural toxins: animal, plant, and microbial. Oxford: Clarendon Press. 353 pp.

Lijinsky, W. 1989. Prepared statement, pp. 567–571. *In* U.S. Congress. Senate. Committee on Labor and Human Resources. Hearing on food safety amendments of 1989, held June 6, 1989. Washington: Govt. Printing Off; 1989; 101st Congress, 1st Session; 572 pp.

Lijinsky, W., R.M. Kovatch, C.W. Riggs, and P.T. Walters. 1988. Dose response study with N-nitrosomorpholine in drinking water of F-344 rats. Cancer Res. **48**:2089–2095.

Litovitz, T.L., B.F. Schmitz, and K.M. Bailey. 1990. 1989 Annual report of the American Association of Poison Control Centers National Data Collection System. Amer. J. Emergency Med. **8**:394–442.

Litovitz, T.L., B.F. Schmitz, and K.C. Holm. 1989. 1988 annual report of the American Association of Poison Control Centers National Data Collection System. Am. J. Emer. Med. **7**:495–545.

Luster, M.I., J.A. Blank, and J.H. Dean. 1987. Molecular and cellular basis of chemically induced immunotoxicity. Ann. Rev. Pharmacol. Toxicol. **27**:23–49.

McDuffie, H.H., D.J. Klaassen, D.W. Cockcroft, and J.A. Dosman. 1988. Farming and exposure to chemicals in male lung cancer patients and their siblings. J. Occup. Med. **30**:55–59.

Morris, P.D., T.D. Koepsell, J.R. Daling, J.W. Taylor, J.L. Lyon, G.M. Swanson, M. Child, and N.S. Weiss. 1986. Toxic substance exposure and multiple myeloma: a case-control study. J. Natl. Cancer Inst. **76**:987–994.

NAS. 1973. Toxicants occurring naturally in foods, 2nd ed. Washington, D.C.: National Academy Press. 624 pp.

NAS. 1982. Diet, nutrition and cancer. Washington, D.C.: National Academy Press. 316 pp.

NAS. 1987. Regulating pesticides in foods: the Delaney paradox. Washington, D.C.: National Academy Press. 272 pp.

NAS. 1989a. Diet and health: implications for reducing chronic diseases. Washington, D.C.: National Academy Press. 749 pp.

NAS. 1989b. Recommended dietary allowance, 10th ed. Washington, D.C.: National Academy Press. 284 pp.

NCI. 1988a. Eating for life. NIH Publ. 88-3000. Washington, D.C.: National Cancer Inst.

NCI. 1988b. Cancer prevention brief: nutrition. Washington, D.C.: National Cancer Inst. 7 pp.

Neal, G.E. 1987. Influences of metabolism: aflatoxin metabolism and its possible relationships with disease, pp. 125–168. *In* D.H. Watson (ed.). Natural toxicants in food: progress and prospects. Chichester, England: Ellis Horwood Ltd. 254 pp.

Olson, L.J., B.J. Erickson, R.D. Hinsdill, J.A. Wyman, W.P. Porter, L.K. Binning, R.C. Bidgood, and E.V. Nordheim. 1987. Aldicarb immunomodulation in mice: an inverse dose-response to parts per billion levels in drinking water. Arch. Environ. Contam. Toxicol. **16**:433–439.

Oser, B.L. 1978. Natural toxicants in foods. New York St. J. Med. **78**:684–685.

OTA. 1988. Pesticide residues in food: technologies for detection. Washington, D.C.: Office of Technology Assessment. U.S. Congress.

Pamukcu, A.M., S. Yalçiner, J.F. Hatcher, and G.T. Bryan. 1980. Quercetin, a rat intestinal and bladder carcinogen present in bracken fern (*Pteridium aquilinum*). Cancer Res. **40**:3468–3471.

Perera, F.P., D.P. Rall, and I.B. Weinstein. 1991. Carcinogenesis mechanisms: the debate continues. Letters. Science **252**:903–904.

Pimentel, D. 1988. Herbivore population feeding pressure on plant hosts: feedback evolution and host conservation. Oikos **53**:289–302.

Pimentel, D. 1990. The potential impact of the withdrawal of 19 pesticides based on the proposed environmental protection initiative: a preliminary assessment. Draft prepared for Natural Resources Defense Council.

Pimentel, D. and L. Levitan. 1986. Pesticides: amounts applied and amounts reaching pests. BioScience **36**:86–91.

Pimentel, D., L. McLaughlin, A. Zepp, B. Lakitan, T. Kraus, P. Kleinman, F. Vancini, W.J. Roach, E. Grapp, W.S. Keeton, and G. Selig. 1991. Environmental and economic impacts of reducing U.S. agricultural pesticide use, pp. 679–718. *In* D. Pimentel (ed.). Handbook of pest management in agriculture, 2nd ed. Boca Raton, FL: CRC Press.

Pitot, H.C., T.L. Goldsworthy, S. Moran, W. Kennan, H.P. Glauert, R.R. Maronpot, and H.A. Campbell. 1987. A method to quantitate the relative initiating and promoting potencies of hepatocarcinogenic agents in their dose-response relationships to altered hepatic foci. Carcinogenesis **8**:1491–1499.

Porter, W.P., R. Hinsdill, A. Fairbrother, L.J. Olson, J. Jaeger, T. Yuill, S. Bisgaard, W.G. Hunter, and K. Nolan. 1984. Toxicant-disease-environment interactions associated with suppression of immune system, growth, and reproduction. Science **224**:1014–1017.

Potashnik, G. and I. Yanai-Inbar. 1987. Dibromochloropropane (DBCP): an 8-year reevaluation of testicular function and reproductive performance. Fertility and sterility. **47**:317–323.

Rosenberg, M.J., P.J. Feldblum, and E.G. Marshall. 1987. Occupational influences on reproduction: a review of recent literature. J. Occup. Med. **29**:584–591.

Schröter, C., W. Parzefall, H. Schröter, and R. Schulte-Hermann. 1987. Dose-response studies on the effects of a-, b-, and g-hexachlorocyclohexane on putative preneoplastic foci, monooxygenases, and growth in rat liver. Cancer Res. **47**:80–88.

See, R.H., B.P. Dunn, and R.H.C. San. 1990. Clastogenic activity in urine of workers occupationally exposed to pesticides. Mutation Res. **241**:251–259.

Sewell, B.H. and R.M. Whyatt. 1989. Intolerable risk: pesticides in our children's food. New York: Natural Resources Defense Council. 141 pp.

Shaked, I., U.A. Sod-Moriah, J. Kaplanski, G. Potashnik, and Buckman. O. 1988. Reproductive performance of dibromochloropropane-treated female rats. Int. Jour. Fert. **33**:129–133.

Sharp, D.S., B. Eskenazi, R. Harrison, P. Callas, and A.H. Smith. 1986. Delayed health hazards of pesticide exposure. Ann. Rev. Public Health **7**:441–471.

Sirica, A.E., C.S. Wilkerson, L.L. Wu, R. Fitzgerald, R.V. Blanke, and P.S. Guzelian. 1989. Evaluation of chlordecone in a two-stage model of hepatocarcinogenesis: a significant sex difference in the hepatocellular carcinoma incidence. Carcinogenesis. **10**:1047–1054.

Smith, A.H. and M.N. Bates. 1989. Epidemiological studies of cancer and pesticide exposure, pp. 207–222. *In* N.N. Ragsdale and R.E. Menzer (eds.). Carcinogenicity and pesticides: principles, issues, and relationships (ACS Symp. Ser. 414). Washington, D.C.: Amer. Chem. Soc. 246 pp.

Stimmann, M.W. and M.P. Ferguson. 1990. Potential pesticide use cancellations in California. Calif. Agric. **44**(4): 12–16.

Strong, F.M. 1974. Toxicants occurring naturally in foods. Nutr. Rev. **32**:225–231.

Thomas, J.A. 1981. Reproductive hazards and environmental chemicals: a review. Toxic Substances J. **2**:318–348.

Thomas, P.T. and R.V. House. 1989. Pesticide-induced modulation of the immune system, pp. 94–106. *In* N.N. Ragsdale and R.E. Menzer (eds.). Carcinogenicity and pesticides: principles, issues, and relationships (ACS Symp. Ser. 414). Washington, D.C.: Amer. Chem. Soc. 246 pp.

Tufts University. 1989. Diet and Nutrition Letter (Medford, MA) **7**(8): 1–2.

USBC. 1990. Statistical Abstract of the United States 1990. Washington, D.C.: U.S. Bureau of the Census, U.S. Government Printing Office.

USDA. 1989. Agricultural statistics. Washington, D.C.: U.S. Government Printing Off. 547 pp.

USDA. 1990. Agricultural statistics. Washington, D.C.: U.S. Government Printing Off.

Warren, V.A., V.N. Hillers, and G.E. Jennings. 1990. Beliefs about food supply safety: a study of cooperative extension clientele. J. Am. Dietet. Assoc. **90**:713–714.

Watson, D.H. 1987. Introduction, pp. 9–10. *In* D.H. Watson (ed.). Natural toxicants in food: progress and prospects. Chichester, England: Ellis Horwood Ltd. 254 pp.

Weinstein, I.B. 1991. Mitogenesis is only one factor in carcinogenesis. Science **251**:387–388.

White, F.M.M., F.G. Cohen, G. Sherman, and R. McCurdy. 1988. Chemicals, birth defects and stillbirths in New Brunswick: associations with agricultural activity. Can. Med. Assoc. J. **138**:117–124.

WHO/UNEP. 1989. Public health impact of pesticides used in agriculture. Geneva: World Health Organization/Nairobi: United Nations Environment Programme. 140 pp.

Whorton, D., R.M. Krauss, S. Marshall, and T.H. Milby. 1977. Infertility in male pesticide workers. The Lancet Dec. 17, 1977: 1259–1261.

Part II

Methods and Effects of Reducing Pesticide Use

7

Socioeconomic Impacts and Social Implications of Reducing Pesticide and Agricultural Chemical Use in the United States

Frederick H. Buttel

Introduction

The struggle over agricultural pesticides, which will soon enter its fourth decade, has long involved conflicting interests, opposing ideologies, and contradictory technical and social science data. This paper will focus on one dimension of these conflicts: debate over the prospective socioeconomic impacts of substantial reductions in the use of pesticides and other agricultural chemicals. Research on this topic is obviously central to pesticide politics and policy, since both sides of the struggle ultimately base their positions on claims that chemicals are or are not integral to the social and economic well-being of agriculturalists and of society as a whole.

We cannot, however, consider the data and knowledge claims of social scientists and others on the socioeconomic implications of reduced pesticide usage in a vacuum. The nature of these data and how they are employed in the policymaking process are shaped by what might be called the "new politics of agricultural chemicals," which will be briefly portrayed as a preface to my discussion of the current knowledge base on the social consequences of reducing pesticide use.

The New Politics of Agricultural Chemicals

Prior to 1962, when Rachel Carson's bombshell, *Silent Spring*, was published, and 1963, when the thrust of Carson's book became the subject of a prime-time documentary on a major television network, the means for control of weeds and agricultural pests and pathogens were of little concern to the American public and agricultural policymakers. While there are now widely recognized environmental and health shortcomings of the 1940s and 1950s generation of industrially developed pesticides, these chemicals were generally less toxic than the arsenicals and copper compounds used prior to World War II. The fact that these chemicals were viewed as relatively benign compared to their predecessors, plus their labor-saving role in the (relatively) high-wage U.S. agricultural economy, contributed to the lax regulatory environment that continued until the Environmental Defense Fund won a case against the use of DDT before an administrative law hearing in the State of Wisconsin in 1969. Shortly thereafter the Federal Environmental Pesticide Control Act of 1972 was passed. This Act, which remains the basic statutory authority for pesticide regulation policy, gave the Environmental Protection Agency (EPA) the authority to regulate pesticides on environmental and health grounds and authorized amendments to the Federal Insecticide, Fungicide, and Rodenticide Act of 1947 that shifted the focus of pesticide regulation from efficacy to health and safety.[1] Even environmentalists were relatively disinterested in agriculture until the late 1960s and early 1970s, as the American conservation movement continued to devote the bulk of its efforts to wilderness and wild area preservation and to wildlife conservation. *Silent Spring*, however, would ultimately prove pivotal in ushering in a new politics of agricultural chemicals, which has had several distinctive characteristics.

First, the agricultural environment has slowly but surely become one of the major foci of the modern environmental movement. While post-Carson agricultural-environmental activism went through a 1970s phase of preoccupation with issues such as loss of agricultural land and the high levels of energy consumption in on-farm production and the food system as a whole, environmental activism over the past decade has become ever more firmly focused on agricultural chemicals. Along with the land-degradation and water-quality implications of soil erosion, tillage practices, and cultivation of fragile lands, pesticides and fertilizer use in agriculture remain cornerstone issues of contemporary environmental activism.

Second, agricultural chemicals, particularly pesticides, involve multiple fora or pressure points for activism. These chemicals have on-site and off-site environmental impacts (particularly on water quality) at the level of agricultural production. Agricultural chemicals also have human health implications for both agricultural workers (who are arguably most directly affected) and consumers. As Sachs notes elsewhere in this volume, public concern with the health and environmental effects of agricultural pesticides has been one of the most prominent and durable aspects of public opinion for two decades. Recent data also suggest growing concern about the health and environmental impacts of pesticides. This concern has translated into

[1]See Bosso (1987) and Reichelderfer and Hinkle (1989) for useful overviews of the evolution of American pesticide policy.

increased attention to agricultural chemicals by the environmental movement.

Third, environmentalists have come to recognize the many ways agriculture affects the environment, and the centrality of agricultural technology in exacerbating or ameliorating these impacts. These groups now consider federal[2] and land-grant agricultural research policy central to their agenda, and to be their business. While land-grant administrators were generally reluctant to admit environmentalists to the table of their major clientele groups in the 1960s and 1970s, they now must do so, grudgingly or not, since active environmental opposition to a state land-grant university's agricultural research program threatens the traditionally cordial relations between land-grant universities and state legislatures. It has not been unknown for a dean of agriculture or an agricultural experiment station director who was recalcitrant in recognizing environmentalists and "alternative agriculturalists" as bona fide or legitimate constituents of colleges of agriculture to be pushed aside by a university president or chancellor concerned with the university's image in state houses and state legislatures.

Fourth, the 1980s are now recognized as a decade during which "green forces" rose to prominence across essentially all of the advanced industrial countries. "Greening," which can be seen as an extension and expansion of the late 1960s and 1970s agenda of environmental organizations, has been strongly focused on agriculture, in both the developed and developing countries. One of the major dimensions of greening has been the rise of "sustainable development" vis-à-vis international development policy. (See, for example, Adams, 1990; Redclift, 1987.) In the advanced industrial countries, the rise of Green has led to a rapid rise in prominence of programs and notions such as low-input sustainable agriculture (LISA). (See, for example, Edwards et al., 1990.) Today most land-grant universities have LISA-type research programs. Though these LISA programs are typically small in nature, sustainable agriculture has come to be enough of a sacred cow so that the length and breadth of the agricultural establishment that once ridiculed alternative and organic agriculture finds it obligatory

[2]Environmental groups constitute one of the major "players" in shaping the past two farm bills, and they have scored a number of major successes (authorizing the LISA—low-input sustainable agriculture—research program, "swampbuster" legislation, and the Conservation Reserve Program) (Reichelderfer, 1990). Why environmental groups have been more successful during the Reagan/Bush years than during previous (more proenvironmental) administrations cannot be explained solely by the increasingly "Green" cast of modern politics. It must be recognized as well that the environmental provisions of these farm bills were to a significant degree a disguised means of supply control that passed muster with the Reagan and Bush administrations, despite their ideological reservations, because these provisions would reduce farm production and raise farm prices with modest federal outlays (*The Economist*, 1990).

to embrace LISA-type symbols. That sustainable agriculture has a very substantial foothold in the agricultural research community is attested to by the NRC's (1989) report, *Alternative Agriculture*, which took a prosustainability and pro-LISA position that would have been unimaginable a decade ago. Also, the rise of Green has led to growing pressure in national political fora to tighten and extend restrictions on agricultural chemical use. These efforts have led to some notable policy breakthroughs in a few countries, particularly some of the countries in Nordic Europe, whose programs for reducing pesticide use are discussed briefly in appendix A.

Fifth, while it was the case as recently as a decade ago that the bulk of the farming community was of a single mind with the agricultural establishment in opposing attempts by environmentalists to restrict agricultural chemical usage, there appears to be growing support among farming circles for low- or no-chemical farming practices. This shift has been accounted for mainly by a number of commercial-scale, and otherwise "conventional," farmers becoming more interested in low-input practices, rather than by the growth of "organic farming," as commonly understood in its relatively pure or uncompromising form (Buttel and Gillespie, 1988; Buttel et al., 1990). The motivations of farmers who express growing interest in lower-input practices are diverse, involving concerns such as farm household health and interest in reducing input costs[3] and preventing consumer resistance to or regulatory scrutiny of pesticide residues on agricultural products. Nonetheless, the combination of a growing cadre of "sustainability researchers" and growth of support in the farmer constituency of low-input practices has given environmental groups even more of a foothold in national and land-grant research policymaking.

The principal implication of the new politics of agricultural chemicals should not necessarily be seen as an exponential accumulation of political pressure and scientific expertise being brought to bear against current pesticide and other agricultural chemical practices. This *may* prove to be the case, but whether this occurs in a widespread way outside of the strongly Green Nordic countries of Europe is unclear at this point. Instead, there has been a trend toward more parity between pro- and antichemical forces within land-grant universities, within the farming community, and within national policymaking apparatuses. There nonetheless remain, within the current motherhood stage of Greening and embracement of the symbol of sustainability, very strong forces for the preservation of traditional chemical practices. The bulk of agronomists, horticulturalists, and entomologists

[3]It is generally estimated, for example, that pesticides account on average for about 20% of total (variable plus fixed) agricultural input costs in the United States (NRC, 1989:175). For corn, for example, fertilizers and pesticides account for about 55% of variable costs and 34% of total costs (NRC, 1989:38).

within land-grant universities are by no means ready to announce the end of the agrochemical age (a particularly dramatic expression of which is CAST [1990a], with the provocative title, *Alternative Agriculture: Scientists' Review*,[4] which is a collection of articles from this wing of the agricultural research community aimed at directly refuting the NRC [1989] *Alternative Agriculture* report referred to earlier). Agricultural chemical manufacturers, the bulk of which are non–consumer-goods companies and are inexperienced in dealing with a public skittish about "chemicals," are seldom shy to roll out "consumer education" programs defending pesticides and appear prepared to protect their chemical markets. The largest and most powerful American farmer organization, the Farm Bureau, remains (cautiously) supportive of chemicals. Many state-level agricultural commodity groups have been more than willing to struggle very actively and to spend large sums to prevent major restrictions on pesticides (e.g., the bulk of agricultural commodity groups in California, which joined the chemical industry coalition against the "Big Green" referendum in California in 1990).[5] There is no evidence that the U.S. Congress, much less the White House or the U.S.D.A., is interested in any policy initiatives such as those adopted in Nordic Europe in the mid- and late 1980s. Further, pesticide usage has increased substantially since the dawn of the new politics of agricultural chemicals in the early to mid-1970s (Conservation Foundation, 1987), though due to the farm crisis, overproduction, and the land-idling impacts of the past two farm bills, pesticide usage has declined in recent years (NRC, 1989:44–45). The outlook is one of a protracted political and intellectual struggle.

One of the major factors that is likely to shape the outcome of the new politics of agricultural chemicals consists of research, data, and knowledge claims relating to the prospective socioeconomic implications of reduced chemical usage. Ultimately, data of this sort will begin to shed light on whether the socioeconomic and environmental benefits of reducing pesticide usage outweigh the disadvantages. It is important to stress, however, that socioeconomic "impact" data were not particularly important in leading to the policies in Nordic Europe that are leading to reduced pesticide usage there, so one should have no illusions that convincing data on this score will alone be the key factor that will shape ultimate policy outcomes

[4]A companion publication to CAST (1990a) is CAST (1990b), which was prepared to address the "Alar controversy" initiated by the Natural Resources Defense Council in 1989.

[5]It is important to note, however, that farmer opposition to regulation of agricultural chemicals and veterinary pharmaceuticals is not accounted for largely, or even mainly, by direct material interests. Survey data, for example, show that farmer opposition to regulation of pharmaceuticals or pesticides is not related to whether farmers actually utilize these substances. The major correlates of opposition to regulation are political, ideology and, secondarily, farm size and assets (Gillespie and Buttel, 1989).

in the United States. For one thing, the socioeconomic issues involved and their policy implications are enormously complex, so neither side is likely to be able to call on unambiguous results that clearly support their policy preferences. Also, as I will stress later, this area of social science research is a somewhat unusual one in that it depends on natural science input for its "technical coefficients," and its own results are subject to the ideological polarities portrayed earlier.

Why do Farmers Use Pesticides?

Ultimately, whether the benefits of reducing chemical usage are greater than the costs depends on why farmers use pesticides in the first place. After all, most farmers, to the degree (which is substantial) that they are able to make production practice decisions without regulatory constraint, choose to employ substantial amounts of pesticides. A good number of them have chosen to go to considerable lengths to preserve their prerogative to use agricultural chemicals. Any social scientist or policymaker who wishes to understand the social and ethical issues involved in reducing chemical usage must therefore confront the fundamental question of why, especially from World War II to the present, farmers have tended voluntarily to use plant protection chemicals.

There are several basic explanations. Each has an element of validity. I will present each in the form of a historical synopsis of the emergence of pesticides and fertilizers as integral components of modern agricultural production practices in the United States and other advanced countries.

The "agrochemical age"[6] in the U.S. has had two principal historical antecedents. First, even well prior to the dawn of the (nitrogen [N] fertilizer) agrochemical age, agricultural research was very strongly oriented to agricultural chemistry (Rossiter, 1975). One of the most visible contribu-

---

[6]There have actually been several different agrochemical ages. Naturally occurring inorganic fertilizers (e.g., guano, bone manures, potash) had been employed for centuries (Goodman et al., 1987), and by the middle of the 19th century superphosphate production facilities had become common in several of the industrial countries. Ever since the pioneering work of the German agricultural chemist Liebig the efficacy of fertilizers in enhancing fertility and yields had been recognized scientifically. But for all practical purposes the first agrochemical age—that of widespread use of industrially produced nitrogen fertilizer—did not come about until the interwar years, following the major technical innovations in synthetic ammonia production (especially the Haber-Bosch process) that occurred during and shortly after World War I. Expansion of N fertilizer use was given particular impetus in the aftermath of World War II, since the U.S. government–subsidized buildup of the munitions industry during the war led to massive overcapacity in ammonia production, which during peacetime reduced N fertilizer prices and encouraged even greater use in agriculture. The next chemical age, that pertaining to industrially produced plant protection chemicals, emerged during the 1940s when DDT, BHC, and 2,4-D were discovered.

tions of early agricultural research in this genre made possible documented successes in the use of fertilizers from off-farm sources, which increasingly were synthetically compounded chemicals produced in factories. These successes established the viability and legitimacy of farmers coming to depend on chemical companies for some of their major inputs and of agricultural scientists working with the chemical industry to help solve farmers' production problems. Put somewhat differently, the sciences that might have undergirded a nonchemical trajectory—ecology in general, and insect ecology and agroecology in particular—were very underdeveloped and generally not represented in college of agriculture faculties.

Second, and equally fundamentally, the formative period of American public agricultural research (the immediate aftermath of the Hatch Act of 1887) and especially the era of the dawn of the agrochemical age were times in which the United States, relative to other major industrial powers, was a high-wage society with labor shortages. It is thus no accident that the early pattern of technical innovation in American agriculture (the bulk of which was essentially unaffected by land-grant research) was largely one of mechanization—initially horse-drawn equipment and later the tractor (Cochrane, 1979). Mechanization thus substituted for expensive labor (and reduced problems of recruiting and disciplining the nonfamily agricultural labor force). In addition, the physical requirements of mechanizing planting, cultivation, and harvest, as well as the major investments in specialized equipment that farmers needed to make, dictated an increased pattern of farm and enterprise specialization. The reinforcement of monoculture, beginning with the horse-drawn equipment phase of the early 19th century and continuing with internal-combustion engine-based equipment in the 20th, began to exacerbate the pest, pathogen, nematode, fungus, and weed problems that are intrinsic to ecosystems with minimal species diversity. Further, biocide chemicals are, in part, labor-saving technologies, which are rational for farmers to pursue in a context of relatively expensive labor.

The reinforcement of monoculture occurred at a time when the chemical industry was becoming increasingly well prepared to provide agricultural inputs. The growth of fertilizer consumption, in fact, was the result of the late teens' developments in synthetic ammonia production and especially of the huge capacity in nitrate production that was built up during World War II. Chemical companies concerned with overcapacity and interested in expanding their chemical markets devoted growing attention to developing agroinput commodities (Goodman et al., 1987). The pattern of farmers increasing their outlays for purchased mechanical, chemical, and seed inputs was also given further impetus through New Deal agricultural commodity legislation. This legislation, which essentially remains in place today, has served to put a floor under prices, to reduce the risk of large-scale capital investments, and to encourage farmers to stress (monocultural) production of the supported commodities. This list of commodities includes all but one (soybeans) that account for the vast bulk of agricultural chemical usage today. Publicly underwritten accumulation in agriculture, and the pattern by which agricultural commodity program benefits were concentrated among the largest farmers, encouraged that massive structural transition of American agriculture in the 20th century. Farms became fewer in numbers, larger in size, more specialized, and so on—and accordingly there was a reinforcement of the agroecological conditions that made agricultural chemical usage logical, if not obligatory (Buttel, 1990). Chemical-company advertising, related extension programs across the nation, the growing public/consumer acceptance of cosmetic standards that pesticides make possible, and the simplification of management that pesticide use may afford also contributed to growth in pesticide usage.

More recently these tendencies have been given even further impetus due to trends in the global economy and in agricultural research. The early 1970s ushered in a phase of rapid expansion of American agricultural exports as a consequence of agricultural trade liberalization, growing demand for imported food and feedstuffs by wealthy and middle-income developing countries, détente, and other factors. This led to massive expansion of American agricultural production capacity later in the decade. American agricultural export growth was focused on three major crops—corn, soybeans, and wheat—these crops caused the most adverse environmental impacts and were the heaviest user of agricultural chemical (NRC, 1989; Conservation Foundation, 1987).[7]

At the same time, the rhythm of American agricultural research was such that the research priorities and technical outputs of the system tended to dovetail with agricultural expansionism. The land-grant system had become increasingly focused on direct service to state-based clientele groups and on applied research geared to their needs; the corresponding imperative of the land-grant system to respond rapidly (in a "putting-out-fires" mode) to the production problems of state-level commodity groups, whose members were investing large sums in anticipation of even further growth in agricultural export levels, tended to make chemical solutions more attractive than the more biologically complex and information-intensive alternatives such as biocontrol. Ecology and agroecology continued to have a very small presence in the land-grant system, which largely remains the case today even though LISA-type research has increased significantly. Despite the growing political controversies over pesticides and other agricultural chemicals, there did not seem to be any persuasive imperative to move beyond chemical rationalization (such as integrated pest manage-

[7] These three crops, plus cotton, account for the lion's share of tons of active ingredients of synthetically compounded biocides used in American agriculture today.

ment, or IPM) in order to respond to growing problems of pest infestations, chemical resistance, EPA and FDA regulation of pesticides, and so on.

While the Greening of colleges of agriculture and state agricultural experiment stations over the past 5 years or so is palpable, and should not be downplayed, it must be stressed as well that this period of time has not been propitious for developing a major land-grant effort, comparable to that which buttressed chemical usage in the post–World War II period, in "low-input" agriculture. In this backdrop of Greening, two forces have emerged that are at least as fundamental as the growing pressure for agricultural research to stress environmentally sound practices. The first is the stagnation—and, in most states, declining real levels—of public funding of agricultural research since 1980, which has the effect of making it extremely difficult to implement a major new program thrust such as agroecology. The second has been the massive expansion of land-grant research in biotechnology,[8] which was originally premised on the expectation of major industrial funding.[9] Biotechnology has not only claimed the vast bulk of agricultural research resources at the margin, but has profoundly preempted the ability of the land-grant system to move into other areas of basic biology—particularly ecology, agroecology, and evolutionary biology—that have great long-term potential to provide the knowledge base necessary to significantly reduce chemical usage. The result is that socioeconomic influences on the research priorities of the land-grant system and the USDA have led to the underdevelopment of the theoretical and applied knowledge base necessary for undergirding a postchemical transition in agriculture. Quite possibly, in fact, a land-grant research effort comparable to that being undertaken in biotechnology right now would make much of the debate, to be discussed shortly, relatively meaningless.

The Socioeconomic Benefits and Costs of Pesticides and Reducing Pesticide Usage: An Overview of Social Science Research

Farmers, probably moreso than agricultural and social scientists, recognize that, in general, insect (and associated mite and nematode) prob-

[8]One of the most prominent issues relating to agricultural biotechnology research is whether it will, on balance, contribute to or detract from "agricultural sustainability." While creating transgenic crop varieties containing herbicide tolerance genes (which increase the effectiveness of chemical weed control by permitting farmers to use higher application levels) has been the most common application of biotechnology to plant agriculture, there are many respects in which biotechnology could contribute to reducing chemical usage.

[9]Large-scale industrial funding of biotechnology has, in general, never materialized, so land-grant-based biotechnology has been funded mainly by the shift of USDA competitive grant funding in the direction of molecular- and cell-level research, by the ability of land-grant universities to obtain new funding streams from state governments, and by reallocating resources from applied research programs such as crop varietal development (Busch et al., 1991).

lems and weed control remain the weak underbelly of "sustainable" or low-input agriculture. (See, for example, Buttel et al., 1990.) Despite the fact that there exist nonchemical or reduced-chemical alternatives for most of these plant protection needs, farmers are less sanguine about their effectiveness, and particularly their reliability,[10] than they are, for example, about decreasing their reliance on chemical fertilizers for fertility. In part, this is a matter of doing a better job to ensure that information on available alternatives is made more widely available to producers (NRC, 1989). However, farmers tend to be reluctant to make major changes in their enterprises (e.g., implementing crop rotation programs and/or adding livestock enterprises, upon which many of the available alternative practices rest), to assume unknown risks, or to jeopardize reducing their base acreages of federally supported commodities (a topic that will be considered later under the broader rubric of chemicals and agricultural policy).

Nonetheless, the available research on the social impact of reducing chemical use essentially amounts to a debate over the degree to which using fewer pesticides[11] and lower levels of N fertilizer will adversely affect agriculturalists and consumers. Beginning with a widely circulated study by Olsen et al. (1982), one prominent group of researchers has done extensive macrosimulation modeling and has concluded that termination of the use of pesticides and N fertilizer would have devastatingly negative implications for farm output, agricultural productivity, export revenues, consumer prices, and the economy as a whole. Interestingly, this initial study by Olsen et al. suggested, somewhat counterintuitively,[12] that farmers as a whole would have higher incomes as a result of declining yields and

---

[10]Olsen (1990), for example, has found that "tinkering policies," such as a tax on herbicides of less than 100%, will not affect farmers' decisions to use chemical weed control. Olsen's study, based on a MOTAD programming model, found that Minnesota farmers' risk-aversion behavior is sufficient so that most farmers will prefer to forfeit the long-term expected benefits of reducing herbicide usage in any given year in order to reduce the variability of crop yields in the current year.

[11]It should be stressed that while this chapter, and much of the social science and related literature on the socioeconomic implications of reduced chemical usage, refer to pesticides in a general and homogeneous fashion, these chemicals vary enormously in their environmental impact, effect on human health, and the extent to which their use can be supplanted with biocontrol or more benign chemicals. Insecticides, for example, are generally more toxic to humans than are herbicides or fungicides. There is, however, great variability among insecticides as to the degree to which they lead to environmental and human health problems. Further, there is comparable variability among crops as to the degree to which various types of synthetically compounded biocides and other agricultural chemicals are employed and to which reduction or elimination of their use would reduce yields and productivity.

[12]This result, however, is not counterintuitive when the tendency toward low price and income elasticities of most agricultural commodities is taken into account. For most such commodities, decreases in output result in disproportionately large price increases, which cause total revenues and net profits to increase.

output. A more recent, more comprehensive—and, as will be briefly discussed later, more controversial—study in this tradition, by Knutson et al. (1990), found a smaller tendency in the direction of increased net income in crop production. Knutson et al. also found that livestock producer income would drop by an amount "that would nearly offset the gain to crop producers" (1990:1) because of increased feedgrain prices. Aside from the benefits to crop producers, all other implications of eliminating chemicals were unambiguously negative.[13] This group of researchers has even contended that eliminating chemicals would actually be adverse for environmental quality, because it is projected that "banning chemicals would lead to a 10 percent increase in cultivated acreage and an associated rise in erosion" (1990:1). This component of the literature will be referred to henceforth as the "prochemical" tradition or perspective. This perspective's views parallel those of the majority opinion represented in CAST's (1990a) *Alternative Agriculture: Scientists' Review*; Gianessi's (1990) article in this collection is a good representative of the social science version of the prochemical position.

An equally prominent tradition in the literature, which will be referred to as the "prosustainability"[14] approach or tradition, was initiated through Oehlaf's (1978) work. More recently this approach has been typified by the social science segments of the NRC (1989) *Alternative Agriculture* report and by Faeth et al. (1991). These researchers have reported that the

---

[13]Some illustrative data from the Knutson et al. (1990) study can provide a feel for their methods and results. First, as will be stressed later, Knutson et al. and other studies in this genre base their results on 100% reductions in chemicals. Second, for example, they project that corn yields will decline by 32% with no use of pesticides and by 53% with no use of pesticides or N fertilizer. Comparable yield reductions for other crops, respectively, are as follows: soybeans (37%, 37%); wheat (24%, 38%); cotton (39%, 62%); peanuts (78%, 78%); rice (57%, 63%); and sorghum (20%, 37%). Per-bushel production costs would generally increase by 100 to more than 200%. Grain and cotton export volumes would decrease by 50%. With no pesticides the average annual household expenditure on food would increase by $228, and with no pesticides or N fertilizer food expenditures per household would increase by an average of $428 (in 1989 dollars).

[14]Researchers from the prochemical and prosustainability traditions might well have some legitimate objections to the use of these categories. They are used here for convenience, and also because, based on personal acquaintance with most of the researchers involved, these categories are a reasonably accurate portrayal of their predispositions, aside from as well as in terms of the results of their research. There has, however, been some mobility of researchers between categories, e.g., compare Olsen et al. (1982) and Olsen (1990). There is also a substantial "nonpartisan" research community working in this area. Some, such as the agricultural economist G. A. Helmers, while known to be quite sympathetic to the sustainability position, is among the more independent minded on these topics. (See, for example, Tweeten and Helmers, 1990.) Others whose work is cited herein whose work steers clear of the poles of opinion include Olsen (1991), Lee (1990), and Reichelderfer and Hinkle (1989). Work by these less partisan observers, however, does not generate the attention that the studies by the more clearly prochemical or prosustainability groups do.

economic implications of reduced use of agricultural chemicals would be benign, the benefits to farmers substantial and the costs minimal, and the implications for environmental quality and productivity in the use of natural resources would be positive. The literature on this topic varies so widely in its conclusions that one is tempted to wonder about whether the researchers are indeed exploring the same reality! Thus, in this section of the paper I will explore some of the major bases of this profound disagreement and examine how these conflicting results can be reconciled.

On-Farm Vs. Controlled Experimental Results

There has been evidence for nearly two decades that many individual organic farmers enjoy relatively high crop yields (typically about 10% lower than under chemical-intensive systems), and because of their low-input costs they tend to have net incomes equal to or in excess of their conventional-farming neighbors. Roughly the same results (yields comparable to those of "conventional" systems, but with lower costs and higher profits) have been noted in studies of agriculturalists who substantially reduce their use of pesticides. Much of this evidence is summarized in NRC (1989), and additional field data collected for purposes of this report are included in the document. These farm-level data have comprised the principal basis of the prosustainability research tradition.

These results, however, have essentially been rejected out of hand by prochemical social scientists and their natural science colleagues who see elimination of pesticides and other agricultural chemicals in a more cataclysmic way. These data, it is contended, should be regarded as anecdotal, since they are neither collected under controlled laboratory or field plot conditions nor have the needed controls (for soil type, agroclimatic regime, managerial ability, the extent of free or low-cost "imported" soil amendments, and so on) been made.

The available on-farm data are accumulating and are increasingly persuasive. Unfortunately, this debate, in the American context, will only be resolved to complete satisfaction with very-large-scale, expensive, on-farm studies in which the most important rival explanations of the apparent success of users who significantly reduce their pesticide use can be assessed through adequate controls.

Micro- Vs. Macrolevel Data

It is widely recognized that many of the more fundamental social consequences of reduced pesticide and overall chemical use will not simply be an extrapolation of on-farm data to the national level. For example, to the degree that reduced-input production systems now rely on "imported, organic" inputs such as rock phosphate or organic wastes, expansion of this type of production will cause the inputs to become scarcer and more expensive. Also, and more fundamentally, for most agricultural commodities reduced output tends to result in disproportionately higher product and consumer prices, and thus in higher farm incomes and reduced "consumer welfare."[15] Rising product prices can be expected to reduce agricultural exports, which moderates the price benefit to farmers and may be adverse for the economy as a whole. A significant component of the higher farm incomes will become capitalized in farmland and other farm asset values, and ultimately will be transferred away from non–land-owning farmers to large-scale owners of agricultural land.

One of the traditional strengths of the prochemical research tradition has been that it was the first to employ sophisticated macrosimulation models (i.e., Olsen et al., 1982) and still involves the most extensive use of large-scale modeling (e.g., Knutson et al., 1990). More recently, however, prosustainability researchers have begun to initiate significant modeling efforts of their own (e.g., Faeth et al., 1991; Azzam et al., 1990), with results consistent with the overall prosustainability position that the various social impacts of reduced use of pesticides and other chemicals would be in the benign-to-positive range.

As noted earlier, there is an acknowledged need to do large-scale modeling of the socioeconomic implications of reducing pesticide and overall chemical usage. This said, there are reasons for caution about taking the scientific imprimatur of these macrosimulation studies too seriously. First, it is widely recognized that the results of these models are essentially determined by the technical coefficients (see below)—such as data, provided by scientists or through the scientific literature, on phenomena such as yield changes, the availability of alternative practices, and so on—that are their fundamental basis. That is, the macrosimulation studies are no more accurate, "scientific," or reliable than the microlevel (biological and economic production function) data that comprise their technical coefficients.

Second, virtually all available macrosimulation models were developed in the "presustainability" era, during which the only agricultural-policy-relevant outcomes of interest were the standard economic ones: productivity, gross farm income, net farm income, average production costs, farmland prices, agricultural export levels, consumer prices, and the like. That is, with the significant, though partial, exception of Faeth et al. (1991), current macrosimulation models are essentially incapable of handling simultaneously the socioeconomic benefits and costs, the environmental ben-

---

[15]It should be stressed, however, that because farm-gate product prices are generally a small and decreasing share of retail food prices, increased agricultural commodity prices will have comparably modest impacts on consumer welfare.

efits and costs, *and* the economic and natural resource productivity trade-offs of alternative policies.

All or Nothing, or Something in Between?

One of the major differences between the prochemical and prosustainability literature is, in a sense, ironic: prochemical researchers have stressed the simulation of the socioeconomic impacts of *100% bans* on chemicals (e.g., Knutson et al., 1990), while prosustainability researchers have stressed the socioeconomic impacts of *fractional reductions* (typically 10–50%) in chemical use.

There are two quite different reasons why this should be the case. First, it is generally accepted that phased-in, gradual, fractional reductions in pesticides and other agricultural chemicals would result in smaller decreases in output and best enable alternative, compensatory practices to come into play. By contrast, total bans on any particular class of chemicals would be far more disruptive. Accordingly, prosustainability researchers elect to employ the assumptions that are relatively more favorable to their case, while prochemical researchers have done likewise. Second, total bans on particular classes of chemicals are, because of the simplicity they afford for calculating technical coefficients, more easy to incorporate in macro-simulation models, which have been the principal method employed in the prochemical tradition.

It should be stressed, however, that total-ban scenarios, while they are viewed sympathetically in the more orthodox organic farming quarters, are not at present a realistic component of the current policy debate on agricultural chemicals. Most sustainability proponents recognize that a total ban on chemicals will not likely be achieved in their lifetimes and that immediate bans would be disruptive. There is growing evidence that the social impacts and implications of pesticide bans are almost certain to be very different from pesticide reduction policies, such as "pesticide-use fees" (Zilberman et al., 1991). Further, even in the countries that have taken the boldest steps to reduce agricultural chemical use, their policies have taken the form of phased-in fractional reductions of the sort summarized in Appendix A.[16]

Obtaining Realistic Technical Coefficients

The most immediate prospective socioeconomic impacts of reduced use of pesticides and agricultural chemicals are those that must be derived

[16]Also, as noted by Lee (1990), the most important U.S. policy issue currently driving chemical reduction issues in the United States is water quality. Lee notes that total bans on chemicals are essentially irrelevant to debates over agricultural contributions to water pollution and water resource degradation, since much more moderate fractional reductions would be adequate to reduce agriculture's contribution to water resource destruction to an acceptable level. See also OTA (1990).

directly or indirectly from microlevel data on how fewer chemicals will affect yields, output, production costs, crop mixes, and the like. That is, social science research on the future impacts of reduced chemical use depends on accurate, realistic natural science data. As noted earlier, even the most sophisticated methods and comprehensive research designs for exploring the socioeconomic consequences of reduced pesticide and chemical use are those whose direction, if not their magnitude, can be inferred from the technical coefficients used.

Ultimately, natural scientists, either through expert opinion or specially designed experimental results, are the source of these technical coefficients. Methodologies vary as to how natural scientists' data and opinions can be summarized into a few numbers to be plugged into models. For example, the widely discussed Knutson et al. (1990) study involved interviews and other consultations with over 140 agricultural scientists at about two dozen land-grant universities.

The three key parameters in assembling reasonable, realistic technical coefficient data are as follows:

1. *Selection of Scientist-Experts*—As noted earlier, the land-grant agricultural science community exhibits an enormous range of opinion on the feasibility of reduced use of pesticides and other agricultural chemicals. The full range of expertise and opinion should be represented.[17] One of the most important reasons for consultation with a range of agricultural scientists is that the technical coefficients to be employed in macrosimulation models are extremely sensitive to the assumptions made about "compensatory practices" and "induced technological change" that will come into play to compensate for chemical use reductions. As persons whose work focuses on these compensatory mechanisms are most likely to be aware of their potentials (and limitations), this expertise is highly germane to estimating such coefficients in a realistic way. There is, however, growing evidence that "expert opinion" is often at variance with field-plot experimental data relating to pesticide regulation decisions and that neither expert opinion nor experimental data alone are sufficient to make reliable judgments (Gianessi and Puffer, 1991).
2. *Providing Realistic Policy Parameters*—The data to be collected from scientist-experts should be based on assumptions that have meaning in the policy process. The current "limits of the feasible"

[17]As noted later, one of the reasons that the Knutson et al. study has become controversial is that the list of scientists they (1990:45–50) consulted did not include any land-grant scientists known to be strongly sympathetic with alternative agriculture, reducing chemical use, and so on.

are that some fractional reduction in pesticide and agricultural chemical usage may be warranted to reduce environmental problems such as surface and subsurface water contamination.

3. *Reconciling Diversity of Opinion*—Representation of the full range of expertise and opinion creates the imperative to have a meaningful process for summarizing the diverse opinions. Ideally, this should be a Delphi-type process, in which scientists with varying opinions are able to express them fully and to discuss the bases of differences of opinion publicly before moving to compilation of a consensus opinion.

European Research on the Impacts of Reduction of Pesticide Use

The U.S. debate on pesticide and chemical use reductions and their socioeconomic impacts has been a largely self-contained one, based on U.S. scientists, U.S.-based studies, and U.S. policy issues. It is important to stress, however, that the U.S. social science, agricultural science, and agricultural policy communities will need to look to Nordic Europe, where pesticide use reduction programs of various types have been in place for several years and where there is a growing literature on these programs' socioeconomic impacts. It should be recognized, of course, that these results cannot be immediately extrapolated to U.S. conditions because of agroclimatic and farm structural variations and differences in agricultural policy.[18] Nonetheless, European research may be helpful for two major reasons: First, in due course, as current programs are implemented and results become available, evidence from Europe can ultimately permit the investigation of the socioeconomic impacts of reduced pesticide usage through ex post, rather than simply ex ante, research designs. As of this time, however, the available data from Europe are largely of the ex ante sort that characterize the U.S. debate. Second, the agricultural systems in the countries of Northern Europe have generally come to be much more chemical intensive than is the case in the United States. Accomplishment of pesticide and other chemical-use reductions in Europe without major social dislocations would suggest that they could be implemented with even less dislocation here.

The most significant European research program on the socioeconomic impacts of reduced chemical use has been that of Alex Dubgaard (1989, 1991), a Danish agricultural economist at the Institute of Agricultural Economics in Copenhagen. Dubgaard has initiated several studies relating to

[18]One of the major differences in the agricultural policy environments of the United States and most European nations concerns output and exports. While American agricultural policy has tended to stress maintaining or increasing output in order to preserve export levels, export levels are not nearly so important in most European countries, especially those that have implemented programs to reduce pesticide usage.

the Danish policy targets of reducing the use of chemical fertilizers and pesticides by 30% and 50%, respectively. As noted in appendix A, the Danish chemical reduction program currently is limited essentially to use-reduction targets, stiffer pesticide regulation and registration procedures, and a redirection of agricultural research priorities to maintain productivity while reducing the use of fertilizers and pesticides. Dubgaard's research program, however, has been premised on the assumption that these policy instruments are likely to be insufficient to achieve the chemical-use reduction targets. His research has therefore focused on several policy instruments, particularly tax levies on nitrogen fertilizer and pesticide applications, that will likely be needed to accomplish these goals.

Dubgaard's research has involved combining micro (farm-level) decision-making models and macrosimulation models. The basic results of his studies are as follows. Dubgaard has found that subsidies of environmentally favorable practices are a very inefficient way of accomplishing environmental goals. Only a "polluter pays" principle can significantly reduce chemical usage. Because neither pesticide chemicals nor the lands on which they are applied are homogeneous, the most widely discussed polluter-pays policy instruments, such as a "flat-rate" tax based on either the purchase price or on the kilos of active ingredient, would be inefficient and inequitable, and not be conducive to reduction of pesticide usage. Dubgaard has found that a flat-rate tax per labeled dosage on all pesticides would minimize these problems. His research shows that a tax of this sort, at a level that would increase the average pesticide price by about 120%, would reduce pesticide usage by between 40 and 45%. About 60% of the pesticide use response was estimated to be due to farmers' reducing their use of pesticides to economic-threshold levels, while the remainder is estimated to come from induced technological improvements in pest management and pesticide application. This tax levy on pesticides, combined with a 150% levy on N fertilizer, was estimated to result in an estimated decline in Danish crop output of about 10%. Since two-thirds of Denmark's land area is already utilized in agriculture and there is virtually no additional land area that can be brought under production, this output decrease is essentially equivalent to an average per hectare crop yield decline.

In the context of the European Economic Community's Common Agricultural Policy, in which there is a common Community-wide price for each agricultural commodity, an output decrease in any one country will not affect product prices in that country.[19] Dubgaard notes, however, that

[19]The policy instruments evaluated by Dubgaard would thus also not involve impacts on consumer food prices. Given the EEC's Common Agricultural Policy framework, in which crop output declines would not result in product price increases, livestock farmers would not be penalized through a program to reduce agricultural chemical use, as has been projected for the United States. In fact, Dubgaard has noted that after tax levies are rebated to farmers based on total acreage, his proposed policy instruments would result in livestock farmers benefiting slightly at the expense of specialized crop producers because of the fact that crop-livestock operations are better able to reduce their chemical usage.

for any one EEC country, decreased aggregate output results in a decreased share of EEC farm commodity program payments that go to that country. Thus, rebating the input levies to farmers would cause their incomes to be comparable to current levels. Denmark, however, would be penalized under current EEC rules because of its being able to collect only 90% of current subsidy levels from the Common Agricultural Policy agricultural price support program. Further, as the EEC currently has an overproduction level of about 15%, 10% reductions in crop output for all EEC countries would serve to reduce EEC agricultural overcapacity and Community outlays on agricultural commodity programs. He has thus stressed the need for a common EEC agroenvironmental policy, along the lines he has suggested for Denmark. A common agroenvironmental policy for the EEC would have the benefits of reducing agropollution levels, thereby reducing the degree to which individual European countries see it as in their interest to maximize output (and to continue using pesticides to that end), and significantly decreasing the current $40 or so billion currently spent on price-support programs.

On Balance

There remain major differences in the assumptions, data, and interpretations of data pertaining to the socioeconomic consequences of reducing pesticide usage in the United States. Nonetheless, we can make some general conclusions based on the available data. First and foremost, the social implications of reducing pesticide usage will rest heavily on the amount of pesticide use reduction to be achieved, the policy instruments used, and the yield declines that result. Second, there is growing evidence that yield reductions will not be so large as has often presumed, so a prospective pesticide usage reduction program would not be accompanied by dramatic dislocations. Third, pesticide use reductions would very likely reduce aggregate crop yields, and reduced crop yields could be expected to have a number of consequences (given the current U.S. agricultural policy context): agricultural product prices would increase, net farm income of crop producers would increase, net farm income of specialized livestock producers would decrease (due to increased costs of feedgrains), farmland prices would increase, and consumer food prices would increase. Consumer price increases would be disproportionately smaller than farm-gate product price increases, though any increase in consumer food prices hurts the poor more than the affluent. Increased product prices would reduce agricultural export earnings (or increase the level of export subsidies required to move a given level of grain in world commerce). To the degree that world market prices are substantially lower than domestic product prices, world price levels would serve to put a ceiling on domestic product price levels (and thus moderate the increase in net farm income) and on consumer food prices.

The effects of pesticide use reductions on farm structure would likely include a decreased level of farm- and regional-level specialization. The distributional consequences would very likely be indeterminate. On one hand, increased profitability of farming and higher land values would make agriculture more attractive for larger investments and to off-farm investors. On the other, the management-intensive practices required to deal with reduced availability of agricultural chemicals would hinder the expansion of large-scale, industrial-type farms that grew in prominence during the agrochemical age.

To the degree that a pesticide and overall chemical-use reduction program reduces aggregate crop output, it would serve as a kind of production control program as well. Assuming that current incentives to retire highly erodable land from crop production remain in place, a pesticide and chemical-use reduction program could be expected to result in modest declines in federal commodity program expenditures.

Agricultural Chemicals and Agricultural Policy

It has long been recognized that agricultural policy (including commodity programs; regulatory policy; agricultural credit, trade, and research policies; as well as nonagricultural policies such as tax policy) and the incentives these policies present to farmers play a vital role in shaping the production practices used by farmers. Agricultural commodity programs that have been invoked to stabilize farm prices and farmer income are arguably the most important in this regard. Approximately two-thirds of U.S. farmers participated in one or more agricultural commodity programs in the late 1980s. It has been estimated that between 80 and 95% of the acreage of program-supported crops was actually enrolled in federal commodity programs at this time (NRC, 1989).

Agricultural commodity programs, which in their current form remain based on the Agriculture Adjustment Act of 1938, have had both long-term and short-term impacts on pesticide-related practices. In a long-term sense, agricultural commodity programs have encouraged farmers to produce the supported commodities, particularly in monocultures. Many of these crops, especially corn, cotton, peanuts, wheat, and rice, tend for various reasons to have significant environmental impacts. These programs have also placed a floor under farm prices and stabilized incomes, and thus have tended to increase the attractiveness of agriculture for large investments and to encourage a more capital-intensive agriculture than would otherwise have been the case. Pesticides, which themselves are capital-intensive inputs, are also highly compatible with the other agricultural practices—mechanization, high-yielding hybrid varieties, fertilizers, and so on—that together have made agriculture an industry with one of the highest ratios of capital investment per worker in the United States.

More recent and more specific technical components of commodity programs also contribute to growth or maintenance of chemical use in agriculture, mainly by creating strong incentives against crop rotations. For example, all price and income support programs for the supported crop commodities are based on the concepts of (1) base acre requirements and (2) proven crop yield for these base acres. Program benefits are limited, or may become limited in the future, if a farmer either (1) employs more extensive crop rotations, and thereby reduces the base acreage of one or more supported commodities and accordingly receives lower base-acre levels for one or more crops, or (2) uses reduced-input practices that reduce crop yields, even if for only a few years. The 1985 Farm Bill (the Food Security Act of 1985) also introduced a cross-compliance provision designed to control government outlays by reducing the production of program commodities. The cross-compliance provision stipulated that to receive any benefits from a given base acreage for a crop, a farmer may not exceed his/her base acreage for any other supported crop. The upshot is that a farmer wishing to move toward crop rotations could not do so, without forfeiting all commodity program benefits, if the rotation would involve growing any acres above the established base acreage for another program crop. Thus, for example, a farmer who currently grows corn on all acres, but who would prefer to introduce another program crop (e.g., wheat or rye) in rotation, could do so only by walking away from all program benefits. These provisions are particularly adverse for farmers who wish to shift from specialized crop operations to systems involving crop rotations and/or to integrated feed and forage operations for adding livestock enterprises.

Many government policies other than federal commodity policies—for example, favorable tax treatment of capital gains and accelerated depreciation and investment tax credit provisions of the tax code, and agricultural credit subsidies—have in the past contributed, or currently provide incentives, to specialized, monocultural, capital-intensive agricultural systems. Most have tended to support capital-intensive, chemically based agriculture.

Agricultural policy is germane to reduced pesticide and agricultural chemical use in several respects. First, to the degree to which policies that buttress chemical use remain intact, it will be more difficult to reduce pesticide usage voluntarily. Second, should there be invoked specific regulatory policies aimed directly or indirectly at reducing pesticide use while current agricultural and agriculturally related policies remain intact, the socioeconomic implications of these reductions may be more adverse than would otherwise be the case, and resistance in the agricultural community can be expected to be stronger.

The past decade has witnessed growing interest in restructuring federal agricultural commodity programs. In the abstract, such an agenda has support from two quarters—the current and predecessor Republican administrations in the United States and many environmental groups—that normally do not have much in common. The recent Republican administrations have favored the phase-out of current commodity programs in order to reduce budget outlays and to place agriculture on a market footing. Groups favoring such a change in agricultural policy also note that a harmonized "liberalization" of government commodity programs across the world would, in addition to increasing American agricultural exports, lead to higher Third World agricultural commodity prices, to more Third World food production, and to a net increase in producer and consumer welfare across the world.

Environmentalists likewise have come to recognize that current agricultural commodity programs tend to subsidize and reinforce chemical usage and farm practices that are environmentally destructive. For this reason environmentalists have become actively involved in formulating the last two Farm Bills, with some successes as noted earlier. They have supported the Reagan and Bush Administration proposals to "decouple" agricultural subsidies and farm product prices. (See, for example, Reichelderfer and Hinkle, 1989.) The basic framework of 1930s farm subsidy programs remains intact, however, mainly because domestic commodity groups and their supporters have been able to muster sufficient political support to frustrate the Administration's efforts to this end. Also, the Uruguay Round of the General Agreements on Tariffs and Trade (GATT), which was heavily premised on a U.S.-backed agenda for working out an agreement on farm trade liberalization, essentially ended in failure due to the recalcitrance of the EEC. An additional reason for the failure of the Uruguay Round was environmental-group resistance to agricultural trade liberalization. Environmental groups opposed farm trade liberalization on the grounds that increased Third World agricultural exports would stimulate chemical-intensive agricultural practices in the developing world and encourage greater production of tropical export commodities in sensitive rain-forest zones.

How is it, then, that environmentalists might be sympathetic to restructuring U.S. agricultural commodity programs, particularly "decoupling," yet have helped to frustrate world agricultural commodity program reforms? There are three basic aspects to this paradox. First, reducing or eliminating subsidies to chemical-intensive practices and creating a market-based world agricultural economy should not be taken as being coterminous. Environmentalists have not objected to state intervention in the agricultural economy in principle, nor do they feel that a market-based agricultural system is intrinsically superior to a subsidized one. Rather, their concerns have been to eliminate provisions of commodity program legislation that provide strong incentives to chemical-based agricultural intensification (such as the base-acreage provisions discussed earlier).

Second, the goal of environmental protection will not necessarily be advanced through agricultural trade liberalization per se. The stimulus that liberalization might provide to Third World agricultural production and exports, which will normally be produced under permissive conditions for the use of fertilizers and pesticides, would very likely yield an increment of global environmental destruction that would more than counterbalance the gains in agricultural-environmental quality achieved in the developed industrial world. Thus, environmental quality will not likely be advanced if liberalization results in increased industrial-country imports of cheap Third World farm commodities that are cheap, in part, because they are produced with environmentally destructive practices and in an absence of environmental regulation. Third, insofar as one of the major impetuses to reduced use of agricultural chemicals has been human health concerns, restricting chemical use in the United States will make little difference if it is accompanied by growing imports of Third World agricultural commodities produced with high levels of pesticide usage.[20]

Most observers consider it unlikely that significant reductions in the use of agricultural chemicals can be achieved solely through reform or restructuring of agricultural commodity programs alone, since the multiplicity of interests that now comes to bear on Farm Bills is such as to virtually guarantee indefinite stalemate (Reichelderfer and Hinkle, 1989). At present it seems most likely that state and local governments, through the authority they have to regulate the quality of surface and subsurface water, will be the arena within which the new politics of agricultural chemicals will move over the next decade (Lee, 1990). There have already been a number of successful initiatives at the state and local levels to accomplish these goals. (See, for example, OTA, 1990.) It is quite possible, however, that growing successes at the subnational levels, and the patchwork of regulation that would result, may become an inducement for chemical companies to agree to uniform national-level regulation.

Even if pesticide and overall chemical regulation becomes a viable national-level policy issue, it should be stressed that elaborating a coherent

[20]Runge and Nolan (1990), for example, have noted that environmental and health regulations on imported agricultural commodities have increased recently due to the growing tendency for imported foods to be produced with recourse to dangerous chemicals and due to the public concerns this has generated. He notes that this type of regulation has become a growing "nontariff" barrier to world trade, which is often justified but sometimes is not.

environmental policy for agriculture is not a straightforward or unidimensional matter. Three observations can be made in this regard. First, while there are several examples of national and subnational policies aimed at reducing pesticide usage by some a priori amount (typically 50%), research in Northern Europe, where such policy goals are most common and ex ante impact research most extensive, shows that such a policy may be ambiguous. Policy to reduce pesticide usage begs the question of what policy levers will be employed to accomplish this goal. This is particularly the case if the main policy instruments are confined to stricter pesticide registration and expansion of research and extension services. Dubgaard's (1989, 1991) work on Denmark has demonstrated that many widely discussed, more proactive policy approaches (e.g., subsidizing environmentally desirable practices) are likely to be ineffective, inefficient, or both; he suggests that effective policies for reducing agrochemical usage must combine a "polluter-pays" principle and substantial taxes on agrochemicals.

Second, it must be kept in mind that policy for reducing the environmental and social costs and risks of pesticides should be conceived of as part of a larger policy for environmental management in agriculture as a whole. For example, there will inevitably be a number of major policy trade-offs in developing an environmentally sound agricultural policy (Reichelderfer, 1990). Major pesticide and chemical fertilizer use constraints may result in bringing more environmentally fragile lands under cultivation, and thus may conflict with the current Conservation Reserve Program. Policy instruments that seek to control groundwater pollution by encouraging use of animal manures as substitutes for chemical fertilizers might exacerbate surface-water pollution.

Third, the most effective policies for reducing the costs and risks of pesticide usage may not necessarily be policies aimed specifically at pesticides. For example, the development of "polluter-pays" policies with respect to water quality or imposition of a major flat-rate tax on N fertilizer may be as or more effective in reducing pesticide usage than pesticide regulation as such. Also, as Reichelderfer (1991:5) notes,

> the Clean Air and Water Quality Acts impose technology-based standards that affect the location, configuration, operating conditions, and costs of virtually all industrial and public utility facilities, yet they place no limits on effluents or emissions from agricultural and other nonpoint sources of air and water pollution. Similarly, industries and municipalities spend an estimated $23 billion to $30 billion annually to comply with the 1972 Federal Water Protection Control Act, yet that act authorizes federal subsidies to help states plan and farmers adopt water quality management strategies for which there are no associated standards. . . . Thus, while the centralized or command-and-control approach to environmen-

> tal policy has been given precedence in nonagricultural sectors, incentive-based and subsidy approaches have predominated in the agricultural sector.

She notes further (1991:7) that "as the U.S. economy grows, new information on the environmental effects of agriculture is made available, and existing environmental legislation is applied to nonpoint pollution sources, the level of environmentally motivated government intervention in agriculture will begin to approach that of other industries." Thus, policy toward pesticides should be viewed in a long-term resource and environmental management sense—that there is a growing likelihood that the environmental standards now expected of virtually all nonagricultural industries will ultimately be extended to agriculture. Accordingly, it is important that public-research and technology-transfer programs recognize this reality, shift their research priorities appropriately, and begin to generate knowledge that will enable farmers to reduce their chemical usage with a minimum of dislocation.

Conclusion

Debates over the desirability of achieving a more environmentally sound agriculture through reduction of pesticide usage will very likely continue to be as politicized in the 1990s as they have been ever since the publication of *Silent Spring*. Research results on the socioeconomic consequences of various prospective policy instruments for enhancing environmental quality will be integral in providing ammunition for both sides of the debate. The literature on this issue, the bulk of which comes from respected quarters of the social science research community, can be construed to provide support for the claims of both.

It is worth noting, however, that the past 2 years have witnessed two interrelated developments within this literature, both of which will undoubtedly serve to sharpen the polarities of opinion. A significant trend in the literature is that represented in the Faeth et al. (1991) monograph. They suggest that the large-scale modeling procedures typically employed by economists and other social scientists to address issues relating to the impacts of environmentally related policy are largely inappropriate to this task, since they are structured so as to address a set of performance parameters (farm income, product prices, exports, and so on) that reflect only one dimension of agricultural policy issues. Faeth et al. suggest the need to develop national and subnational performance measures of natural resource productivity, in addition to the more traditional measures such as labor productivity and average levels of production costs.

A second development has been engendered by one of the studies noted earlier, that by Knutson et al. (1990). The Knutson study, insofar as it has been the most wide-ranging of its genre, seemingly provided the most comprehensive and convincing evidence thus far for the prochemical position. This study, however, has become subject to a number of damaging critiques and to unflattering commentaries because of its methodological procedures and ethical posture. The Knutson et al. study, rather than having been received as conclusive evidence in support of the prochemical position, has wound up being seen in the eyes of many agricultural economics colleagues as a prime case of the questionable ethics of conducting research paid for by industry sponsors with a direct interest in the outcomes. In a vociferous exchange of articles and letters in *Choices* (4th Quarter, 1990, and 1st Quarter, 1991), a policy-oriented magazine published by the American Agricultural Economics Association, Knutson et al. have been forced to defend themselves against the charge that the consistency of the methodological procedures with the proprietary interest of the sponsoring corporations was not accidental.

Regardless of the specifics of the controversy over the Knutson et al. study, the very fact that controversy emerged reinforces a number of points stressed in the foregoing. Neither the agricultural/natural sciences, which ultimately must provide the most important inputs into social science research on technical and policy choices in agriculture, nor the agricultural/social sciences themselves, are immune to social and political influences on their research designs and results. Neither has a monopoly on the information needed to make product policy decisions in this area. The manner in which these issues will ultimately be resolved, however, will be based, in part, on which groups will be able to mobilize the authority of science behind their claims.

Acknowledgments

The author wishes to acknowledge the assistance of Stacy Stephans in preparing this paper and financial support of this research through Hatch Project #438 at the Cornell University Agricultural Experiment Station.

Appendix A: Summary of Major National Policy Initiatives Aimed at Reducing Pesticide Usage: The Cases of the Netherlands and Denmark

This appendix reports information on national pesticide use reduction programs in two Northern Europe countries, the Netherlands and Denmark. Note that comprehensive discussions of pesticide reduction policies in Sweden and Ontario are reported elsewhere in this volume (by Pettersson and by Surgeoner and Roberts, respectively).

Netherlands

Eutrophication of lakes by nitrate and phosphate, acidification of lakes, and soil and groundwater contamination have all been determined to be greatly affected by pesticide use in the Netherlands. In 1984 the Department of Agriculture and Fishery published a report announcing government initiatives to achieve a reduction in the use of chemicals. Formulation of guidelines for registering pesticides in the Netherlands is based upon the same type of soil and microbial studies that the EPA uses in the USA to set up their guidelines. Government regulations in the Netherlands require that agricultural chemicals be submitted for registration to the Pesticide Bureau. The actual conduct of field experiments for registration of pesticides is being delegated to the industry. Submitted chemicals must meet certain requirements, and the applicant has to provide the information needed to make this evaluation possible. A procedure has been established between the chemical industry and the Dutch government in which the applicant may perform field trials supervised and checked by the Plant Protection Service in order to ensure that they obtain all relevant data. However, since that time, research input has been small, and governmental services have not been very active in promoting the introduction of integrated pest management practices.

A long-term project regarding the ecological sustainability of the use of chemicals involves four bills in the Ministry of the Environment. They are the National Environmental Policy Plan, Nature Conservation Policy Plan, Third Water Management Plan, and Fourth Physical Planning Strategy Plan. Source restrictions, effect-mitigation measures, and steps to rehabilitate polluted ecosystems are all involved in the plans. Strict ecological standards with regard to agrochemicals will be implemented. A bill on criteria for rescreening "old pesticides" has just been brought to Parliament. The acceptable criteria involve the half-life of the pesticide and the amount of it that leaches into groundwater. Criteria for the effects of pesticides on air pollution are also presently being developed. The nation's current risk assessment of existing chemicals takes steps, including establishing a list of 400 risky chemicals; profiles of calculated parameters for each chemical; risk scores; ranking of risks based on characteristics of the chemicals and modes of production, distribution, and uses; basic policy documents for 50 priority chemicals; and 15 top priorities selected for direct action.

Denmark

Because of a considerable increase in the use of pesticides in Denmark in the period from 1981 to 1984 and published studies showing negative impacts of herbicides on bird fauna, soil leaching, and links between pesticides and cancer, there has been much public debate and concern over the issue and a demand for reduction of pesticide use.

The present Danish legislation (Consolidation Act of 1987 and its two Statutory Orders) was implemented with the intention to prevent health and environmental problems caused by chemical substances and products. Laws concerning registration are as follows: registration will not be granted if the product is considered to be especially dangerous or harmful to health or the environment; registration will not be granted if there is an existing alternative method that is less harmful; a time limit of 5–8 years is allowed for registrations; and registrations must be renewed for all substances and products which have been registered before 1/10/80. The Minister of the Environment has also published a plan of action to aid in the reduction of pesticide use. Pesticide consumption must be reduced by at least 25% by January 1, 1990 (relative to average usage in 1981–1985), and a further reduction of 25% has to be met before January 1, 1997. These reductions are based on the quantity of active ingredients sold as well as on the spraying intensity, i.e., the yearly frequency of treatments with recommended dosage.

Projects are being initiated to provide information to farmers and also to discover new alternatives and methods. Following input from farmers, a decision model is being developed in order to suggest the most appropriate herbicide and its proper dose. Research into mechanical control of annual weeds has been reintroduced. Encouragement of farmers to grow crops without any input of agrochemicals is also being suggested as a possible option

References

Adams, W.M. 1990. Green Development. London: Routledge.

Azzam, A., G.A. Helmers, and M.F. Spilker. 1990. "U.S. agriculture under fertilizer and chemical restrictions, part 2." Report No. 163, Department of Agricultural Economics, University of Nebraska-Lincoln.

Bosso, C.J. 1987. Pesticides and Politics. Pittsburgh: University of Pittsburgh Press.

Busch, L., W.B. Lacy, J. Burkhardt, and L.R. Lacy. 1991. Plants, Power, and Profit. Oxford: Basil Blackwell.

Buttel, F.H. 1990. "Social relations and the growth of modern agriculture." Pp. 113–146 in C.R. Carroll et al. (eds.), Agroecology. New York: McGraw-Hill.

Buttel, F.H., and G.W. Gillespie Jr. 1988. "Preferences for crop production practices among conventional and alternative agriculturalists." American Journal of Alternative Agriculture **3** (Winter):11–18.

Buttel, F.H., G.W. Gillespie Jr., and A. Power. 1990. "Sociological aspects of low-input agriculture in the United States: A New York case study." Pp. 515–

532 in C.A. Edwards et al. (eds.), Sustainable Agricultural Systems. Ankeny, IA: Soil and Water Conservation Society.

Cochrane, W.W. 1979. The Development of American Agriculture. Minneapolis: University of Minnesota Press.

Conservation Foundation. 1987. State of the Environment. Washington, DC: Conservation Foundation.

Council for Agricultural Science and Technology (CAST). 1990a. Alternative Agriculture: Scientists' Review. Ames, IA: Council for Agricultural Science and Technology.

Council for Agricultural Science and Technology (CAST). 1990b. Pesticides and Safety of Fruits and Vegetables. Ames, IA: Council for Agricultural Science and Technology.

Dubgaard, A. 1989. "The need for a common environmental policy for EC agriculture." Pp. 35–48 in A. Dubgaard and A.H. Nielsen (eds.), Economic Aspects of Environmental Regulations in Agriculture. Kiel (Germany): Wissenschaftsverlag Vauk Kiel KG.

Dubgaard, A. 1991. "Pesticide regulation in Denmark." In N. Hanley (ed.), Farming and the Countryside. Oxon, U.K.: C.A.B. International.

The Economist. 1990. "Good times are back on the farm, for a bit." The Economist (10 March):25–26.

Edwards, C.A. et al. (eds.). 1990. Sustainable Agricultural Systems. Ankeny, IA: Soil and Water Conservation Society.

Faeth, P., R. Repetto, K. Kroll, Q. Dai, and G. Helmers. 1991. Paying the Farm Bill: U.S. Agricultural Policy and the Transition to Sustainable Agriculture. Washington, DC: World Resources Institute.

Gianessi, L.P., and Puffer, C.A. 1991. "Inadequacy of scientific and economic data in pesticide benefits analyses." Resources No. **104**:14–17.

Gillespie, G.W. Jr., and F.H. Buttel. 1989. "Understanding farm operator opposition to regulation of agricultural chemicals and pharmaceuticals: the role of social class, objective interests, and ideology." American Journal of Alternative Agriculture **4**:12–21.

Goodman, D., B. Sorj, and J. Wilkinson. 1987. From Farming to Biotechnology. Oxford: Basil Blackwell.

Knutson, R.D., C.R. Taylor, J.B. Penson, and E.G. Smith. 1990. Economic Impacts of Reduced Chemical Use. College Station, TX: Knutson & Associates.

Lee, L.R. 1990. "Relationships among groundwater quality, agricultural production, and consumer food prices." Manuscript, Department of Agricultural and Resource Economics, University of Connecticut.

National Research Council (NRC). 1989. Alternative Agriculture. Washington, DC: National Academy Press.

Oehlaf, R. 1978. Organic Farming. Montclair, NJ: Allanheld, Osmun.

Office of Technology Assessment (OTA). 1990. Beneath the Bottom Line. Washington, DC: OTA.

Olsen, K.D. 1990. "Modeling farm-level interactions between policy and a farmer's choice of weed control method." Manuscript, Department of Agricultural and Applied Economics, University of Minnesota.

Olsen, K.D., J. Langley, and E.O. Heady. 1982. "Widespread adoption of organic farming practices: estimated impacts on U.S. agriculture." Journal of Soil and Water Conservation (January–February):41–45.

Redclift, M. 1987. Sustainable Development. London: Methuen.

Reichelderfer, K.H. 1990. "Environmental protection and agricultural support: are trade-offs necessary?" Pp. 201–230 in K. Allen (ed.), Agricultural Policies in a New Decade. Washington, DC: Resources for the Future.

Reichelderfer, K.H. 1991. "The expanding role of environmental interests in agricultural policy." Resources No. **102**:4–7.

Reichelderfer, K.H. and M.K. Hinkle. 1989. "The evolution of pesticide policy: environmental interests and agriculture." Pp. 147–173 in C.S. Kramer (ed.), The Political Economy of U.S. Agriculture. Washington, DC: Resources for the Future.

Rossiter, M.W. 1975. The Emergence of Agricultural Science. New Haven: Yale University Press.

Runge, C.F., and R.M. Nolan. 1990. "Trade in disservices: environmental regulation and agricultural trade." Food Policy (February):3–7.

Tweeten, L.R., and G.A. Helmers. 1990. "Comment on Alternative Agriculture Systems." Pp. 134–138 in Alternative Agriculture: Scientists' Review. Ames, IA: Council for Agricultural Science and Technology.

Zilberman, D., A. Schmitz, G. Casterline, E. Lichtenberg, and J.B. Siebert. 1991. "The economics of pesticide use and regulation." Science **253**:518–522.

8

Swedish Pesticide Policy in a Changing Environment

Olle Pettersson

Introduction

As in most industrial or postindustrial countries, Swedish agriculture during past decades, as well as today, is being subjected to major technological, environmental, economic, and social changes. The dominating trends in this development are fairly uniform; higher yields and more efficient production by utilization of better plant and animal varieties, together with technical and chemical aids.

Accompanying the changes in technology, intensity, and cultivation as well as in the industrial structure, agriculture and horticulture have been the subject of increased environment impact and the object of intensified environmental conflicts and discussion. The conflicts and controversies, as well as the political results, differ in time and character between countries due to sociological, ethical, and political differences.

In this respect, some specific Swedish (or maybe Scandinavian) characteristics and outcomes can probably be identified which have their origin in the role of agriculture in Swedish history, sets of values, or political experiences. The result of these sociological, ethical, and political factors is that pesticides as well as other environmental agents have been put on the agenda of politics and legislation in Sweden at an earlier stage and more emphatically than in many other countries.

These aspects of environment policy will be discussed on the basis of the Swedish decision to reduce pesticide use by 50%, a decision that has attracted interest also in other countries. What is the cultural and political background? Why is it that the opinion and the demand for a decision of this kind first takes place in Sweden, where the problems concerning pesticide use are much less pronounced than in many other countries and agricultural areas? Does the Swedish policy imply a new approach with completely different conditions for pesticide use, or should it preferably be described as an adaptation to what modern pesticide and agricultural technology can achieve?

Forces and Factors in Pesticide Use

In the conventional discussion on pesticides, the approach is frequently dominated by static and egoistic considerations. The whole problem is converted into "What are the consumers prepared to pay?" or "What does society require of the farmers?" These are relevant concerns but at the same time they attend to only part of the truth. This section attempts a more complete and dynamic explanation of the driving forces and changes involved. (Pettersson, 1989, 1990). See also Figure 8.1.

All technological development influences and is influenced by the natural conditions and the values of society. Climatic and cropping conditions are important in problems concerning agricultural production hindered by weeds and pests. Pesticides are used when available to protect crops from pests.

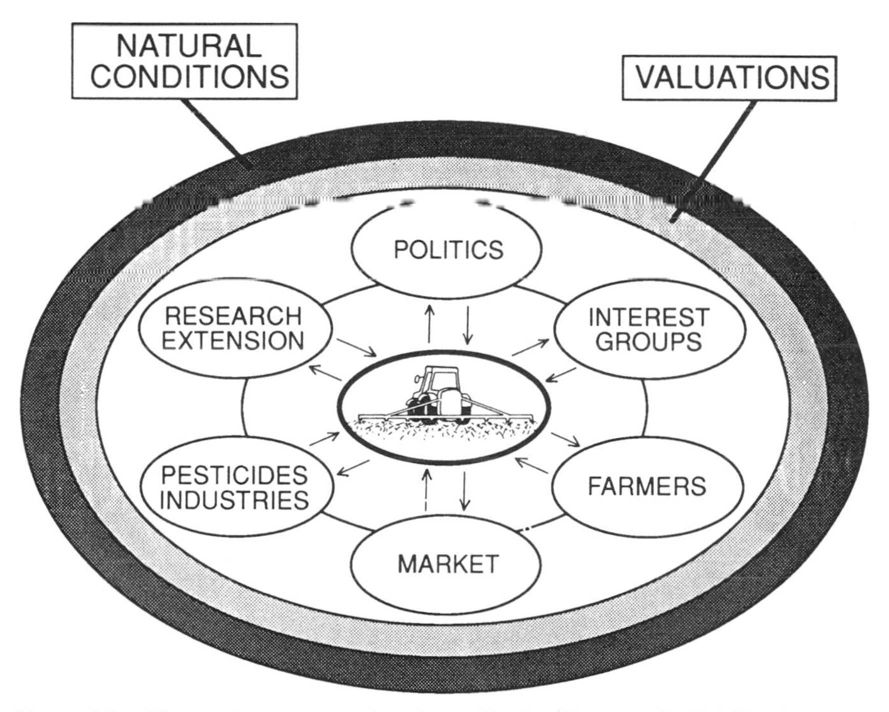

Figure 8.1. Factors in nature and society affecting the use of pesticides in agriculture.

Our cultural and political heritage determine what is "right" or "wrong" in our relationship to Nature. This situation influences the technologies used by society.

Within this framework, several factors play a role for technological development. In this context, *the technological opportunities* for pest control are important. *The commercial structure*, includes companies and the *market* politics influence regulations and restrictions in pest control.

These driving forces *together* influence the actual application of pest control technology. For many years, the technological opportunities have been the most important driving force, whereas today we appear to be in a phase where restrictions from society are playing an increasing role. Hitherto, public policy had little impact on *how much* pesticide was used, but has been important determining *which pesticides* may be used. This has modified the assortment of pesticides used but had little influence on the total use of pesticides.

Pesticide Industry

The pesticides appearing on the market today are the results *research and development (R&D) work* within the chemical companies. The early synthetic substances were largely by-products of war-time laboratories. Developmental work specifically directed at producing suitable substances for different agricultural purposes was limited.

This has changed over the years. Pesticides for use in cultivations have, in themselves, become commercially interesting to develop, and R&D work has become more target-directed. The substances produced during recent years with relatively well-known modes of action reflect this development. In general, we can say that R&D has provided different opportunities for society and the farmers to consider.

The pesticides actually used depend on what society accepts and what its framework of regulations permits, but they also depend on *the decisions of the chemical companies and the commercial commitments of agricultural enterprises*. If a pesticide is to become commercially interesting, it must have a sufficiently large market. Ideally, it must be used in several crops and in several climatic areas.

The demand that health and environmental effects must be tested for also leads to relatively high costs, which contribute to there being a lowest critical size as regards the market for a given pesticide. *Thus, we may expect that only a small number of all the chemical, biological, and technical possibilities of developing pesticides are in fact utilized.*

The Agricultural Industry and the Farmers

As regards the *agricultural enterprises*, both the farmers' attitudes and their profitability calculations play an important role. The relationships between the cost of an application and the economic result will probably be the governing factor for the development in the long-run. The prices obtained for marketed products will also depend on which control inputs are profitable.

Market and Consumers

The demand on the market may work in different directions from time to time. Certain technical and cosmetic demands on foodstuffs have encouraged the use of pesticides. At the same time, restrictive use of pesticides is a criterion that can be recognized, at least among certain groups of consumers. This is expressed most clearly in the demand for "organically grown" products.

The Importance of Technology and Public R&D

The public R&D is important in utilizing the pesticides available in an optimal way on individual crops and by individual enterprises. It may encourage the use of pesticides by identifying more situations where it is profitable to use such pesticides that are available. R&D by the public can also function as a counterweight which eliminates an overoptimal use of certain pesticides.

Politics and Legislation

The political aspect consists of the regulations that determine if, when, and how pesticides may be used. Hitherto, they have largely played a restrictive role in the inputs used in pest control. Changing facts, sets of values, and opinions are reflected in changed criteria that include health and environmental concerns.

The demands placed on the *individual pesticide* reflect a risk/benefit assessment from the state or controlling authority with regard to the use of just this pesticide. General environmental tariffs on pesticides have principally different functions and may be regarded as an attempt *on the social* level to balance advantages against disadvantages.

Sets of Values Among Citizens, Farmers, and Consumers

The sets of values that characterize society also influence the political approach as well as the action and scope for action of individual actors. To some extent, the values are based on direct experience and knowledge of the subjects to which they are applied. Sets of values and modifications of these values with regard to, e.g., pesticides, are governed, however, also by certain "megatrends" which are linked to more superficial changes in society.

Consequently, other sets of values are predominant in the urbanized "postindustrial" society than in the agrarian society or the earlier industrial society. Groups of citizens living far from cultivated areas, agricultural production, and the struggle against the forces of nature have different sets of values with regard to agriculture and its methods than the farmers themselves have. Environmental destruction and the environmental debate not only give rise to a specific aversion to specific environmental impact factors but also to more general attitudes concerning what is permitted in relation to Nature.

Characteristics of Swedish Pesticide Use

Depending on natural conditions and the structure of agriculture, pesticide use varies between different countries and regions. Sweden and Scandinavia are characterized by a relatively moderate degree of specialization and nondiversification in agriculture in comparison with the most intensive agricultural areas in Europe and the United States and in comparison to areas in developing countries where cash crops are grown.

The nature of Swedish agriculture also implies that the environmental impact and other side effects of pesticide use will not be overwhelming. For example, herbicide resistance problems primarily are studied in the international literature and not in the field. The fact that certain environmentally questionable substances were removed early from the range of permitted pesticides in Sweden also contributes to this. Sweden was also in the forefront of this earlier phase of more stringent pesticide policy (Pettersson, 1989).

Some Aspects of Swedish Agricultural Development

The dominating trends in the agricultural development in Sweden as well as in other industrialized countries are well known and fairly uniform; higher yields and more efficient production by utilization of better plant and animal varieties together with technical and chemical aids.

Today a smaller area is utilized for the crop production than previously. The reduction in cropping area decreases the amount of land in open countryside. At the same time, however, several negative environmental effects of cultivation have increased, most markedly in the shape of nutrient (and to some extent pesticide) leaching.

This is connected with the Swedish climate, with increasingly concentrated production, and with the increased intensity. When calculated per kilogram wheat or per kilogram milk, the leaching is perhaps not much greater than formerly, but locally it has increased considerably per hectare and per liter of water passing through the agricultural land on its way to lakes, waterways, seas, and to the groundwater.

The actual content of agricultural production in terms of area, crops, and different habitats is an environmental factor in itself. The influence of cultivated land on what surrounds it is another such factor. Between these two factors there is an interaction to the extent that a certain mix of production will lead to a certain type of disturbance. Figure 8.2 illustrates the utilization of the Swedish agricultural land and its changes with time.

Natural Conditions: Climate and Soils

All Scandinavia has a relatively humid climate. In the agricultural areas of Sweden, the annual rainfall varies from around 600 to more than 1,000 mm, half of which is passing through the soil to the drainage system. The

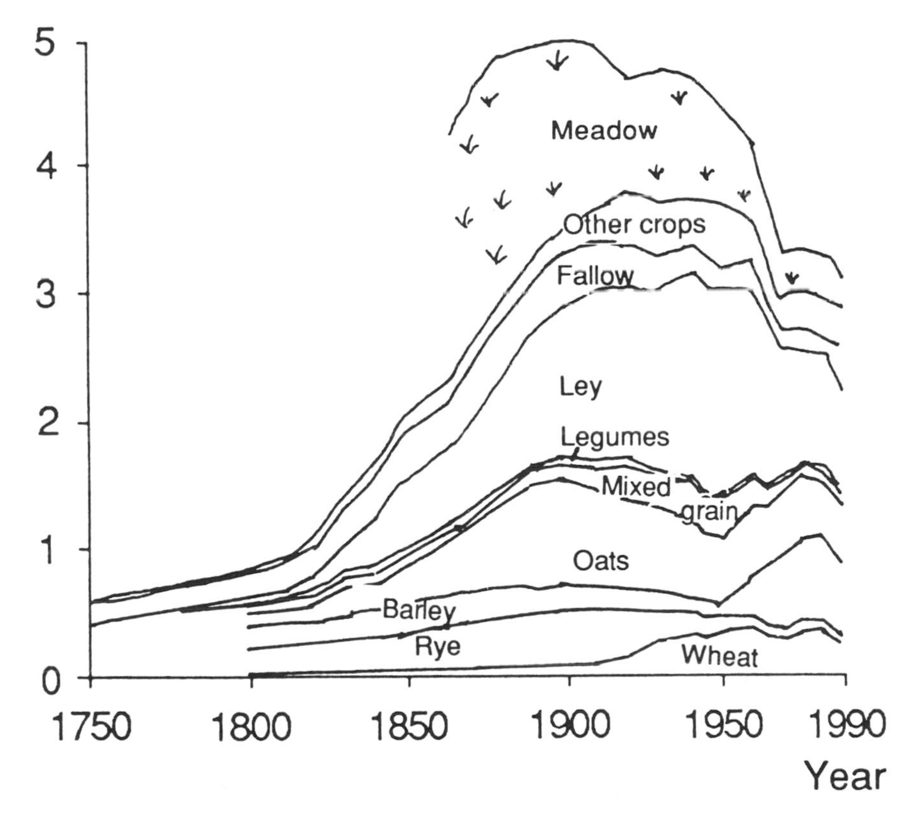

Figure 8.2. The breakdown of Swedish farmland in different periods.

large precipitation and runoff are preconditions for the diffuse, area-associated water pollution caused by agriculture. In addition, the distribution of precipitation throughout the year is to some extent counter to a distribution that would be optimal for agriculture. Comprehensive precipitation and runoff during the part of the year when the soil is not covered by vegetation also provides conditions for leaching and erosion of nitrogen and phosphorus.

Sweden is not a uniform agricultural area. It includes fertile plainlands in southern Sweden with an agricultural structure and yields that are similar to the granaries of Central Europe. It also includes agriculture in sparsely populated areas along the borders of the deep forests of northern Sweden. When described in terms of soil types, Sweden varies from heavy clay soils with relatively little leaching of nutrients and almost negligible soil erosion to sandy soils exposed to wind erosion, even though the dimensions are small in a global perspective.

The relatively humid and cold weather implies that weed problems in annual crops may be greater than the problem of pests. Compared to the European continent, Sweden uses more herbicides and less fungicides and insecticides. The use of pesticides in Sweden expressed as area sprayed is shown in Figure 8.3.

Structure of Swedish Agriculture

Sweden covers many forms of agriculture and associated environmental problems. They extend from the environmentally fairly harmless grassland and cattle-dominated farms in central and northern Sweden to the more intensive grain- or pig-dominated units in the southern provinces, similar to what is found in Denmark and Germany.

In terms of economic and social structure, Swedish agriculture consists primarily of family farms. This is true generally, even though some sectors have been dominated by titled landowners.

The Impact of Agricultural and Environmental Policy

Today, the main issue of agricultural policy is to make adjustments *between agriculture and other sectors* and forces of society. For several decades the agricultural policy of Sweden as well as of many other countries has aimed at an adjustment of the social, technological, and economic forces working for structural changes *within the agricultural sector*. This includes price regulations as well as acquisition of land and support of research and technological innovations. At the same time, a withdrawal of labor from agriculture has been stimulated.

The result of the policy has been a more orderly and socially acceptable revolution than a more market-oriented policy probably would have pro-

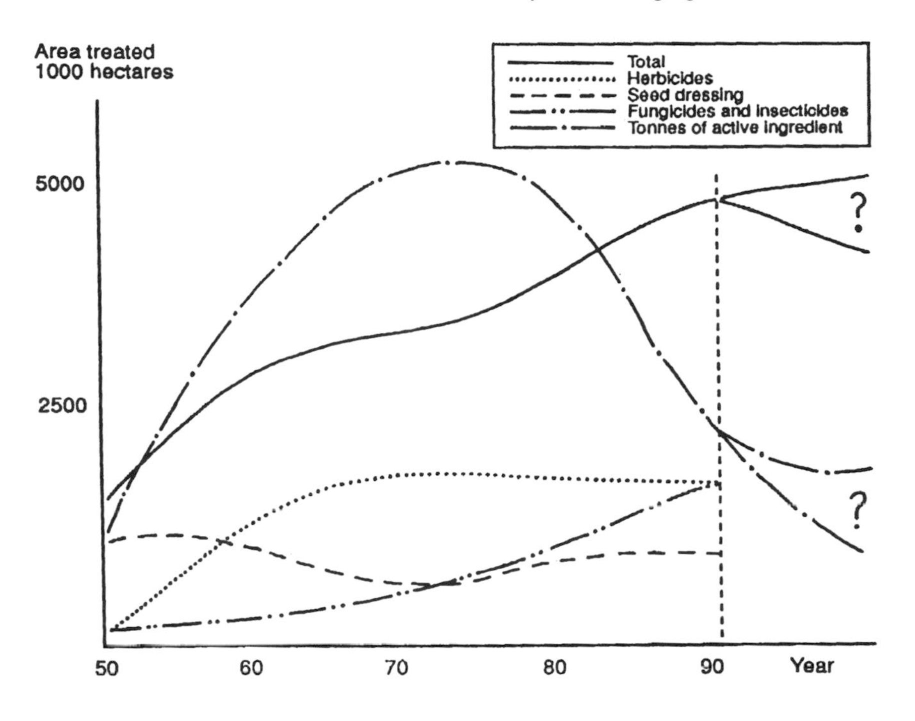

Figure 8.3. The use of pesticides in Swedish agriculture expressed as the area treated and as tons of active ingredient. Source: Swedish National Chemical Inspectorate and others.

duced. The combination of "push" and "pull" factors in this social transformation has at least avoided the creation of areas of rural poverty. Those who left the agricultural sector were stimulated to do so and those who stayed were subsidized in order to be able to make a decent living. Pure production efficiency was compromised in order to achieve other social objectives.

At the same time, the farmers have not been exposed to the capitalist market economy to the same extent as other parts of the industry. Add to this the fact that the input and processing industries in the agriculture and food sector to a high degree have been governed by the farmers cooperative and you will find some explanations of the mentality and culture of the people on the farms.

Politically, reflecting the social structure of agriculture, the farmers have played an important role in the center of the Swedish political arena. Cooperation and also coalitions between the Swedish Labor Party and the Farmers Party were important characteristics of the unique political stability during the 30s, 40s, and 50s.

The side effects of agricultural policy and protectionism are well known in most industrial countries. The production volume and the production intensity have been stimulated too much and a surplus with low value on the world market has been the result. This has also meant that the inputs of fertilizer and pesticides have been overstimulated with increased environmental impact as one of the results.

During the 1970s and 1980s, agricultural policy has been changed so that different kinds of environmental objectives have been included. At the same time, the environmental policy in general has formulated different demands on the production methods and environmental influence of agriculture. Part of this modified political environment is made up of the program for reduced use of pesticides in agriculture which is presently being adopted.

Differences Between Sweden and Other Countries

Even though the characteristic features of agricultural conditions, structure, and changes are of the same *kind* in several other countries as in Sweden, I believe that we can identify a number of specific features with regard to *their degree*. Certain natural conditions such as soil quality and a relatively diversified and widespread agricultural landscape offer advantages from the environmental viewpoint. However, the wet climate leads to leaching problems.

Despite the fact that agriculture has been exposed to the same structural forces as in other countries, the result, seen as a lack of diversification in cultivation—monocropping—and a concentration of livestock-keeping, has not been so pronounced. This inertial resistance depends both on agricultural policy and also on the uniformity of farmers as a class and upon their dominating cooperative enterprises. Large production units for livestock have not been established around the cities, something a freer market would probably have led to. It probably also to some extent explains why pesticide use in Sweden is lower than in many other European countries.

Also when considering social and political aspects, we can identify a difference in comparison with many other countries. The "Swedish model" of agriculture is dominated by independent peasant proprietors through history, strong political organization and action, and an active food processing industry.

Sweden is a young industrial country where the agricultural way of life has long had hegemony over the philosophy of the inhabitants. Most Swedes have long had contacts with agriculture either directly or via relatives. Consequently, therefore, in the dispositions and sets of values among Swedes, agriculture has had a greater dominance than reflected in its importance for gross national product (GNP).

Changes in the Environmental Impact

Both production and environmental problems in agriculture can be found to arise from the original *conflict between cultivation and nature*. Also, the positive environmental values and biological productivity of agriculture are associated with how this contradiction is handled (Pettersson, 1990).

The efforts to produce food, feed, and fiber for the survival or well-being of mankind have always been in conflict with the balance and recycling of nature. The self-regulation of the natural ecosystem is replaced by something which is more controlled and targeted. This control has always been associated with problems. The solutions have always led to new problems, although to varying extents. One of these problem areas has been how to cope with weeds, insect pests, and plant pathogens.

The types of questions placed with regard to chemical pesticides are, therefore, not principally new. Secondary effects and conflicting objectives are to be found around every type of production technique. *Food quality, the internal problems of the agrarian ecosystem with regard to its sustainability and productive ability*, effects on *the surrounding nature*, and questions with regard to *resource management*, are all influenced by cultivation methods and technology (Fig. 8.4).

Food Quality

In the present context, *food quality* concerns both product quality as well as health impact. This includes several issues of toxicological, tech-

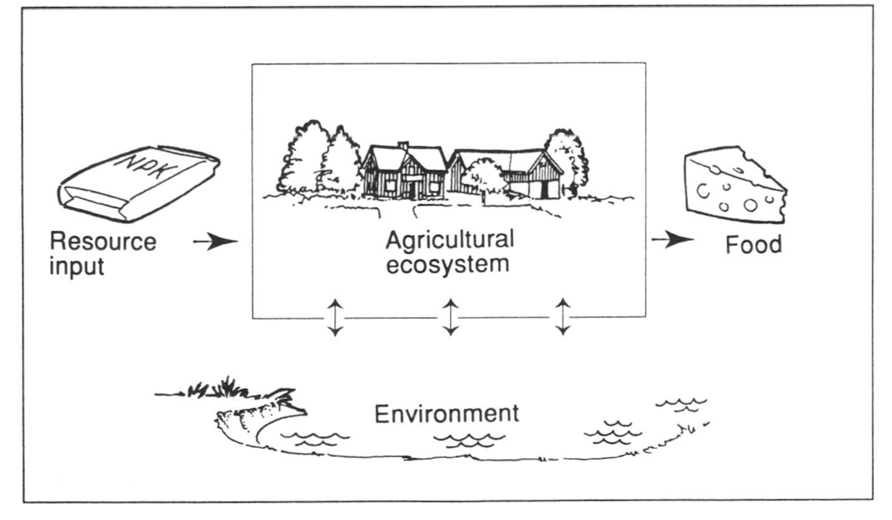

Figure 8.4. Cultivation and its relation to the environment.

nical, and nutritional character. As regards the chemical pesticides, this mostly concerns the risks connected with residues in foodstuffs and the importance of the pest control input for the technical and nutrient quality of the products.

The chemical pesticides largely have positive effects as regards *product quality*. This mainly depends on the fact that efficient pest control improves certain general properties of the product and makes the cultivation more effective. Smaller amounts of weeds give better harvesting conditions and thus opportunities to achieve better technical quality, faster drying, etc. Control of insects and fungi on the crop has the same effect. The absence of attacks leads to fewer technical and hygienic problems.

The general assessment with regard to *health effects* suggests that pesticide residues in food play a minor role in our health. This assessment is based on the type of assessments made when tolerance values are established, on epidemiological studies, as well as on a more qualitative discussion. The latter contains, e.g., the awareness that the natural occurrence of risky substances in food, including carcinogenic substances, for a "normal" consumer is several orders of magnitude larger than those represented by pesticide residues. From ecological, evolutionary, and social points of view it is no surprise that ordinary natural food contains numerous substances with unknown or hazardous long-term effects. Biological evolution and human experience have only selected away the natural toxins in foodstuffs which have more acute effects.

Production Sustainability

Problems within the *agrarian ecosystem* imply, in principle, risks to the soil and the crop plants when using pesticides. Influence on the soil life and resistance in weeds, insect pests, and plant pathogens to pesticides are important aspects.

Availability of and dependence on chemical pesticides have, in themselves, caused problems. The clearest example is resistance formation in pests against individual chemicals. This has both economic and ecological consequences, and it is also a problem that may become more common with some of the modern pesticides and their specific modes of action. Continuous use of a certain substance will lead to a more intensive selection pressure which may "hit back."

With the amounts of pesticides normally applied in Swedish agriculture, it is unusual to find permanent adverse effects on the soil ecosystem. The ecological impact of pesticides is small compared to other kinds of impact caused by agricultural technology. Cultivation measures such as crop rotation, application of organic material, and mechanical tillage of the soil are of greater importance for the biological activity and for the composition of the soil flora and fauna than modification of pesticide levels.

Resource Management and Economy

Resource management in the present discussion mainly concerns utilization of nonrenewable resources. The most important of these is fossil energy, e.g., fuels and pesticide manufacturing.

As regards pesticides and *resource management* in agriculture, we can note that chemical control is frequently a technique with good resource economy. This is most clearly seen in the weed sector. Mechanical tillage and other alternative methods of weed control often require a considerably larger energy input in the form of fuel than the manufacture and application of pesticides requires.

Impact on Surrounding Ecosystems

Environmental effects *on surrounding Nature* include direct secondary effects caused by pesticides that leave the agrarian ecosystem as a result of drift, evaporation, or leaching. The modification of the agrarian landscape for which the use of pesticides is an important factor, as well as the influence of pesticides on the flora and fauna in this part of the countryside, are also included here.

The more obvious *environmental effects* are both direct and indirect. The influence on seed-eating birds of mercury and DDT is among the classical examples of direct effects from the 1950s. This type of problem has decreased in most advanced industrialized countries, even though it has not disappeared. The decrease is largely a result of changes in the assortment of pesticides used.

Pesticide use reduces biological diversity in agricultural ecosystems. Monocultures and a less-diversified landscape have reduced the number of species in agricultural systems. About 75% of the 400 species of vascular plants in Sweden are affected by agricultural practices. The number of partridges have decreased considerably due to the lack of insect diversity in the grain fields due in part to reduced number of weeds and their diversity.

From Specific Side Effects to General Ecological Effects

These different problems can be discussed for each method and aid used within agriculture. "Modern" and "traditional" technologies have advantages and disadvantages when it comes to satisfying demands for product quality, sustainability of the agroecosystem, the well-being of the environment, and good management of nonrenewable resources.

At the same time, there are numerous *conflicts of objectives* which become apparent when effects are divided into problem sectors and which have been discussed previously. A sustainable agriculture is not necessarily the one which has the *fewest negative* and *most positive* effects on the environment. Low-input agriculture need not be the most environmentally optimal and simultaneously lead to the best product quality. We need improved relations between agriculture and Nature, but we cannot expect more than short moments of harmony. The conflicts should be handled and not denied in deference to some Eco-Utopia.

Depending on the pesticides available and the development of technology and knowledge, the dominating problems during different eras in chemical pesticide control will vary. Superficially, we might describe the past and ongoing development in Sweden as follows (Pettersson, 1989). Of course, this description could be true for most of the industrialized countries with slight modifications.

During *the 1950s*, the "first-generation" pesticides dominated against insects (DDT), weeds (phenoxy acids), and fungal pests (inorganic chemicals). Direct effects on the flora occurred as a side effect of the intensified weed control. The influence of insecticides and seed-dressing fungicides on the bird fauna was a prominent problem. The debate over measures to be taken culminated, however, in the 1960s. The restructuring of agriculture has changed the agrarian landscape.

During *the 1970s*, or perhaps earlier, weed control had reached its highest level. New herbicides were introduced: Chemical weed control had become a routine measure in agriculture. Phenoxy acids dominated. The availability of new fungicides and insecticides, together with increased awareness of what the pests actually implied with regard to the yield, led to increased utilization, particularly in cereals. This process was still continuing in Swedish agriculture during the 1980s. Increasingly large areas were treated, although with lower doses.

The restrictions placed on pesticide use as a consequence of health and environmental concern have led to some substances being banned. Problems with damage to apiaries are, e.g., less common today than earlier due to the changes that have been made in technology and assortment.

"Second-generation" pesticides were introduced or were already available. Examples of these are the synthetic pyrethroids for control of insects and glyphosate for control of weeds. The possible health and ecotoxicological effects of most of these new pesticides are better known than were those of the first-generation pesticides.

The result of these change is that *the amount of active ingredient of pesticide used in agriculture in Sweden has decreased since around 1980 whereas the area treated has increased continuously up to the present time.*

The changes in the future will probably mainly depend on the price relationships between pesticide and crop.

During *the 1990s*, we may expect that technological development, together with increasing demands placed on the pesticides, will have led to the zone between second- and "third-generation" chemicals. Low-dose pesticides of different kinds and compounds whose modes of action are better known will be used. Insecticides and fungicides will be introduced that are based on a more fundamental knowledge of mechanisms behind fungus and insect ecology and biochemistry.

The precision in different application measures will increase. At the same time, there is a risk that additional problems will occur in agriculture having to do with resistance, residual effect, and disturbances to the balance between different pests and their enemies. This type of problem is more commonly found in countries or regions which are dominated by more nondiversified crop rotations than are common in Sweden.

The price relationships between pesticides and crop will be decisive for the development of control inputs as regards the area treated. Decreased grain-crop prices can be expected as a part of the deregulation taking place within agriculture. This suggests that certain fungicide and insecticide inputs which today have low profitability will decrease in extent. If the agricultural land simultaneously decreases, this may lead to a reduction in the treated area.

On the other hand, it is not probable that changed conditions for agricultural enterprises and the use of pesticides will lead to completely new "alternative" cropping systems becoming competitive on a large scale. Chemical pesticides will remain an important component of agricultural methods and aids, even though their quantity and quality will vary with the external conditions of agriculture.

Figure 8.5 gives a survey of how the problem has changed with time. The ecological secondary effects and the health risks associated with the earlier pesticides could be eliminated by technological improvements in the pesticides. This suggests not that the secondary effects disappear, but that the type of *specific secondary effects* associated with the toxicity of the chemicals decreases, whereas the *general ecological effects* remain just as large or increase with a more intensive control input. This refers to the indirect secondary effects on both the cropping ecosystem and surrounding ecosystems, which are a result of the intended effect of the pesticide—to decrease the occurrence of weeds and the pests.

In summary and in general terms, we may thus state that as a result of modern technology, including pesticides, *food products from agriculture and the basis for production in agriculture are in good shape whereas the environment is in less good shape and that agricultural production takes place with the support of resource inputs* from outside agriculture.

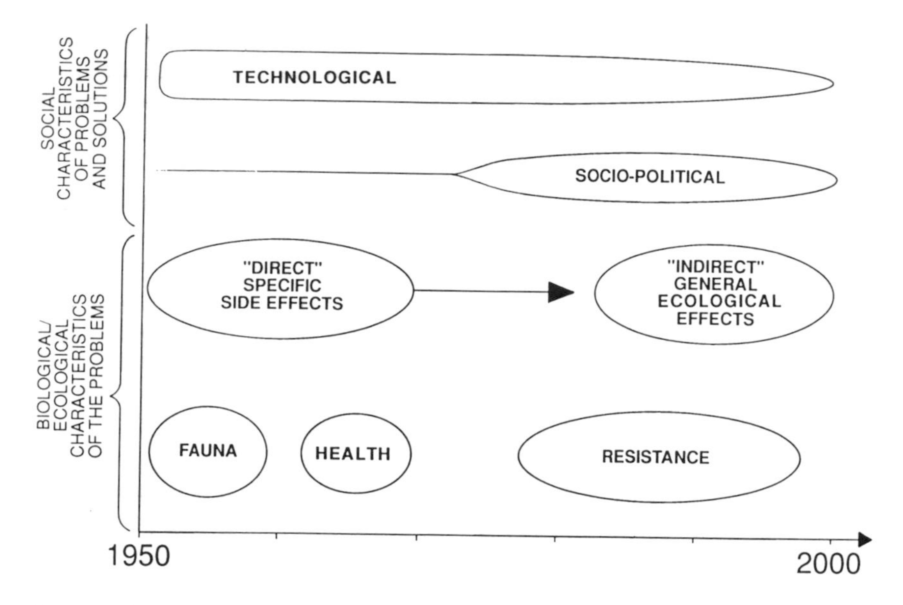

Figure 8.5. The problems, solutions, conflicts, and characteristics of pesticide use and development during different periods.

The changes during past decades can also be interpreted as agriculture becoming increasingly better in its original role as a foodstuffs producer, but today it produces fewer positive but more negative secondary effects. The demands for foodstuffs and the requirement placed on their quality have been dominating whereas the demands for environmental values have not been as concrete and as governing for the development, either directly or indirectly via political measures.

Sociological, Ethical, and Political Aspects

If it were only the absolute level of the environmental problem which decided the extent to which pesticides appear on the political agenda, we would not expect that the discussion on reduced use would first start in Sweden. Instead, the explanation must be sought in the sociopolitical sphere (Fig. 8.1). Politics and political measures emanate from the events within different social and democratic structures (Carlsson et al., 1989; Pettersson, 1990). *Consequently, there is no reason to look for relationships between the absolute level of environmental problems in a certain country, on the one hand, and the political discussion and political measures, on the other.* Instead, one must discuss which political, historical, demographic, and social conditions are prevailing in the country concerned.

The values and attitudes prevalent in society directly or indirectly influence the discussion, the policy implemented and—more or less—the methods used in agriculture. The accepted and permissible vary because of historical and cultural differences. In order to clarify this point, we may consider the situation concerning which animals and plants are considered correct to eat. Such attitudes are deeply embedded in tradition and values which vary between individuals and groups, very few of which are based on strictly "rational" considerations.

"The Green, Green Grass of Home"

Ethics and values are thus extremely important in attitudes to changes in food, agriculture, and in the cultivated landscape. If certain values are more prominent in Sweden and in the Scandinavian countries, then this will characterize the discussion on and assessment of the changes in agriculture, including pesticide policy.

As regards Sweden, we must not forget the exceptional place of agriculture in the culture and in the consciousness of the people. A common factor for all the Scandinavian countries is that the more comprehensive industrial revolution came late for many of the inhabitants, even though some aspects of industrialism have long traditions in these countries, e.g., mining.

Postindustrial Sets of Values and Preindustrial Experiences

Until fairly recently, the Scandinavian countries were peasant countries, and there has never been a pronounced "big-city culture," accompanying dominating changes in sets of values, such as may be found in the United Kingdom and in Central and Southern Europe. At the same time, Sweden is, in many respects, an advanced welfare state and a postindustrial society. Altogether, this will result in a unique combination of *typically preindustrial experiences and also typically postindustrial sets of values.*

Most people have relatively close contacts and relationships with the farming culture. This sector was also characterized to a greater extent than in many other countries by free and independent farmers and to a smaller extent by nobility and large estate owners. Consequently, it is probable that more people feel affected by the changes in agriculture and in the countryside in Scandinavia than in many other parts of Europe. It is their own cultural inheritance that is being changed—not just that of other people.

The deep concern about pesticide use in agriculture among the Swedish people has also a background in the history of forest herbicide use in the 1960s and 70s. This practice had impact only on minor areas of forest land, but after a long period of controversy, herbicide use for control of brush vegetation in forestry was restricted, decreased, and finally almost prohibited by national and local authorities. (Exceptions are herbicide use in plant nurseries.)

Part of this process also had its specific Swedish characteristics. Probably because of the historically based unique public access to private land, this conflict arose earlier and became more intense in Sweden than in other countries. This customary right of the public is of specific relevance for forest land because, among other things, it allows every citizen to pick mushrooms and berries on private forest land. Thus the changing technology of forestry, including herbicide use, was not regarded as an issue only for the forest companies but also for the general public.

The Welfare State and the Confidence in Political Solutions

In Sweden and in Scandinavia there has also traditionally been a high degree of *confidence in political measures* and solutions to different problems in society. This is probably connected with a relatively stable political situation and also with numerous successful examples of the art of "social engineering."

Consequently, it is natural to expect that both disappointment and worry about the development as well as demands for changes and the creation of better times would lead to greater political demands and expectations than in other countries. The dreams of Utopia are more often directed toward the political arena and not toward the market.

From Technological Solutions to Ethical and Political Adjustments

Many questions concerned with environmental influence have, in addition to scientific aspects, both esthetic and ethical dimensions. The "permissible" or "suitable" in connection with food, livestock management, and cropping are closely connected with traditions and sets of values that are only partly based on rational considerations. The possibility itself of causing fundamental influences on Nature also leads to new questions. Chemical pesticides are an example of a technology that enables the transformation of both the landscape as well as Nature in a more comprehensive manner than has been possible earlier.

The new situation therefore perhaps defines not the character of the problem but possibly its dimensions. A more powerful technology, which the chemical pesticides in agriculture frequently represent in comparison with the more traditional methods, will lead to more revolutionary changes ecologically, economically, and socially than those that result from a less powerful technology. In the same way as, e.g., parts of biotechnology do, the chemical pesticides therefore give rise to *a new question: How successful should we be in our manipulations and controls?*

The increasingly perfectly "designed" cultivation is productive; it yields products of high quality and is resource efficient, but at the same time it leads to a more uniform countryside. The positive environmental values of agriculture decrease. This occurs at the same time as the *shifts in social sets of values* have also resulted in a more general questioning of *technology and manipulation*. The advanced pesticides and biotechnology are exposed to the same kind of doubt. *It is no longer the actual or possible magnitude of the secondary effects that are decisive. Instead, it is the actual objective which is questioned.* Technological efficiency gives in itself birth to new doubt.

This leads to *a true conflict between the most original aims of agriculture and the requirements and sets of values found in society. There are no technological solutions to this problem. Instead, the problems and the conflicts have entered the sociopolitical arena.*

Depending on the modifications to sets of values that will occur in the future, this conflict may increase or decrease. It is possible that increasing numbers of people will become increasingly skeptical about the increasing perfection of agriculture and control inputs, but it is also possible that fewer people will focus their doubt and anxiety on just this part of technology and that the symbolic value of pesticides will be replaced by something completely different, e.g., something which involves biotechnology.

Pesticide Policy in a Changing Environment

Political decision-making suggests that opinions and values should be formalized within the rules of society, in this case concerning the use of pesticides. The demands of the citizens as regards safety, information, and political influence should be expressed in a reality that nonetheless only partly is influenced by political decisions and is more influenced by technological development, by the market (in the widest sense), and by the actors within these spheres.

More powerful technological aids with a comprehensive influence on the landscape and its ecosystems are, in principle, of importance for the individual not only as a consumer in the market but also as a member of society.

Thus, at the same time as there is something extremely reasonable in the growing demands on environmental policy—of which pesticide policy is part—from a democratic viewpoint, there are clear difficulties in finding the functioning political instruments and in actually influencing the development within the sector.

The shifts in sets of values that have occurred in society during recent decades have, among many other things, also implied moments of general questioning with regard to technology as such and manipulation of Nature. This skepticism is sometimes well justified but sometimes it is based on less-than-solid foundations. The changes in values, where Nature is increasingly regarded as being the standard, are tending to become a source of conflict, particularly in matters concerning agriculture and its methods, since cultivation must be defined as a manipulation of the ecosystems.

I believe that it is somewhere within this perspective of modified sets of values and difficulties in actually influencing the development and use of technology within a sector of great concern to the people that we should recognize the decision in Sweden to reduce the use of pesticides by 50% between 1985 and 1990 and subsequently to reduce it by 50% again. To some extent this reflects general political and public opinion trends in the industrialized countries. To another extent the reduction programs have specifically Swedish—or Scandinavian—characteristics.

From the Quality of Pesticides to the Quantity of Pesticide Use

In Sweden, as also in most other developed countries, different systems of regulations have developed and become formalized during the 1950s and 1960s with regard to the demands placed on individual pesticides concerning health and ecological risks. As a result of this and of new technological and chemical opportunities, the individual compounds have gradually been improved. *Policy-making has largely, so far, suggested restrictions for which economically interesting control inputs will also be utilized.*

The risk/benefit assessment of society is reflected in changed demands with regard to health and environment viewpoints which result in rules on which compounds may be used, together with when and how. However, at the same time, the use of pesticides has increased as a result of growing knowledge of the economic importance of pests and with access to new and more efficient means of controlling pests which earlier could not be dealt with. In addition, there is also an increased need for control measures in the nondiversified and intensive cropping systems that have become possible as a result of (or because of) access to new pesticides. *The very hegemony of pesticides in agricultural technology gives rise to questions where the answer is, More pesticide use.*

Policy and public opinion directed at chemical aids used in agriculture during recent years in Sweden, and which have resulted in different reduction programs, can be seen in this context. Public opinion and the political system do not accept the total effects of a technology that has hitherto been approved and accepted piece by piece and compound for compound.

The Impact of the "New" Pesticide Policy in Sweden

If we try to evaluate what actually happens with pesticide use in Sweden as a result of the programs for reduced use that are now in force, we will find that the use of pesticides measured in kilograms of active ingredient has decreased considerably since 1980 (Figs. 8.3 and 8.6). From 1981 to 1990, the reduction is about 50%. The official goal of 50% refers to the reduction during the 5-year period of 1986–1990 *compared to the average of 1981–1985*. Because this procedure raises the starting point compared to the trend value of 1986, the solution of the calculation also gives approximately a 50% reduction for the 5-year period ending in 1990.

This reduction during the 1980s is the result of several factors affecting pesticide use in agriculture discussed in the earlier section, Forces and Factors in Pesticide Use. (See also Bernson and Ekström, 1991 and Pettersson, 1989.) It has been achieved by the introduction of new pesticides that are active in lower doses. For example, the herbicides TCA and sodium chlorate have been replaced by glyphosate and phenoxy acids by sulfonylurea derivatives.

Another element of the pesticide reduction program is the use of lower doses in general. This is also of specific relevance for herbicides. The weed pressure has decreased in the grain crops as a long-term effect of herbicide use. Thus, it is possible today to use lower doses of herbicides with less

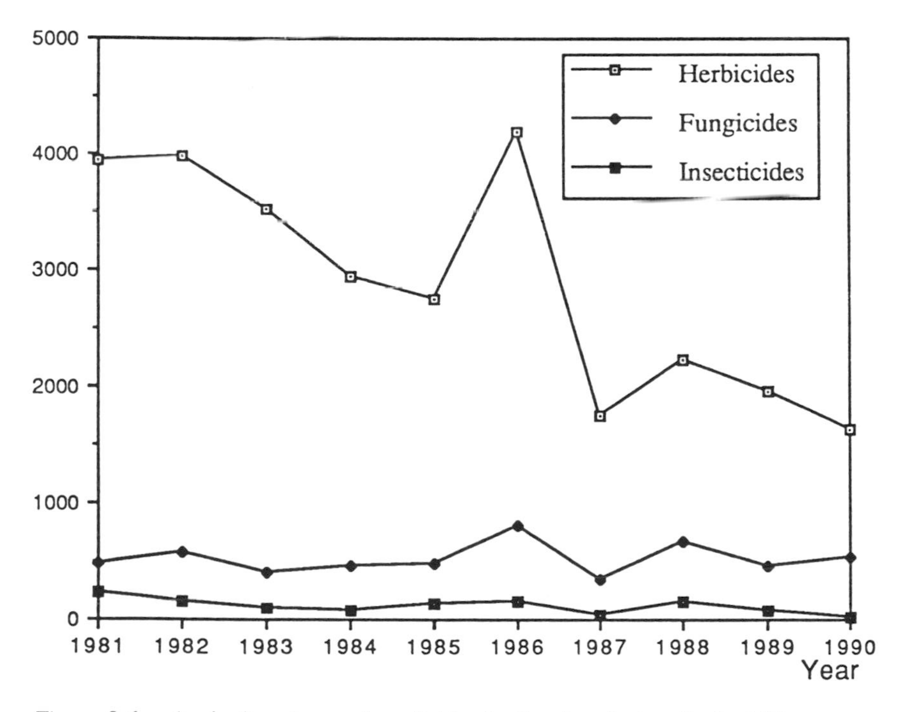

Figure 8.6. Agricultural use of pesticides in Sweden during the last 10 years—tons of active ingredient. Source: Swedish National Chemicals Inspectorate.

herbicidal efficacy and still maintain weeds at an acceptable level without creating any increased problems in the future.

With lower doses of pesticides, the spraying technique becomes more important. A voluntary test of sprayer condition has been in operation since 1988 with a subsidy covering 75% of the test cost. The extension service has also been strengthened in order to provide better tools for pest prognoses.

Another important part of the new legislation of 1986 is a stricter regulation of individual compounds, thus reducing the risk. The new regulations include "cutoff criteria" for pesticides that are clearly unacceptable from health and/or environmental protection points of view. Pesticides can be approved for a maximum of 5 years. During 1990, 450 of around 600 pesticide products were due for reconsideration. As a consequence of this, the manufacturers did not apply for continued approval for 100 products and asked for withdrawal of another 50. The Chemicals Inspectorate denied reregistration for another 50 products which did not match the new criteria. All together, the data from this reregistration procedure imply that the number of pesticide products decreased from 600 to 400 and that the number of active ingredients used in Swedish agriculture today is about 100.

However, at the same time as the kilograms of active ingredient have decreased considerably, the pesticide-treated area has continued to increase. The use of insecticides and fungicides in cereals has intensified in Sweden during the 1970s and 1980s as in many other countries. My conclusion is that *so far, the "new" pesticide policy in Sweden may be described more as an adaptation to what is technically and economically possible rather than as a break* in the trend. Thus, it does not yet imply completely new priorities and new conditions for commercial agriculture in Sweden.

If we look to the future, both technology and economy speak for reduced use, both as regards kilograms and the area treated. More low-dose pesticides are introduced; spraying equipment and methods will become more effective. Swedish agricultural industry is becoming more market-oriented and less regulated, which should lead to lower grain prices. The final outcome of this will probably be that some of the control inputs that today are profitable will become unprofitable.

Thus, several factors suggest that the use of pesticides in Sweden can be reduced by a further 50% without altering the basic conditions for their application. However, it is also clear that there are, at the same time, political expectations for more drastic and actual reductions in pesticide use. At present, it is too early to be able to hazard a guess as to what this will mean.

Discussion and Conclusions

During the relatively short period of agrarian history in which the chemical pesticides have played a dominating role they have been evaluated in different ways in different periods and by different people. They have come to symbolize the *technological dream* that eliminates the wear and tear of everyday life. Others have regarded them as *a necessary evil*, an attitude that is also predominant today. Some people assert that they are an *unnecessary evil*, whereas *the more pragmatic attitude*—that these aids in cultivation, in the same way as many others, have their problems but at the same time can be improved and developed—is finding difficulty in being understood at present.

Changes in pesticide use are taking place with and without political measures. Policy-making always includes an element of adaptation as well as the implementation of public opinion. The result and the political action that freely occurs vary from one area of society to the other. For those topics with a long political tradition, for example, economic and social policy, the aims and means are fairly clear. For topics with a short political history like environmental policy, the policy-making has not yet matured in its aims and means. This is certainly true for the area of pesticide policy.

Pesticides as an Environmental and Ethical Subject

A recurrent question in the discussion of agricultural methods and with regard to the use of chemical pesticides is whether, as a result of their properties and possible effects, they give rise principally to new environmental or ethical questions, or whether they can be regarded as aids in cultivation in the same way as other, "traditional," methods. Is there a *difference in type or a difference in degree* between these and other aids used in agriculture?

My conclusion is that when *considering the individual pesticide, there are no principally new questions arising.* Pesticides can be evaluated and compared with other aids and cultivation methods. It is a matter of differences in degree when discussing the impact of pesticides compared to other means in agriculture. In some cases, the pesticides represent the optimal solution to the problems in cultivation; in others, the "nonchemical" methods do.

Thus *the use of chemical pesticides is also possible in an agriculture where environment, quality, and resource economy are given a higher priority in comparison with the present levels of priority given to "efficiency" and "production." Methods and cultivation systems which exclude chemical pesticides are not generally regarded as more environmentally considerate, quality-minded, or resource-efficient. A rational consideration thus does not deal with "chemical agriculture" or "organic" (nonchemical) agriculture. Instead, considerations should include the priority of environmental objectives in a wide context regardless of the means* used to attain these *objectives.*

I cannot see why individual pesticides *in themselves* should provoke any different questions than those arising from other cultivation inputs and aids. It appears to have been, more than anything else, historical coincidence that only pesticides have come to have the symbolic value that they do today in the environmental and agricultural debate. The problems around pesticides are, in principle, capable of being handled. However, this does not always mean that they are handled in the correct way.

On the other hand, *when considering the total use and effects of pesticides, the differences compared to "traditional" cultivation methods are more than a matter of degree.* Thus, it is at this superficial level that one might argue that pesticide use gives rise to the essentially new questions of ethical and political character mentioned earlier.

The Need for Public Regulation of Pesticide Use

Already the differences in degree between pesticides and other aids in cultivation raise important questions as to their utilization and the ecological and social limitations for such utilization. Pesticides are one of several factors which can be used to make an impact on Nature in a comprehensive manner.

Even the ideal and optimal pesticides without direct secondary effects cause impacts on Nature and the society outside agriculture. In this way, they concern the general public and are not simply and internal problem for the people who use or utilize pesticides in production. In addition, most pesticides used are neither ideal nor optimal. The need for public regulation of pesticide use is justified by both these factors. The task of regulation is both to balance risk against benefits and to contribute to eliminating substances with unacceptable risks and secondary effects.

Thus, it is obvious that regulations are necessary to determine when, where, and how the individual pesticide may be used, as well as to determine which pesticides are permitted. *Already the fact that the pesticides have external effects which the market cannot handle requires public and social assessment and control.* The same conclusion can, both pragmatically and empirically, be reached from the historical experience of pesticide use in industrial countries as well as from their use today in many developing countries, which lack the social infrastructure which can make their use both ethically and socially acceptable. The consequences of free and uncontrolled experimenting with pesticides and other technologies are frightening.

The questions arising from, e.g., the Swedish decision to reduce the total use of pesticides, involve whether, apart from a regulation of individual pesticides, there is *also need for public control over the total use of pesticides*. Control of this kind may be justified if the total negative effects of pesticide applications are greater than the sum of what the individual pesticides achieve. It can, however, also be justified if the total impact of pesticide use is of qualitatively different character than that of individual pesticides.

Such an impact may be of a purely environmental nature—changing the agricultural landscape and ecosystems to a degree that is considered unacceptable—or of a socioeconomic nature. The utilization of pesticides may, by means of its own dominance, create ecological problems that result in an increased need for pest control. This may restrict the development and utilization of other techniques and methods. This could also result in society becoming extremely dependent on pesticides and thus technologically/economically vulnerable.

Personally, I believe that this is a problem and that there is *justification for public regulation both as regards individual compounds and regarding the total use of pesticides*. On the other hand, I find it rather ironic, both as an agricultural scientist and as a Swedish citizen, that this question first surfaces on the political agenda of one of the more successful countries and regions, where it is least justified, whereas the development continues unabated in those parts of the world where the problem is more serious.

References

Bernson, V. and Ekström, G. 1991. *Swedish Policy to Reduce Pesticide Use*. Pesticide Outlook **2**:3. August 1991. 33–36.

Carlsson, M., Pettersson, O. and Jeffner, A. 1989. *Livsåskådning, miljö och odling* (Conception of Life, Environment and Cultivation). Fakta Mark/Växter Nr 10, Uppsala 1989.

Pettersson, O. 1989. *Bekämpningsmedel, hälsa, miljö* (Pesticides, Health and Environment). Aktuellt från lantbruksuniversitetet 378, Uppsala 1989.

Petterson, O. 1990. *Alternativ i odlingen* (Alternatives in Agriculture). Aktuellt från lantbruksuniversitetet 388, Uppsala 1990.

9

Reducing Pesticide Use by 50% in the Province of Ontario: Challenges and Progress

G. A. Surgeoner and W. Roberts

Introduction: This Farm Ontario

The Province of Ontario, in central Canada, has the largest agriculture industry of any province in the nation. Approximately, one-half of the class 1 agricultural land for Canada lies within its boundaries. In 1986, Ontario farmers sold food and other agricultural products worth 5.5 billion dollars. About 2 billion dollars of this production was exported, principally to the United States (O.M.A.F. 1988). A wide diversity of agricultural commodities are produced, from tobacco and peanuts in the south, to fruit and grapes in the Niagara Peninsula, to vegetables and pasture lands throughout the province.

Approximately 60% of the gross farm income is derived from livestock or livestock products. Consequently, the major crops are grain corn (704,000 ha), silage corn (154,000), soybeans (518,000), winter wheat (279,000), and hay (1,040,000). There were approximately 3.2 million hectares of field crops grown in 1988 (O.M.A.F., 1988). Our field-crop commodities are similar to many of the Midwestern U.S. states, e.g., Illinois and Ohio. We do have, however, 76,000 ha of vegetable production and 34,000 ha of fruits and vineyards.

Ontario is the most populated province in Canada, accounting for 9.5 million of a national population of approximately 26 million (O.M.A.F., 1988). The 1986 Canadian census reported 72,713 farms, an 11.8% decline from 1981. These farms totaled 13.9 million ha in 1986, a decline of 6.5% from 1981 (O.M.A.F., 1988). Farmers make up less than 4% of the provincial population.

Most of the population is confined to large urban centers like Toronto, Hamilton, Ottawa, and London. In counties surrounding large urban centers like Toronto the majority of land under cultivation is not owned by farmers but rather by speculators and urbanites who hope to eventually retire to the country. Indeed, just over 15% of those living in rural Ontario are farmers (Davies and Penfold, 1990). Many in the rural landscape are commuters who work in the cities and live on small parcels of land. In 1986, the farm population of Canada represented only 14.6% of the actual rural population (Davies and Penfold, 1990). We estimate that there are approximately 40,000 full-time farmers in the province. Public-opinion polls indicate that our farm population is well respected. A recent multiple-choice, public-opinion poll of Ontario citizens asked, "What positive things come to mind about Ontario farmers?" They had the following responses: (1) 36%, hardworking/dedicated/heritage, (2) 18%, provide food, (3) 13%, vital to economy, and (4) 16%, don't know. Nevertheless, the political voice of farmers does not match that of the urban or most rural populations. The farmers are, however, well organized with a variety of commodity marketing boards and joint associations to deal with common problems such as international trade, animal welfare, and public misperceptions about farming. A consortium of 15 marketing boards has formed an association known as AGCARE to address environmental issues facing agriculture such as pesticides.

Food Systems 2002: An Ecological Systems Approach to Sustainable Agriculture

History

It would be a triumph of virtue if one could say that the program to reduce pesticides by 50% in the Province of Ontario was based on a consultative process between farmers, agriculture researchers, politicians, and the general public. As in many endeavors the truth is more revealing. In the summer of 1987, there was a provincial election to choose the ruling party for our provincial parliament and our premier, who is the equivalent of a U.S. state governor. Various political parties were seeking issues on which the public could make decisions. Public-opinion polls in Ontario, as in most jurisdictions throughout North America, had indicated that the public in general was concerned about pesticides from both a human health and environmental perspective. The electorate made their choice in the late summer of 1987. Quite simply, the party which won the election and became the ruling party had as one of their platforms a promise to reduce pesticides by 50%.

Prior to the election in 1985, there had been an ad hoc committee of pest-management specialists including provincial, federal, and university personnel who had attempted to determine where the province should be going relative to pest management. Six commodity areas where pest management was currently in place were examined and attempts were made to analyze how much these IPM programs had reduced pesticides. Estimates varied between 40 and 70% dependent on commodity. There was an overall concensus that a 50% reduction across the province was feasible and briefing papers were submitted to senior management outlining this possibility and mechanisms to achieve it. Details were lacking but obviously this caught the eye of some politicians.

In essence, the political platform would not have been developed without a preliminary agenda having been developed by pest-management personnel. In one sense, our program to reduce pesticides by 50% was a master pest-management plan for the next 15 years. The public would have shown little interest in it as a simple pest-management strategy, but as a program to reduce pesticides by 50% it was readily accepted.

Why 50%? It had a nice advertising ring to it, as did other programs, such as that to reduce garbage by 50% through recycling programs. The public doesn't want terms like 37% or 63% which may be defined as logical reductions determined by scientists. As indicated, however, pest-management personnel, based on previous IPM programs in place, felt that it was an attainable goal. The point is that a populace can choose whatever level of reduction it wants. The politicians could have called for a complete ban on all pesticides, which would have had catastrophic impacts on agriculture in the province. A major and sometimes ignored caveat in the 50% pesticide reduction was that this would be done while maintaining economically competitive food production. There had been brief consultation with agriculturalists, who indicated that a 50% reduction would be possible under certain conditions, with certain resources, and over certain time frames. The role of personnel within the Ontario Ministry of Agriculture and the Food and Advisory Boards to the Ministry was to

1. Devise a conceptual frame work under which this goal could be met
2. Identify the resources required to meet the objectives
3. Identify the time frame required to complete the task

The time frame was identified as 15 years, which meant, because the program was conceived in 1987, that the objectives would be completed by the year 2002. The program was thus entitled, Food Systems 2002—A Program to Reduce Pesticides in Food Production by 50%.

The Agriculture Perspective

While the origins of the program were a political response to the public's perception about pesticides, many in the agriculture community concurred with the objectives, but for reasons very different than those of the general public. Obviously, we were concerned about environmental issues and human health issues, but perhaps not to the same degree as the public. We obviously had a better understanding of the regulatory programs available to minimize environmental and health impacts (Archibald, 1990). We also recognized that there were significant economic benefits derived from crop-protection compounds without which our farmers would be at a competitive disadvantage. We felt that the program, in the long term, was absolutely critical to the competitiveness of our farmers and to rural society for a number of reasons.

First, despite the fact that Canada uses only 2% of the world's pesticides, we have national registration demands in terms of health and environmental data as stringent as those of the U.S. E.P.A. (Archibald, 1990). Often Canada requires data submissions that major international companies cannot justify based on market potential. Consequently, Canadian producers often do not have the choice of products that producers in other countries have. Registrations, if they do come, often lag several years behind the United States. Obviously, international companies will attempt to secure the large U.S. market before turning to Canada. Even if Canada is deemed to have an economically attractive market to pesticide companies, registrations often lag 3–5 years behind the United States.

Because of added regulatory costs and smaller markets, Canadian farmers will typically have to pay more for products than their U.S. competition. Many pesticides in Canada cost from 7 to 44% more than in the United States (Martin et al., 1990). The pesticide input costs for growing an acre of tomatoes in Ontario vs. Ohio are outlined in Table 9.1. Simply stated, many believe that we will always be at a competitive disadvantage with regard to pesticides. One way to overcome this is to reduce our reliance relative to other nations.

Second, we believed that costs of pesticides would increase at rates significantly above those of commodities, so these input costs would take

Table 9.1. Comparison of crop-protection chemical costs in Canadian dollars for field tomatoes grown in Ontario to those in Ohio.

Pesticide	Price ($/L)		Number		Cost	
	Ontario	Ohio	Sprays	L/acre	Ontario	Ohio
Guthion 240	9.77	6.45	2	0.91	17.78	11.74
Bravo 500	8.30	8.17	6	1.6	79.68	78.43
Ethrel	23.60	14.27	1	1.5	35.40	21.41
TOTAL					132.86	111.58
Difference					$21.28	

Source: Submitted Scot Makey by H.J. Heinz (in Martin 1990:17).

a greater component of a farmer's capital resources. Costs would increase rapidly because (1) pesticides are typically fossil-fuel derived, (2) regulatory data requirements will only get worse, and (3) ideas of establishing environmental taxes (e.g., as in Iowa) will likely occur. We believed that any reduction in pesticides while maintaining control would thus have returns in excess of inflation.

Third, we had major concerns with respect to resistance of pests to pesticides. In closed environments, e.g., greenhouses and flies in barns, there were often no viable products for control of insects (Harris et al., 1982). Resistance to herbicides and to fungicides continues to increase. At the same time, we are experiencing a net loss in the number of insecticide and fungicide active ingredients in Canada. With greater reliance on fewer products the probability of resistance continues to escalate. The prospect of not having any viable pesticide option for control of a pest in Canada is not all that hypothetical.

Fourth, many in the farm community were concerned for their own health and with how they were viewed by society. The farmers read the same media reports that urban consumers do. They recognized that there was concern in the general public relative to pesticides (Sachs et al., 1987). Rightfully, they wondered, if the public was so concerned about residues in the parts per million, what of their own health, because they were using products in concentrated form and spraying large volumes over a lifetime.

A fifth factor for some was what we call the pariah syndrome. The public and indeed farmers themselves questioned whether they were poisoning the environment. Some felt that the high esteem they previously had with nonfarming neighbors was being eroded because of their use of pesticides. Farmers don't spray because they want to but rather out of perceived necessity. Anything that could be done to reduce reliance on pesticides was appreciated by the farm community as long as it was economically viable and entailed minimal risk of losses.

The final and perhaps ultimately the most significant concern was that if the farm community didn't begin to reduce pesticides, legislation would evolve that would force reduction and perhaps ban pesticides. In Ontario, we see many communities concerned about application of pesticides on lawns and a few attemps to develop bylaws to ban pesticides. The goodwill that farmers have with the general public and specifically with their rural neighbors is critical to continued viability. By being proactive, by reducing reliance of pesticides, there is a tremendous opportunity to affect public opinion so that farmers are seen as stewards of the land with full public support.

50% Reduction of What?

A 50% reduction in pesticides can be measured different ways by different people. What is your benchmark? Tons of active ingredient, number of applications, or something nebulous like reliance? In Ontario, every 5 years, a survey is conducted by the Economics and Policy Branch of the Ontario Ministry of Agriculture and Food which measures pesticide use in conjunction with the International Joint Commission on Pollution in the Great Lakes Basin. The last survey had been conducted in 1983; the results are displayed:

Metric Tonnes of Pesticides (by Active Ingredient) Used in Ontario in 1983.

	Tonnes	*Percentage*
Herbicides	*5,921*	*67.9*
Fungicides	*1,727*	*19.8*
Insecticides	*610*	*7.0*
Nematocides	*462*	*5.3*

These data have been identified as our benchmark. In Ontario, we are talking about a 50% reduction in tons of active ingredient as measured by weight. We recognize that there are problems with this methodology. It does not measure reliance, actual use pattern, or biological activity of the pesticides in the past or currently used; nor indeed does it measure new products being added to the system. One can argue that introduction of newer herbicides and insecticides that work at grams of active ingredient per hectare vs. kilograms per hectare will achieve our objectives. It can also be argued that as certain high-pesticide-use patterns such as that of nematocides on tobacco are reduced because of lower acreage, not reliance, we will achieve our objectives. We would concur with these concerns but stress that the philosophy of the program managers is to reduce reliance and not simply play semantic games. A benchmark was available and it was chosen. We don't make apologies because every system that was considered had inherent problems. We would also emphasize that we would predict that newer products, although biologically more active against target organisms, will have much better health and environmental data bases than many of our older products and be more target specific.

The Master Game Plan

To most in Ontario agriculture the concept of a 50% reduction was radical. None felt that it could be achieved over a single year or even within 5 years without significant sacrifice by the farmers. It was critical that farmers be involved in the consultative process and agree not only with the objectives but the methodologies and time frame. Facetiously, some stated that a time frame of 15 years was accepted because most felt that by the year 2002 they wouldn't be held accountable for the success or failure of the program.

More pragmatically, the program evolved into three 5-year time periods each of which would be evaluated at the end; reassessments would be made accordingly. We, therefore, report on the resources and methodology being used in the first 5-year installment of the program. The next 5-year plan (1992–1997) has not been formulated, and the political party in power has changed. The new party, which ran on a strong social and environmental platform, has recently publicly endorsed Food Systems 2002.

A total of $10 million (Canadian—Cd) was made available for the first 5 years. Three broad areas of effort were identified and addressed; these were (1) education ($1 million), (2) research ($5.6 million), and (3) infrastructure of pest-management and education personnel ($3.4 million). The dollars were made available at the rate of $2 million per year.

Some Novel Concepts

The senior administrators of the program allowed a few of what we consider simple revolutionary management concepts. One of the best was that you didn't have to spend your allocated budget each year. In fact, the allocated budget was placed in trust and actually allowed to collect interest at current deposit rates, which was then available to the program. All too often there is the concept that you must spend your budget or it will be cut. As this program developed in the first year or 2 researchers were often committed to other programs or the program hadn't filtered down to non-agricultural research institutions. There was never the urgency that we must spend, but rather the understanding that we would fund quality, and that if we didn't spend all the money, or hire quality personnel, the dollars would carry over to the next year along with the interest accumulated. In addition it was made clear that we could fund expertise from outside the province if that was where the best value for the dollar was to be found.

Another important concept was that, because this was a 15-year program, personnel hired for our pest-management infrastructure were made permanent employees rather than individuals on temporary funds who might have wondered whether at the end of 5 years they would have jobs. Previous experience from other program initiatives had shown that starting about year 3, people would begin to seek new jobs because of uncertainty about contract renewal. Logically, the best people were hired away first. Replacements were difficult to get because you could only offer 2-year instead of 5-year contracts. Retraining of pest-management scouts and grower confidence with new personnel each year would present problems. From the start there were some new management concepts. Importantly, for all involved, the sense of committment rather than political expediency and reacting from crisis to crisis was evident.

Education

With the cooperation and in full consultation with our grower associations, the Ontario Ministry of the Environment and the Agriculture Chemical Manufacturers Association (CPIC), the Ontario Pesticide Education program was expanded. This program had been in existence since 1986. Originally, the program was voluntary; producers took an intensive 1-day course and wrote an examination at the end of the day upon which they had to achieve a 75% average to receive certification.

The program was expanded and an excellent 120-page training reference manual was developed. Instructors for the course are often fellow farmers or people retired from the ag-chem industry. The key was that instructors be practical individuals all of whom could identify with farmers. The obvious areas of chemical safety, storage, calibration, and spray-equipment maintenance are covered. The cost of the course is $35, which includes the cost of the manual, which producers are sent prior to participating in the course. Training sessions are scheduled across the province, often partially sponsored by equipment dealers and suppliers. Meetings are always scheduled from November to April with a maximum enrollment of 25 individuals. The program is extremely successful but there was still a need for some mechanism to make education mandatory. This was accomplished by some superb political maneuvering by our grower associations. It was the growers who pushed hard for mandatory training. Responsible growers recognized that those who did not take the course and didn't use pesticides responsibly would negatively impact on the availability of products to themselves. The public doesn't discriminate between the pesticide being at fault vs. misuse and abuse by the applicator.

Since 1987 there have been over 1,500 training sessions with about 28,000 producers having participated. Their response has been extremely positive; 92% have passed the course on the first attempt. Evaluation sheets from producers are filled with comments like, "Why wasn't this available 25 years ago?" As most agriculturalists realize, the best salesperson you have is a satisfied farmer. In Ontario, the farmers spread the word. At this writing, winter 1991, there are about 5,000 farmers still to be certified. Many of these are not of the same caliber as the first students. They often have a lower educational background, have major language-translation problems, and in some cases difficulty in reading. A great deal of effort involving fellow farmers as tutors—private training for those worried about exposing their lack of reading skills to their peers—is being made. Every attempt is being made so that this will be a positive experience rather than a negative one. If the farmer fails there is no charge for taking the course again.

By April 1991, all farmers will have to show a license which certifies that they have successfully completed the course before they can purchase agricultural pesticides. All vendors will also be required to have licensed personnel on the premises. There are fines for selling or applying product without a minimal level of pesticide training. Food Systems 2002 has within its conceptual framework that farmers and vendors will have to be recertified every 5 years. What we hope to see is more specialized certification. At present, there is a special program available for greenhouse and mushroom growers. One concept is to design programs to go beyond pesticide training to Integrated Pest Management (IPM) Training Packages which start to provide a more holistic approach to pest control. Such a well-developed manual is available for our apple producers, who already have been involved in pest management. Obviously, there would be significant resources required to customize manuals on a crop-by-crop basis. In fact, it appears more logical to work at least initially on an animal-commodity basis, e.g., dairy, beef, swine, poultry, etc. Because most of these groups produce fields crops such as corn and soybeans, we can start to develop IPM strategies from a farm-management level rather than strict crop-by-crop management. Crop rotation, the economics-associated with low-input production, and proper sequencing of crops will obviously play a bigger role.

In summary, the administrators of the program, and more importantly the grower associations, believed that a certain level of competency was required before pesticides could be applied. A general awareness of the risks associated with pesticides and a mind-set of responsibility for your actions with pesticides was instilled. Hopefully, there will be some reduction in product used as farmers calibrate their equipment. It is continually emphasized that there are significant economic incentives to only apply the proper amount and that poorly calibrated application equipment is an economical and environmental liability.

Research ($5.6 Million)

From the onset it was recognized that there were no simple solutions and that considerable research would be required to develop alternative pest-control strategies and to make better utilization of existing pesticides. In the first year a number of literature reviews were funded related to topics such as application technology, conventional vs. organic pest-management approaches, etc. The objective of the literature reviews was to ensure that we didn't reinvent the wheel. Key focus areas for research effort were identified. These included (1) sprayer technology; (2) non-chemical control alternatives through biological control, cultural practices, crop rotations, and other means; (3) a systems approach to crop production including such concepts as critical periods for weed control; (4) plant breeding (pest-resistant cultivars); and (5) evaluation of newer pesticides for crop protection. The concept of funding research on more pesticides would appear contradictory. However, we strongly believe that newer products, which are generally of lower volumes, have modern health and environmental data bases, are often more target specific, and have lower residual activity, are important tools to be used in reducing potential impacts to farmers and the environment.

The call for research proposals was circulated widely to both Ontario universities and the agribusiness community. Each year $800,000 was made available for research. Typically, proposals were funded for 2–4 years and most are still ongoing. In addition to general submissions there was a conscious effort to direct research to meet the goals of the program. For example, after the second year, approximately 80% of the funds had been directed toward entomologically oriented proposals and a partial moratorium on entomological research was instituted. Concurrently, the weed scientists of the province were encouraged to make submissions, because if major reductions of pesticides were to occur, most of that had to come from herbicides. The overseeing committee also recognized where the major pesticide-use patterns were and would often turn down proposals of excellent scientific merit and encourage resubmission using different crop systems, e.g., reject biocontrol of weeds on pastures for a proposal on biological control of weeds in row-crop systems like corn and soybeans where most of our herbicides were used. Another concept was to encourage team approaches to control, where weed scientists, entomologists, and plant pathologists simultaneously would attempt to research a cropping system along with good economists and agronomists. Breaking down discipline barriers has been difficult but progress is occurring.

The expectations of immediate payoffs in research were never foremost in evaluators' minds. We recognized that implementation of research findings will often be well down the road and that many research projects, although of good science, will not lead to workable solutions at the farm level. Research benefits are often intangible, but we definitely have a research fraternity that is beginning to interact between disciplines. They now meet at a common time once a year and make oral presentations on research funded by this program. Personal interactions, respect for the other disciplines, and respect for others' abilities as good scientists will hopefully lead to more interactive research.

Pest-Management Infrastructure

When Food Systems 2002 began there were only six IPM programs for horticultural commodities. While these crops relied heavily on insecticides and fungicides, they contributed relatively little to the overall tonnage of product in the province because of small acreages involved. Corn rootworm insecticides constituted more insecticide use than all combined in fruit production. Importantly, the IPM programs that were in place had demonstrated a 25–40% reduction in pesticides with net savings of about $100 per ha. There had already been widespread acceptance of IPM by the growers involved such that grower organizations were actively funding some IPM research and hiring the necessary scouts that would be used by the IPM specialists.

An additional 11 IPM specialist have been hired and placed at strategic locations around the province in the vicinity of major concentrations of their particular commodity. Most have an M.Sc. in one of the crop protection disciplines; two are Ph.Ds. Again each year only two or three individuals were hired because there was concern that there wouldn't be enough qualified personnel available and the logistics of getting personnel on stream in 1 year presented risk in terms of making faulty advice to producers.

Producer confidence in the pest management specialists takes time to develop and can be quickly eroded if the farmer's crop is unduly damaged. As some farmers were known to say, "I spray so that I can sleep at night." At this point they have more confidence in the pesticide than the information provided by the IPM scouts. Farmers are often first concerned that all IPM specialists do is recommend not spraying. They do, however, see an additional function of the IPM specialists when they see them recommending sprays under certain conditions. We now have 19 IPM specialists established in the province and commodities such as corn, soybeans, and potatoes (and intensive cereal management) are coming into workable IPM programs covering larger acreages than the intensive fruit and vegetable production.

Principles

A few key points should be emphasized.

1. The mechanism of reducing pesticides involves strong grower support.
2. Education; research into alternative methods of pest control and into newer, more effective technologies; and an infrastructure to provide farmers timely pest-management advice all play a role. In the final analysis what we are attempting to do is to substitute knowledge for product. That knowledge base must be reliable and added slowly to the system to ensure grower confidence.
3. The program is based on good science rather than on satisfying political masters and the latest public crisis for today.
4. A 50% reduction in pesticides is a goal we honestly intend to achieve. If we don't quite make it we will still have succeeded in improving significantly our agriculture community. The pesticide education program and its enthusiastic acceptance by producers constitute one example.

5. There will be detractors who want greater or less reductions and disagree with the methods of implementation, methods of assessment, and semantic definitions. They have a legitimate voice, but the program must not be allowed to get "hung up in details." Its people and their enthusiasm that ultimately make the difference.

Where Are We in 1991

The last pesticide survey was conducted in 1988 and there has been a 17.4% reduction in pesticide use. (See Tables 2a–j). Pesticide issues such as food safety, groundwater contamination, and human exposure continue to have widespread media coverage. This program is cited as one ahead of its time and in response to critics of agriculture production methods. In 1990, in an Angus Reid Poll by Agriculture Canada, famers were asked, "Have you changed farming practices in the last year as a result of environmental concerns"? In Ontario, 34% said, Yes, they had reduced crop chemicals. Unfortunately, although obviously widely known in the farm community, this program isn't well known in the urban populace. We must make an effort to convey positive initiatives so as to have the general public on our side and not just the farm community. Finally, there has been a change in the governing party. The new government, within 3 months, has already announced that it agrees with the principles of "2002" and will continue to support it in the future. All political parties have shown support for this initiative.

Table 9.2a. Pesticide use summary of results comparing 1973, 1978, 1983, and 1988 for corn.

	Survey year			
	1973	1978	1983	1988
Area grown in ha (thousands)	758.8	1,019.8	1,052.0	858.0
Area sprayed	N/A	969.8	1,039.4	832.3
Herbicides				
Triazine	1,228.2	1,918.2	2,150.6	1,224.0
Phenoxy	29.6	36.8	81.5	162.4
Other	535.0	1,136.0	1,421.2	1,104.9
Fungicide	—	3.7	—	0.3
Insecticides	118.3	61.1	145.2	93.8
TOTAL pesticide (tons)	1,911.1	3,155.8	3,798.5	2,585.4
% TOTAL pesticides		47.4	43.3	35.9

Table 9.2b. Pesticide use summary of results comparing 1973, 1978, 1983, and 1988 for soybeans.

	Survey year			
	1973	1978	1983	1988
Area grown in ha (thousands)	190.2	285.3	364.0	518.0
Area sprayed	—	271.3	361.1	499.9
Herbicides				
Triazine	—	53.8	192.1	235.6
Phenoxy	—	—	0.2	8.0
Other	355.7	467.4	1,089.6	1,451.2
Fungicides	—	1.7	—	—
Insecticides	1.8	4.6	—	3.4
TOTAL pesticides (tons)	357.5	527.5	1,281.9	1,698.2
% TOTAL pesticides	—	7.9	14.6	23.6

Table 9.2c. Pesticide use summary of results comparing 1973, 1978, 1983, and 1988 for small grains.

	Survey year			
	1973	1978	1983	1988
Area grown in ha (thousands)	907.7	815.4	852.0	842.7
Area sprayed	—	471.3	535.1	566.3
Herbicides			(Growth regulators 1.3)	
Triazine	—	—	0.3	6.8
Phenoxy	349.1	259.0	333.4	409.6
Other	13.0	10.6	42.5	75.0
Fungicides	0.5	—	0.1	3.2
Insecticides	2.7	0.2	—	—
TOTAL pesticides (tons)	365.3	269.8	376.3	495.9
% TOTAL pesticides	—	4.1	4.3	6.9

Table 9.2d. Pesticide use summary of results comparing 1973, 1978, 1983, and 1988 for hay.

	Survey year			
	1973	1978	1983	1988
Area grown in ha (thousands)	1,153.4	1,153.4	1,720.0	1,477.0
Area sprayed	—	40.4	32.7	62.0
Herbicides				
Triazine	—	6.0	—	2.5
Phenoxy	8.1	8.0	15.1	16.1
Other	2.8	2.8	9.6	25.8
Fungicides	0.2	—	—	—
Insecticides	1.0	2.4	0.2	—
TOTAL pesticides (tons)	12.1	13.8	24.9	44.4
% TOTAL pesticides	—	0.2	0.3	0.6

Table 9.2e. Pesticide use summary of results comparing 1973, 1978, 1983, and 1988 for fruit.

	Survey year			
	1973	1978	1983	1988
N[a] 3.7				
GR[b] 1.3				
Area grown in ha (thousands)	N/A	28.4	28.4	29.0
Area sprayed	—	27.6	N/A	N/A
Herbicides				
Triazine	—	1.8	2.0	6.5
Phenoxy	—	—	0.1	0.4
Other	—	1.9	5.8	12.6
Fungicides	—	187.9	411.0	429.6
Insecticides	—	157.8	143.8	144.4
TOTAL pesticides (tons)	—	349.4	562.7	598.5
% TOTAL pesticides	—	5.3	6.4	8.3

[a]N = nematocides.

[b]GR = growth regulators.

Table 9.2f. Pesticide use summary of results comparing 1973, 1978, 1983, and 1988 for vegetables.

	Survey year			
	1973	1978	1983	1988
N[a] 28.9				
GR[b] 3.2				
Area grown in ha (thousands)	N/A	71.0	72.9	65.9
Area sprayed	—	68.1	N/A	N/A
Herbicides				
Triazine	—	24.3	19.1	18.2
Phenoxy	—	12.9	6.5	2.9
Other	—	95.8	75.2	68.2
Fungicides	—	184.3	154.7	196.5
Insecticides	—	104.6	100.5	78.5
TOTAL pesticides (tons)	—	421.9	356.0	396.4
% TOTAL pesticides	—	6.3	4.0	5.5

[a]N = nematocides.

[b]GR = growth regulators.

Table 9.2g. Pesticide use summary of results comparing 1973, 1978, 1983, and 1988 for dry beans.

	Survey year			
	1973	1978	1983	1988
Area grown in ha (thousands)	49.4	67.0	32.0	53.4
Area sprayed	—	66.2	31.5	49.6
Herbicides				
Triazine	—	1.4	0.6	12.6
Phenoxy	—	—	—	0.2
Other	97.2	160.2	56.8	206.8
Fungicides	—	0.9	1.1	—
Insecticides	6.8	4.3	1.3	0.8
TOTAL pesticides (tons)	104.0	166.8	59.8	220.4
% TOTAL pesticides	—	2.5	0.7	3.1

Table 9.2h. Pesticide use summary of results comparing 1973, 1978, 1983, and 1988 for tobacco.

	Survey year			
	1973	1978	1983	1988
Area grown in ha (thousands)	43.0	43.0	40.5	24.3
Area sprayed	—	43.0	40.5	24.3
Herbicides		N[a] = 1156.6	N = 1610.6	N = 766.8
		GR[b] = 403.5	GR = 590.8	GR = 323.6
Triazine	—	—	—	—
Phenoxy	—	—	—	—
Other	14.7	16.3	17.8	7.4
Fungicides	5.1	3.3	—	—
Insecticides	1,719.4	85.3	39.9	21.2
TOTAL pesticides (tons)	1,739.2	1,665.0	2,259.1	1,119.0
% TOTAL pesticides	—	25.0	25.7	15.5

[a]N = nematocides.

[b]GR = growth regulators.

Table 9.2i. Pesticide use summary of results comparing 1973, 1978, 1983, and 1988 for pasture.

	Survey year			
	1973	1978	1983	1988
Area grown in ha (thousands)	892.0	675.8		
Area sprayed	—	6.1		
Herbicides				
Triazine	—	—	Include as hay and pasture	
Phenoxy	11.4	3.5		
Other	7.7	0.4		
Fungicides	—	—		
Insecticides	0.3	—		
TOTAL pesticides (tons)	19.4	3.9		
% TOTAL pesticides	—	0.1		

Table 9.2j. Pesticide use summary of results comparing 1973, 1978, 1983, and 1988 for total all crops.

	Survey year			
	Field crops only:	Field/horticulture/ roadsides		Omit roadsides:
	1973	1978	1983	1988
Area grown in ha (thousands)	3,994.5	4,159.1	4,161.8	3,894.6
Area sprayed	—	2,018.0	N/A	N/A
Herbicides				
Triazine	1,228.2	2,000.2	2,364.7	1,538.9
Phenoxy	398.2	394.7	493.7	599.6
Other	1,026.1	1,899.8	2,719.7	2,954.9
Fungicides	5.8	381.8	566.9	629.6
Insecticides	1,850.3	420.3	430.9	349.6
Nematocides	—	1,156.6	1,610.6	799.4
Growth regulators	—	403.5	590.8	329.4
TOTAL	4,508.6	6,656.9	8,777.3	7,201.4

References

Archibald, B.A. 1990. Review of data requirements for the registration of pesticides in Canada and the United States. 38 pp. Submission to Federal Pesticide Registration Review Team, Ottawa, Ont.

Davies, E. and G. Penfold. 1991. Exploration of a framework for evaluating sustainable development in a rural context. Paper presented American Planners Association Conference: Rural Planning, Visions of the 21st Century. Univ. of Florida; Orlando, Florida, Feb. 1991.

Harris, C.R., S.A. Turnbull, J.W. Whistlecraft, and G.A. Surgeoner. 1982. Multiple resistance shown by field strains of the house fly *Musca domestica* (Diptera: Muscidae), to organochlorine, organophosphorous, carbamate and pyrethroid insecticides. Can. Ent. **114**:447–454.

Martin, L.J. 1990. Ed. Growing Together. Report to Ministers of Agriculture Task Force on Competitiveness in the Agri-Food Industry, Appendices Working Group Reports June 1990. 125 pp.

Sachs, C., D. Blair and C. Richter. 1987. Consumer pesticide concerns: A 1965 and 1984 comparison. J. Consumer Affairs **21**:96–107.

10

Environmental and Economic Impacts of Reducing U.S. Agricultural Pesticide Use*

David Pimentel, Lori McLaughlin, Andrew Zepp, Benyamin Lakitan, Tamara Kraus, Peter Kleinman, Fabius Vancini, W. John Roach, Ellen Graap, William S. Keeton, and Gabe Selig

Introduction

Several studies suggest that it is technologically feasible to reduce pesticide use in the United States 35–50% without reducing crop yields (PSAC, 1965; OTA, 1979; NAS, 1989; Palladino, 1989). Two recent events in Denmark and Sweden support these assessments. Denmark developed an action plan in 1985 to reduce the use of pesticides 50% before 1997 (Mogensen, 1989). Sweden also approved a program in 1988 to reduce pesticide use by 50% within 5 years (NBA, 1988). The Netherlands is developing a program to reduce pesticide use 50% in 10 years (Süddeutsche Zeitung, 1989). These proposals, along with Huffaker's (1980) assessment that the United States overuses pesticides, prompted us to investigate the feasibility of reducing the annual use of synthetic organic pesticides by approximately one-half.

Farmers use an estimated 320 million kg (700 million lb) of pesticides annually at an approximate cost of $4.1 billion (Table 10.1). These figures do not reflect the "indirect costs" of pesticide chemical use, such as human pesticide poisonings, reduction of fish and wildlife populations, livestock losses, destruction of susceptible crops and natural vegetation, honeybee losses, destruction of natural enemies, evolved pesticide resistance, and creation of secondary pest problems (Pimentel et al., 1980a).

Investment in pesticidal controls has been shown to provide significant economic benefit through increased crop yields. Dollar returns for the direct benefits to farmers have been estimated to range from $3 to $5 for every $1 invested in the use of pesticides (PSAC, 1965; Headley, 1968; Pimentel et al., 1978). However, these benefits are calculated using current agricultural practices, some of which actually increase pest problems. Clearly, the direct and indirect benefits and risks of using pesticides in agriculture are highly complex.

---

*Reprinted with permission from CRC Press, from Handbook of Pest Management in Agriculture, Volume I (2nd ed., 1991).

Table 10.1. U.S. hectarage treated with pesticides (modified from Pimentel and Levitan, 1986).[a]

Land-use category	Total hectares ($\times 10^6$)	All pesticides		Herbicides		Insecticides		Fungicides	
		Treated hectares ($\times 10^6$)	Quantity ($\times 10^6$ kg)	Treated hectares ($\times 10^6$)	Quantity ($\times 10^6$ kg)	Treated hectares ($\times 10^6$)	Quantity ($\times 10^6$ kg)	Treated hectares ($\times 10^6$)	Quantity ($\times 10^6$ kg)
Agricultural lands	472	114	320	86	220	22	62	4	38
Government and industrial lands	150	28	55	30	44	—	11	—	—
Forest lands	290	2	4	2	3	<1	1	—	—
Household lands	4	4	55	3	26	3	25	1	4
TOTAL	917	148	434	121	293	26	99	5	42

[a]Total for hectarage treated with herbicides, insecticides, and fungicides exceeds total treated hectares because the same land area can be treated with several classes of chemicals and several times.

The objective of this investigation is to estimate the potential agricultural and environmental benefits and costs of reducing pesticide use by approximately 50% in the United States. To obtain a "best estimate" of the costs and benefits, this study (1) examines current pesticide use patterns in about 40 major U.S. crops; (2) quantifies current crop losses to pests; (3) estimates the agricultural benefits and costs of reducing pesticide use by substituting currently available biological, cultural, and environmental pest-control technologies for some current pesticide control practices; and (4) assesses the public health and environmental benefits associated with reduced pesticide use.

Extent of Pesticide Use

Of the total estimated 434 million kg of pesticides used in the United States, 69% are herbicides, 19% insecticides, and 12% fungicides (Table 10.1). The 320 million kg of pesticides used in agriculture are applied at an average rate of about 3 kg/ha to about 114 million ha—62% of the 185 million ha that are planted to crops (Pimentel and Levitan, 1986). Thus, a significant portion (38%) of crops receives no pesticide.

The application of pesticides for pest control is not evenly distributed among crops. For example, 93% of all row-crop hectarage, like corn, cotton, and soybeans, is treated with some type of pesticide (Pimentel and Levitan, 1986). In contrast, less than 10% of forage-crop hectarage is treated. Herbicides are currently being used on about 90 million ha in the United States—greater than half of the nation's cropland—but nearly three-quarters of these herbicides are applied to just two major crops, corn and soybeans. Field corn alone accounts for 53% of agricultural herbicide use.

The case is similar for insecticide use. About 62 million kg of insecticides are applied to 5% of the total agricultural land (Table 10.1). Approximately 25% of all insecticides are used on cotton and corn. Fungicides are primarily used on fruit and vegetable crops (Pimentel and Levitan, 1986).

Insecticide use also varies among geographic regions. Warmer regions of the United States often suffer more intense pest problems. For example, while only 13% of alfalfa hectarage in the United States is treated with insectides, 89% of the alfalfa area in the Southern Plains states is treated to control insect pests (Eichers et al., 1978). In the Mountain region, where large quantities of potatoes are grown, 65% of the potato cropland receives insecticide treatment, while in the Southeast, where only early potatoes are grown, 100% of the potato cropland is treated (USDA, 1975). Cotton insect pests such as the boll weevil are also more of a problem in the Southeast than in other regions (USDA, 1983). In the Southeast and Delta states, 84% of the cotton cropland receives treatment, while in the Southern Plains region less than half of the crop (40%) is treated. Also, a crop hectare can be treated 20 times per season (e.g., apples and cotton), whereas other crop hectares may be treated only once (e.g., corn and wheat).

Crop Losses to Pests and Changes in Agricultural Technologies

Since 1945 the use of synthetic pesticides in the United States has grown 33-fold. The amounts of herbicides, insecticides, and fungicides used have changed with time due in large part to changes in agricultural practices and cosmetic standards (Pimentel et al., 1977a). At the same time, the toxicity to pests and biological effectiveness of some of these pesticides have increased at least 10-fold (Pimentel, 1989). For example, in 1945 DDT was applied at a rate of about 2 kg/ha. Today, similar effective insect control is achieved with pyrethroids and aldicarb applied at 0.1 kg/ha and 0.05 kg/ha, respectively.

Currently, an estimated 37% of all crop production is lost annually to pests (13% to insects, 12% to plant pathogens, and 12% to weeds) in spite of the use of pesticides and nonchemical controls (Pimentel, 1986). While pesticide use has increased over the past four decades, losses from pests have not shown a concurrent decline. According to survey data collected from 1942 to present, losses from weeds have fluctuated but declined slightly from 13.8% to 12% (Table 10.2). This is due to improved chemical, mechanical, and cultural weed control practices.

Over that same period, losses from plant pathogens, including nematodes, have increased slightly, from 10.5% to about 12% (Table 10.2). This is partly due to reduced sanitation, higher cosmetic standards, and abandonment of rotations.

The share of crop yields lost to insects has nearly doubled during the last 40 years (Table 10.2) despite more than a 10-fold increase in the amount and toxicity of synthetic insecticide used (Arrington, 1956; USBC, 1971, 1988). The increase in crop losses due to insects per hectare has been offset by increased crop yield obtained through the use of higher-yielding varieties and greater use of fertilizers and other inputs (USDA, 1986; Pimentel and Wen, 1989).

The increase in crop losses despite increased insecticide use can be explained by some of the major changes that have taken place in agricultural practices. These include the planting of some crop varieties that are more susceptible to insect pests; the destruction of natural enemies of certain pests, thereby creating the need for additional pesticide treatments (van den Bosch and Messenger, 1973); the increase in pests resistant to pesticides (Roush and McKenzie, 1987); the reduction in crop rotations; the increase

Table 10.2. Comparison of annual pest losses (dollars) in the USA for the periods 1904, 1910–1935, 1942–1951, 1951–1960, 1974, and 1986.

		Percentage of pest losses in crops				Crop value	
Period	Source	Insects	Diseases	Weeds	Total	$\$ \times 10^9$	Source
1986	Pimentel 1986	13.0	12.0	12.0	37	150	USDA 1986
1974	Pimentel 1976	13.0	12.0	8.0	33.0	77	USDA 1975
1951–1960	USDA 1965	12.9	12.2	8.5	33.6	30	USDA 1961
1942–1951	USDA 1954	7.1	10.5	13.8	31.4	27	USDA 1954
1910–1935	Hyslop 1938	10.5	NA[a]	NA	NA	6	USDA 1936
1904	Marlatt 1904	9.8	NA	NA	NA	4	Marlatt 1904

[a]Not available.

in monocultures and reduced crop diversity (Pimentel, 1961; Pimentel et al., 1977b); the lowering of FDA tolerance for insects and insect parts in foods and the enforcement of more stringent "cosmetic standards" by fruit and vegetable processors and retailers (Pimentel et al., 1977a); the increased use of aircraft application technology; the reduction in sanitation, including less attention paid to the destruction of infected fruit and crop residues (Pimentel, 1986); the reduction in tillage with more crop residues left on the land surface; the culturing of crops in climatic regions in which they are more susceptible to insect attack; and the use of pesticides that have been found to alter the physiology of crop plants, making them more susceptible to insect attack (Oka and Pimentel, 1976). These factors will be explored further in the discussion of alternatives to pesticide use.

Estimated Agricultural Benefits/Costs With a Reduction in Pesticide Use

The reduction of U.S. pesticide use would require substituting nonchemical alternatives for chemical pest control and improving the efficiency of pesticide application technologies. Such changes might increase control costs slightly. In some cases, however, these costs might decrease. The costs and benefits of alternative controls are examined below:

Crop Losses to Pests

Estimates of losses by pests for the 40 major crops grown with pesticides were made by examining data on current crop losses, reviewing loss data based on experimental field tests without treatment, and consulting pest control specialists. Combining these data, however, was often difficult. For example, data based on published experimental field tests usually emphasize the benefits of pesticide use; thus, loss data associated with pesticide treatments usually emphasize benefits over costs (Pimentel et al., 1978).

In addition, such studies often exaggerate total crop losses since assessments of insect, disease, and weed pests are carried out separately and then combined. For example, on untreated apples, insects were reported to cause a 50–100% crop loss, disease 50–60%, and weeds 6% (Glass and Lienk, 1971; Pimentel et al., 1978; Stemeroff and George, 1983; Ahrens and Cramer, 1985). This approach yields an estimated total loss of about 140% from all pests combined! A more accurate estimate of losses in the absence of pesticides ranges from 80 to 90% based on current "cosmetic standards" (Ahrens and Cramer, 1985). Exactly how much overlap exists in the loss figures for apples and for the other crops is not known.

Our analysis has other important limitations. The figures for current crop losses to pests, despite pesticide use, are based primarily on USDA data and other estimates obtained from specialists. We emphasize that these are estimates. For certain crops, little or no experimental data are available concerning yields with pesticide use and various substitute alternatives. In addition, in some cases recent data were not available. With these crops, our estimates were generally extrapolated from data on closely related crops.

In summary, while we fully recognize the limitations of the data used in this analysis, we believe the need exists to assemble available information in order to provide a first approximation of the potential for reducing pesticide use by one-half. We hope that better data will be available in the future so that a complete analysis of pesticide costs and benefits can be made.

Reduction of the risks associated with pesticides is in itself a complicated issue, particularly because environmental and health-related trade-offs are often associated with changes in technology. Because of the complexity of these trade-offs, they could not be included in the analysis. One example involves the conflict between reducing pesticide use and promoting soil conservation through the use of no-till and reduced tillage. Although no-till and reduced-till significantly reduce soil erosion (Van Doren et al., 1977), they also significantly increase the use of herbicides, insecticides, and fungicides (Taylor et al., 1984). Reducing pesticide use may require reducing the use of some no-till systems. However, highly cost-effective soil conservation alternatives to no-till exist. These include ridge-till, crop rotations, strip-cropping, contour planting, terracing, wind-breaks, mulches, cover crops, green-mulches, and others (Moldenhauer and Hudson, 1988). Ridge-till in particular is rapidly growing in popularity as an effective replacement for no-till and can be implemented for most row crops. Ridge-till allows for no-till crop culture without the disadvantages of no-till. In addition, ridge-till can be employed without the use of herbicides (Thompson, 1985; Russnogle and Smith, 1988).

Techniques to Reduce Pesticide Use

The increases in crop losses associated with recent changes in agricultural practices suggest the existence of some alternative strategies that might be used to reduce pesticide use. These strategies will be discussed below. Two additional important practices that apply to all agricultural crops include greater use of scouting and improved application equipment. Currently, a significant number of pesticide treatments are applied unnecessarily and at improper times due to a lack of treat-when-necessary programs. Furthermore, an unnecessary amount of pesticide is lost during application (e.g., only 25–50% of the pesticide applied by aircraft actually reaches the target area [Ware et al., 1970; ICAITI, 1977; Ware, 1983; Akesson and

Yates, 1984; Mazariegos, 1985; Pimentel and Levitan, 1986]). These strategies and others are discussed below on a crop-by-crop basis.

The added costs of implementing alternative pest controls to reduce pesticide use ranged from −$10 to +$15 depending on the alternative technology and crop. For example, some studies have demonstrated that weed control in plums using mechanical cultivation for weed control reduces the use of herbicides and saves the farmer about $10/ha (Weakley, 1986). For scouting in general we increased alternative pest control costs by $5/ha. Most often, however, scouting reduces pesticide use and saves the farmer money (NAS, 1975; NAS, 1989; OTA, 1979; other chapters in this book). To be cautious, however, in most cases we estimated an increase of $5/ha instead of a decrease in pest control costs.

Employing crop rotations with corn for control of the rootworm was estimated to increase control costs about $10/ha. In this case we did not consider the benefits from rotations that result for weed control, disease control, soil erosion control, and improved corn yields (Helmers et al., 1986; Cramer, 1988). In fact, we also added $15/ha by employing mechanical cultivation for pest control.

Insecticides

Possible changes in agricultural insect control methods that would help reduce total chemical insecticide use are as follows:

Corn

During the early 1940s, little or no insecticide was applied to corn, and losses to insects were only 3.5% (USDA, 1954). Since then, insecticide use on corn has grown more than 1,000-fold while losses due to insects have increased to 12% (Ridgway, 1980). This increase in insecticide use and the 3.4-fold increase in corn losses to insects are primarily due to the abandonment of crop rotation (Pimentel et al., 1977b). Today about 40% of U.S. corn is grown as continuous corn with 11 million kg of insecticide applied (Table 10.1). By reinstituting crop rotation, great reductions in pesticides could be achieved. Rotating corn with soybeans or a similar high-value crop has little or no impact on net profits; rotating corn with wheat or other low-value crops, however, reduces net profits per hectare.

The rotation of corn with other crops, however, has several added advantages that include reducing weed and plant pathogen losses as well as decreasing soil erosion and rapid water runoff problems (Helmers et al., 1986; Cramer, 1988).

By combining crop rotations with the planting of corn resistant to the corn borer and chinch bug, it would be possible to avoid the use of 80%

of the insecticides on corn while at the same time reducing insect losses (Schalk and Radcliffe, 1977; Lockeretz et al., 1981). Such a move is estimated to increase the costs of corn production by $10 per hectare. This cost increase would affect 40% of the corn hectarage on which corn is grown continuously (Table 10.3A).

A new approach using an attractant combined with insecticides for rootworm control has been reported to reduce insecticide use 99% (Paul, 1989). This would be further enhanced by planting corn in rotation for rootworm control.

Still assuming that a significant portion of the hectarage now planted to continuous corn was returned to rotation, the overall cost would be an added $10/ha (Table 10.3A).

Cotton

The potential for reducing pesticide use in U.S. agriculture is well illustrated by insecticide use in cotton production in Texas. Since 1966 insecticide use on Texas cotton has been reduced by nearly 90% (OTA, 1979). In priority order, the technologies adopted to reduce insecticide use were as follows: "scouting" or monitoring pest and natural enemy populations to determine when to treat, biological control, host-plant resistance, stalk destruction (sanitation), uniform planting date, water management, fertilizer management, rotations, clean seed, and tillage practices (OTA, 1979; King et al., 1986).

Currently, a total of 29 million kg of insecticide is applied to cotton, and it is estimated that this could be reduced by approximately 38% through the use of readily available technologies (Table 10.3A). By effectively using a treat-when-necessary or "scouting" program, one might reduce insecticide use by an estimated 20%. Through the use of pest-resistant cotton varieties and the alteration of planting dates in most growing regions, we could reduce insecticide use by another 3% (Frans, 1985; Frisbie, 1985). An additional 10% reduction in insecticide use could be achieved by replacing the price-support program with a "free-land market" for cotton production (Pimentel and Shoemaker, 1974; NAS, 1989). The result of this change would be to allow cotton to be grown in regions with fewer insect pests, thus reducing the need for insecticide use. The current price-support program will have to change if society is to gain from these benefits. At present, this is a politically unattractive proposition.

Giving greater care to the type of application equipment employed, especially reducing the use of ULV (ultra low volume) application equipment on aircraft, would increase the amount of insecticide reaching the target area from 25 to 50%. The amount of insecticide reaching the target area could be increased to 75% if ground-application equipment were used

Table 10.3A. Field-crop losses from insects with current insecticide use and estimated costs if insecticides were reduced and several alternatives were substituted.

Crop	Ha × 10³ [a]	Total (kg × 10⁶) insecticide use		Insecticide treatment			Current crop pest loss, %	Added alternative control cost $/ha [g]	Total added alternative control cost $ × 10⁶
		Current	Reduced [g]	Ha treated %	Cost $/ha	Total cost $ × 10⁶			
Corn	30,419	14 [c]	2	41 [c]	15 [e]	187	12 [e]	10	92.5
Cotton	3,852	8 [c]	5	61 [c]	118 [j]	320	8 [k]	0	0
Wheat	26,208	1.0 [c]	0.64	7 [c]	23 [i]	42	6 [e]	5	9.2
Soybeans	24,933	5 [c]	2.5	3 [c]	8 [n]	6	3 [e]	5	3.7
Rice	1,013	0.2 [k]	0.1	16 [k]	49 [m]	7.9	4 [e]	5	0.4
Tobacco	279	1.6 [b]	0.26	85 [d]	20 [q]	4.7	11 [f]	5	0.4
Peanuts	594	0.5 [c]	0.25	80 [o]	35 [h]	16.6	10 [f]	5	2.6
Sorghum	6,750	1.1 [c]	0.50	17 [c]	30 [k]	34.4	7 [n]	0	0
Sugar beets	446	0.1 [g]	0.05	30 [o]	18 [p]	2.2	12 [f]	5	0.67
Other grain	8,267	0.1 [g]	0.05	3 [o]	16 [g]	4.0	10 [f]	3	0.4
Alfalfa	10,394	1.0 [g]	0.5	7 [d]	16 [g]	11.6	15 [f]	3	1.1
Hay	14,121	0.05 [g]	0.025	0.5 [o]	16 [g]	1.1	15 [f]	0	0
Pasture	241,564	0	0 [g]	0	0	0	15 [f]	0	0
TOTALS		32.65	11.85			637.5			110.97

[a] USDA, 1986. [b] Pimentel and Levitan, 1986. [c] USDA, 1988. [d] Duffy, 1982. [e] USDA, 1965; Ridgway, 1980. [f] USDA, 1965. [g] Estimated. [h] McArthur et al., 1985. [i] Johnston and Bishop, 1987. [j] Frans, 1985. [k] Teetes et al., 1986. [m] USDA, 1982b. [n] Douce and McPherson, 1988. [o] USDA, 1975. [p] Swenson and Johnson, 1983. [q] Jones et al., 1988.

Table 10.3B. Vegetable crop losses from insects with current insecticide use and estimated costs if insecticides were reduced and several alternatives were substituted.

Crop	Ha × 10³[a]	Total (kg × 10⁶) insecticide use		Insecticide treatment			Current crop pest loss, %	Added alternative control cost $/ha[g]	Total added alternative control cost $ × 10⁶
		Current	Reduced[g]	Ha treated %	Cost $/ha	Total cost $ × 10⁶			
Lettuce	91	0.06[b]	0.04	77[k]	60[g]	4.9	7[i]	10	0.81
Cole	111	0.4[b]	0.20	62[b]	30[g]	2.1	13	10	0.34
Carrots	36	0.02[b]	0.013	44[b]	15[g]	0.2	7[i]	5	0.03
Potatoes	550	2.0[g]	1.2	96[b]	59[h]	31.2	6[c]	10	5.3
Tomatoes	177	0.8[b]	0.01	90[b]	64[d]	10.2	7[i]	0	0
Sweet corn	241	0.6[b]	0.02	85[b]	40[g]	8.2	19[i]	10	2.4
Onions	47	0.09[b]	0.072	95[b]	79[f]	1.4	4[c]	5	0.04
Cucumbers	47	0.02[g]	0.013	43[b]	30[g]	0.61	21[i]	5	0.61
Beans	150	0.5[b]	0.33	76[b]	14[d]	1.7	12[i]	5	0.57
Cantaloupe	45	0.02[b]	0.02	88[b]	30[g]	1.2	8[i]	0	0
Peas	143	0.05[b]	0.033	71[b]	30[g]	3.0	4[i]	5	0.17
Peppers	23	0.01[g]	0.005	77[k]	30[g]	0.50	7[i]	5	0.09
Sweet potatoes	43	0.26[m]	0.13	100[m]	30[m]	1.29	16[j]	5	0.22
Watermelons	83	0.03[b]	0.02	54[b]	30[g]	1.3	4[i]	5	0.07
Other vegetables	100	0.01[g]	0.006	40[g]	30[g]	1.2	13[i]	5	0.1
TOTALS		4.87	2.012			69.00			10.75

[a]USDA, 1986. [b]Ferguson, 1984. [c]Stemeroff and George, 1983. [d]Based on amount applied (Gianessi and Greene, 1988) and price of pesticide used (USDA, 1985). [e]Estimated. [f]USDA, 1982a. [g]See text for details. [h]Warner, 1985. [i]USDA, 1965. [j]Douce and Suber, 1985. [k]USDA, 1975. [m]Schalk, J. M., 1989, personal communication.

Table 10.3C. Fruit and nut crop losses from insects with current insecticide use and estimated costs if insecticides were reduced and several alternatives were substituted.

		Total (kg × 10^6) insecticide use		Insecticide treatment					
Crop	Ha × 10^3 [a]	Current	Reduced	Ha treated %	Cost $/ha	Total cost $ × 10^6	Current crop pest loss, %	Added alternative control cost $/ha [g]	Total added alternative control cost $ × 10^6
Apples	198	7.0[m]	4.5	96[b]	225[f]	42.8	4[c]	0	0
Cherries	54	0.5[m]	0.1	86[b]	40[g]	1.9	10[h]	0	0
Peaches	94	0.7[m]	0.35	96[b]	135[d]	12.2	12[e]	0	0
Pears	34	1.0[m]	0.05	99[b]	40[g]	13	10[h]	0	0
Plums	57	0.7[g]	0.25	72[j]	40[g]	1.6	10[h]	0	0
Grapes	308	2.0[p]	1.0	67[j]	13[k]	2.7	6[h]	0	0
Oranges	276	8.5[i]	4.7	54[b]	50[g]	7.5	6[h]	0	0
Grapefruit	98	3.0[i]	1.6	94[b]	50[g]	4.6	5[h]	0	0
Lemons	32	1.0[i]	0.55	82[b]	50[g]	1.3	6[h]	0	0
Other fruit	100	0.2[g]	0.1	44[j]	50[g]	2.2	25[h]	0	0
Pecans	155	0.01[g]	0.005	60[j]	30[o]	2.8	14[e]	0	0
Other nuts	170	0.01[g]	0.005	60[j]	160[n]	16.3	25[h]		
TOTALS		24.62	13.21			108.9			0

[a]USDA, 1986. [b]Suguiyama and Carlson, 1985. [c]Stemeroff and George, 1983. [d]Folwell et al., 1981; Warner, 1985. [e]Douce and McPherson, 1988. [f]Gerling, 1986; Kovach and Tette, 1988. [g]Estimated. [h]USDA, 1965. [i]Haydu, 1981. [j]USDA, 1975. [k]Warner, 1985. [m]Shwu and Webb, 1981. [n]Calkin, 1987. [o]Perry and Saunders, 1982. [p]Dennehy, 1989.

instead of aircraft application equipment (Ware et al., 1970; ICAITI, 1977; Ware, 1983; Akesson and Yates, 1984; Mazariegos, 1985; Pimentel and Levitan, 1986). In addition, covering the spray boom with a plastic shroud can further reduce drift 85% (Ford, 1986), and thereby allow for an additional reduction in pesticide use (Table 10.3A).

Insecticide use on cotton might be reduced by another 6% if other alternative pest control techniques were implemented. These techniques include cultivation of short-season cotton, improved fertilizer and water management, improved sanitation, rotations, clean seed, and tillage practices (OTA, 1979; Bieber et al., 1981; Cochran, 1985; Grimes, 1985). Depending on the particular environment, insecticide use on cotton might be reduced much more than suggested. For example, Shaunak et al. (1982) reported that insecticide use in the lower Rio Grande Valley of Texas could be reduced 97% by using short-season cotton under dryland conditions. This practice also resulted in a two-fold increase in net profits over conventional methods.

Thus, by using combinations of the above nonchemical pest controls for cotton, insecticide use might be reduced about 40% (Table 10.3A). These alternative controls should pay for themselves through reduced insecticide and application costs (Table 10.3A).

Wheat

The major insect problems (Hessian fly, green bug, wheat stem fly, Russian wheat aphid, and armyworm) in wheat are minimized by host-plant resistance, manipulation of planting date, tillage, use of vigorous lines, sanitation, and rotations (PSAC, 1965; Schalk and Radcliffe, 1977; USDA, 1982a; Hatchett et al., 1987). As a result only about 7% of the wheat hectarage is treated with insecticides (Table 10.3A). Still, a reduction in insecticide use might be possible through the use of scouting and treat-when-necessary programs (Nissen and Juhnke, 1984). These programs might enable farmers to reduce the quantity of insecticides used in wheat production by about 20% (Table 10.3A).

Soybeans

While only 3% of the soybean hectarage is treated with insecticides, a total of 4.2 million kg of insecticide is applied to this crop (Table 10.3A). Note, the percentage of soybean hectares treated has declined from 12% in 1982 to 3% today (Szmedra, 1989). In U.S. soybean production during the past 4 years, insecticide use for control of the Mexican bean beetle, the pea moth, and caterpillars has been reduced 63% through the use of scouting, trap crops, and other alternatives (McPherson, 1983; Flanders, 1985; Wilcox, 1987). Growers prefer scouting over the other techniques

(Greene et al., 1985). In addition, *Bacillus thuringiensis* (Bt) and some viruses can be used for control of several caterpillar pests in soybeans. Soybean insects can also be controlled by large vacuum suction devices (Street, 1989). These insect control techniques have proven successful and are becoming increasingly widespread. Despite a large insecticide reduction in the past, it might be possible to reduce insecticide use by at least one-third through greater use of nonchemical techniques at an estimated cost of $3/ha (Table 10.3A).

Rice

Approximately 16% of the rice hectarage is treated with insecticides to control three primary arthropod pests: the rice water weevil, tadpole shrimp, and rice stink bug (USDA, 1973; UC, 1983). Losses from arthropods in rice are relatively low—4% (Table 3A). Alternatives to reduce arthropod pest problems include scouting, intensive seedbed preparation, manipulation of water levels, and high seeding rates (COPR, 1976; UC, 1983). By employing all or a combination of these alternatives, it might be possible to reduce insecticide use about one-half at an estimated cost of $5/ha (Table 10.3A).

Tobacco

The principal insect pests of tobacco are the tobacco flea beetle, spittlebug, and tobacco budworm and hornworm. About 85% of tobacco hectarage receives insecticide treatments for these pests (Table 10.3A). Alternatives for the control of flea beetle and spittlebug include improved biocontrol, sanitation, tillage, and destruction of all stalks and suckers (Metcalf et al., 1962; Liapis, 1983). Budworm and hornworm control alternatives include sanitation, virus, Bt (Johnson, 1978), and scouting (Liapis, 1983). Using a combination of these techniques, it might be possible to reduce insecticide use by two-thirds at an estimated cost of $5/ha (Table 10.3A).

Peanuts

The primary pests of peanuts are the lesser cornstalk borer and the corn rootworm (Smith and Barfield, 1982). Scouting can be employed against both pest species, and its effectiveness was clearly demonstrated in Texas, where treatments were reduced 70% when farmers adopted scouting or treat-when-necessary programs (Smith and Barfield, 1982). Sanitation, winter plowing, and rotations will also help control the rootworm, and Bt can be used for control of borer larvae. Using a combination of these alternatives, it might be possible to reduce insecticide use by one-half at an estimated cost of $5/ha (Table 10.3A).

Sorghum

Only about 17% of the sorghum hectarage is treated with insecticides, and the two primary pests are the greenbug aphid and the sorghum midge (FAO, 1979; Kramer, 1987). For the greenbug, the alternatives include resistant hybrids, tillage practices, and scouting (Schalk and Radcliffe, 1977; Teetes and Johnson, 1978; FAO, 1979; Burton et al., 1987). For the sorghum midge the alternatives include early, uniform planting and the use of resistant varieties (FAO, 1979; Teetes et al., 1986). In addition, wireworm, white grub, and southern corn rootworm injury is lessened by clean cultivation, rotation, and good sorghum growth (Teetes, 1982, 1985). Weed and crop refuse destruction aids against cutworms. Early planting lessens the infestation level of corn earworm, sorghum webworm, fall armyworm, and some panicle feeding bugs, as well as sorghum midge. Water management (as well as timely irrigation, if available) lessens the severity of spider mites. Employing several of the alternative control techniques might help reduce insecticide use by 50% (Table 10.3A).

Sugar Beets

The two principal insect pests of sugar beets are the beet leafhopper and the sugar beet webworm. The leafhopper can be controlled by destroying alternative host plants like Russian thistle and by eliminating all beet vegetation from the field after harvest (Metcalf et al., 1962). The alternatives for webworm control include improved sanitation, tillage, scouting, and Bt. By employing combinations of these alternatives for the insect pests, it might be possible to reduce insecticide use about one-half at an estimated cost of $5/ha (Table 10.3A).

Alfalfa

The principal pest of alfalfa is the alfalfa weevil. Several alternatives are available for control of this pest, including strip cutting, biological controls, Bt, scouting, and resistant varieties. Resistant varieties are also available to control pea aphid and spotted alfalfa aphid (Schalk and Radcliffe, 1977; Armbrust et al., 1980; Ruesink et al., 1980; Wilson et al., 1984; Davis, 1985). By employing a combination of these alternatives, it might be possible to reduce insecticide use on alfalfa about one-half at an estimated cost of $3/ha (Table 10.3A).

Lettuce

The principal insect pests of lettuce are aphids and the cabbage looper. The looper can be controlled with a nuclear polyhedrosis virus (not yet cleared for use), Bt, and scouting. For aphids, the alternative is scouting. Employing a combination of these techniques, it might be possible to reduce insecticide use about one-third at an estimated cost of $10/ha (Table 10.3B).

Cole

The primary pests of cole crops include the cabbage maggot, cabbage looper, cabbage butterfly, and the diamondback moth (Kirby and Slosser, 1984). By incorporating granular insecticide into the potting soil of seedlings, it has been demonstrated that the quantity of insecticide used could be reduced by over 50% for control of early season pests (Straub, 1988). However, about 70% of insecticide use is for control of caterpillar pests. Scouting is an important means of increasing the effectiveness of sprays and eliminating needless treatments (Kirby and Slosser, 1984).

Both the cabbage looper and cabbage butterfly are highly susceptible to virus diseases, and viruses could be effectively used against these pests (Falcon, 1976; Jaques, 1988). To date, these viruses have not been approved for use on food crops, but there is no evidence of risks to public health or the environment (Summers and Kawanishi, 1978; Pimentel et al., 1984). Bt can be used against all three caterpillar species (Jaques, 1988). Thus, it might be possible to reduce insecticide use in cole crops an estimated two-thirds (Table 10.3B). The cost of these alternatives was estimated to be $10/ha (Table 10.3B).

Carrots

Losses to the carrot-fly, carrot beetle, and carrot-weevil are estimated to be about 7% (Table 10.3B). By implementing a sound scouting program, it might be possible to reduce insecticide use in carrot production about one-half at an added estimated cost of $5/ha (Table 10.3B).

Potatoes

The principal insect pests of potatoes are the Colorado potato beetle, aphids, the potato flea beetle, and the potato leafhopper. The dominant pest is the potato beetle, and alternative control methods include area-wide rotations, early maturing varieties, short-season potatoes, and scouting (Shields et al., 1984; Wright, 1984; Wright et al., 1986; CR, 1987; Radcliffe et al., 1989). The use of short-season potatoes and scouting may reduce insecticide use by 33% (Shields et al., 1984). Radcliffe (1989, personal communication) reports that insecticide use on potatoes could be reduced 75% with effective scouting. Also, Bt has been found to be effective in controlling the potato beetle (Cantwell and Cantelo, 1984; Jaques and Laing, 1989). Using the fungus *Beauveria bassiana* for Colorado beetle control demonstrated the potential for an 80% reduction in insecticide use (Roberts et al., 1981); however, a recent study reported that the fungus was ineffective (Jaques and Laing, 1989). By employing pest control combinations, like rotations, short-season potatoes, scouting, and Bt, it might be possible to reduce insecticide use 40% at an estimated cost of $10/ha (Table 10.3B).

Tomatoes

Tomatoes are a high-value crop (about $7,500/ha). To protect this valuable crop, about $64/ha in insecticide is applied and 90% of the hectarage is treated (Table 10.3B). The primary insect pests are the tomato fruitworm and tomato hornworm; potato beetles and aphids are also occasional pests (Farrar et al., 1986; Zehnder and Linduska, 1987). Insecticide applications to control these pests can be reduced an estimated 20% through scouting and another 60% to 80% by substituting Bt and other natural enemies (Krishnaiah et al., 1981; Antle and Park, 1986; Farrar et al., 1986; Hoffman et al., 1986; Horn, 1988; Jimenez et al., 1988). By employing these techniques, it might be possible to reduce insecticide use on tomatoes by 80% (Table 10.3B). Although additional labor is needed for scouting, total control costs remain the same because of the savings from reduced insecticide applications (Antle and Park, 1986; Jimenez et al., 1988).

Sweet Corn

The primary insect pest in sweet corn production is the corn earworm (McLeod, 1986). Several alternative techniques exist for controlling this pest. These include a highly effective nuclear polyhedrosis virus (Oatman et al., 1970), Bt, mineral oil treatment of ears (Barber, 1942; Johns, 1966), early maturing varieties (Huffaker, 1980), and rotations. More reasonable cosmetic standards could also greatly reduce the need for high pesticide use (Straub and Heath, 1983). Using a rotation sequence of sweet corn and soybeans in Georgia with effective management practices, Tew et al. (1982) reported that, at a minimum management level, pesticide costs decreased 17-fold and net profits increased significantly compared with conventional methods. Employing a combination of several of these alternative technologies, it might be possible to reduce insecticide use in sweet corn by more than three-quarters (Table 10.3B). Because the results of Tew et al. (1982) were for a particular rotation system, we estimated that the added costs for the alternatives would be $10/ha (Table 10.3B).

Onions

The primary insect pests of onions are the onion maggot and the onion thrips (Ritcey and McEwen, 1984; Edelson et al., 1986). Losses of onions where insecticide treatments were made averaged about 4% (Stemeroff and George, 1983) contrasted with losses of 39% for untreated onions (Tolman et al., 1986).

Alternatives for control of these pests include rotations, scouting, and sanitation (Cadoux, 1984; Mayer et al., 1987). Recently, D. Haynes (personal communication, Department of Entomology, Michigan State University, 1988) reported that onions could be produced without insecticides by raising cattle adjacent to the onion field and mulching the onions with straw. A parasitic wasp species used the maggots in the cattle manure as an alternative host. The straw protected a predaceous beetle that preys on the onion maggot. Onion losses to maggots and other insects in the alternate system were only 2–3% compared with the average of 4% in insecticide-treated plots (Table 10.3B). Also, it has been reported that preventing injuries to the onion bulbs during the growing season will help reduce maggot attack (Cadoux, 1984). Employing several of these alternatives in combination, it might be possible to reduce insecticide use about one-third at an added estimated cost of $5/ha (Table 10.3B).

Beans

About 76% of the bean hectarage is treated with insecticides for two primary insect pests, the Mexican bean beetle and the pea moth (Table 10.3B). Insecticide use might be reduced by one-third by using scouting, planting short-season varieties, using resistant cultivars, applying Bt, and employing a vacuum apparatus to remove pests (Krishnaiah et al., 1981; Karel and Rweyemamu, 1985; Karel and Schoonhoven, 1986; Mahrt et al., 1987; Stockwin, 1988; Street, 1989) (Table 10.3B). Scouting and the vacuum techniques were estimated to increase control costs an estimated $3/ha (Table 10.3B).

Cucumbers and Watermelons

The primary pests of cucumbers and watermelons are the cucumber beetle and pickleworm (Douce and Suber, 1985). The most practical substitutes for insecticides are scouting, rotations, and reflective mulches (Schalk et al., 1979). Using these two alternatives, it might be possible to reduce insecticide use by about one-third at an estimated cost of $5/ha (Table 10.3B).

Peas

The major insect pests of peas are the pea aphid and the pea moth (Metcalf et al., 1962). Scouting can help reduce insecticide treatments to about one per season for the aphid (Maiteki and Lamb, 1985). The pea moth can be controlled by deep plowing and early threshing (Metcalf et al., 1962). Through scouting and the improved targeting of insecticides, insecticide use can be reduced 50% (Cranshaw and Radcliffe, 1984; Radcliffe, 1989, personal communication). By employing a combination of the various alternatives, it might be possible to reduce insecticide use about one-third at an added estimated cost of $5/ha (Table 10.3B).

Sweet Potatoes

The sweet potato weevil is reported to be the most serious pest of this crop. Rotation, sanitation, pheromone traps, and scouting are suitable alternative control methods that can be employed (Metcalf et al., 1962; Mullen and Sorensen, 1984; Heath et al., 1986). A complex of other root-feeding larval species, including wireworms, corn rootworms, flea beetles, and white grubs, also cause significant losses. The most effective method to prevent damage from these pests is to use cultivars with multiple insect resistance (Mullen and Sorensen, 1984; Schalk and Jones, 1985; Jones et al., 1987). The reduction in insecticide cost through the use of these cultivars would amount to about $138/hectare. Using these methods, it might be possible to reduce insecticide use about one-half at an estimated $5/ha (Table 10.3B).

Apples

A wide array of insects attack apples, and thus a large number of sprays are currently applied to orchards. In orchards about $80/ha is invested for insecticide treatments (Table 10.3C). Through the use of scouting and more selective insecticides and miticides, several studies have demonstrated that it might be possible to reduce insecticide and miticide use by 50% (EPA, 1975; Asquith et al., 1980; Tette et al., 1987; Kovach and Tette, 1988; Prokopy, 1988) (Table 10.3C).

Additional reductions in pesticide use might be possible if airblast sprayers were replaced by sprayers that apply more spray directly on the trees and less in the surrounding environment. It is estimated that about 35% of the pesticide applied by airblast sprayers is lost to the environment (Byers and Lyons, 1985). A newly designed nozzle and sprayer allows the amount of insecticide and miticide to be reduced by 50% while maintaining the same effective control of the pests (Van der Scheer, 1984). Also, Prokopy et al. (1978) demonstrated that alternate middle-row spraying reduced insecticide and miticide use 50% with no decrease in yield. Recently Seiber (1988) modified a sprayer so that it reduced spray use about 40%.

Peaches

Peaches also have a wide array of insect pests that require several sprays each season. About $100/ha is spent for insect and mite control on peaches

(Table 10.3C), and this amount may reach $338/ha (Kirchner et al., 1987). Hall (1985) reported that in 60% of the cases airblast sprayers were not placing sufficient spray on the trees. Thus, by using more efficient application equipment and switching from routine calendar spraying to scouting and a treat-when-necessary program, it might be possible to reduce insecticide and miticide use by about one-half (Meyer, 1986a) with no added cost (Table 10.3C).

Grapes

The most serious insects on grape include the grape berry moth, grape leafhopper, and the Japanese beetle in some locations. Scouting and fertilizer management were reported to be highly effective in reducing insecticide use by at least 50% (NAS, 1989). Bt and pheromone traps have proven effective against the grape berry moth (Dennehy, 1989; Dennehy et al., 1989). The potential exists to develop a recirculating sprayer that would enclose the rows with a cover and capture all excess spray for recirculation, further reducing the amount of insecticide applied (Dennehy, 1989). Thus, by employing a combination of these alternatives, it might be possible to reduce insecticide use at least 50% at no added cost (Table 3C).

Oranges, Grapefruit, and Lemons

Depending on the crop, from 54% to 94% of the citrus hectarage is treated for insect pests (Table 10.3C). By using biological, cultural, mechanical, and genetic controls, as well as scouting, pesticide use might be reduced 45–50% (Burrows, 1981). Employing a combination of these controls, we estimated that insecticide use in citrus might be reduced by 45% with no added cost (Table 10.3C).

Pecans

An estimated 60% of the pecan hectarage is treated with insecticides (Table 10.3C). Key pests of pecan are pecan nut casebearer, pecan weevil, and hickory shuckworm. Prophylactic sprays are commonly applied for diseases. Use of degree-day models, scouting, and other decision-making aids would result in "as necessary" treatments and reduce insecticide and fungicide use by one-half (Harris et al., 1978; Harris, 1989, personal communication). With a multipest management program including the use of black-light traps for moth suppression, Gentry et al. (1982) reported that the amount of insecticide applied could be reduced while significantly raising net profits. Bt can also be applied for control of some moths. By employing a combination of these alternatives, it might be possible to reduce insecticide use about one-half at no added cost (Table 10.3C).

Other Nuts

For other nut production, including walnut and hazelnut, about 60% of the total hectarage is treated with insecticides. Major insect pests in these crops include the codling moth, navel orange worm, and mites (Metcalf et al., 1962; Headley and Hoy, 1986, 1987). Insecticide use for control of these pests can be reduced about 50% by scouting and introduction of natural enemies (Culver, 1978; Calkin, 1984, 1985, 1987; Headley and Hoy, 1986, 1987). No additional costs were incurred because savings in reduced insecticide use more than covered the scouting costs (Culver, 1978; Calkin, 1985).

Herbicides

In evaluating the economic cost of nonchemical weed controls, we have generally assumed that mechanical and cultural weed controls cost more than herbicidal controls. However, variability in weather often determines the effectiveness of herbicides and mechanical cultivation for weed control. Under wet, rainy conditions, herbicides are more effective, whereas under dry conditions mechanical controls tend to achieve better results (Wilcut et al., 1987). Other factors that may reduce the effectiveness of herbicides include the toxicity of some herbicides to certain crops, which may reduce yields from 2% to 50% (Chang, 1965; Elliot et al., 1975; Akins et al., 1976), and an occasional increase in insect pest and disease problems in crops whose physiology is altered by herbicides (Pimentel, 1971; Oka and Pimentel, 1976).

A total of 220 million kg of herbicides is applied to U.S. crops annually (Table 10.1). The total cost of weed control has been estimated to be $3.1 billion with herbicides and $3.1 billion with tillage and cultivation costs included (Chandler, 1985). Opportunities exist to substitute a variety of ecological alternatives for herbicides in most crops (Altieri and Liebman, 1988).

Corn

More than half (53%) of the herbicides used on crops are applied to corn (Table 10.4A). Over 3 kg of herbicide is applied per hectare of corn, and more than 90% of the corn hectarage is treated. By avoiding total weed elimination in some cases, herbicide use can be reduced 75% (Schweizer, 1989). At present 91% of the land is also cultivated to help control weeds in corn (Duffy, 1982).

The average costs and returns per hectare to no-till, reduced-till, and conventional-till have actually been found to be quite similar (Duffy and Hanthorn, 1984). For example, added labor, fuel, and machinery costs for

Table 10.4A. Field-crop losses from weeds with current herbicide use and estimated costs if herbicides were reduced and several alternatives were substituted.

Crop	Ha × 10³[a]	Total (kg × 10⁶) herbicide use		Herbicide treatment			Current crop pest loss, %	Added alternative control cost $/ha[g]	Total added alternative control cost $ × 10⁶
		Current	Reduced[g]	Ha treated %	Cost $/ha	Total cost $ × 10⁶			
Corn	30,419	111[s]	45	96[e]	50[d]	1,460	10[c]	15	438
Cotton	3,852	8.2[e]	3.0	94[e]	40[r]	145	15[c]	13	33.4
Wheat	26,208	8.2[e]	6.5	50[e]	44[q]	576	13[c]	7	19
Soybeans	24,933	58[s]	25	96[e]	19[g]	450	15[c]	10	148
Rice	1,013	6[h]	4.0	98[f]	90[i]	89	19[c]	10	9.9
Tobacco	279	0.5[b]	0.2	71[j]	15[o]	3.0	8[k]	0	0
Peanuts	594	2.5[e]	1.6	93[h]	62[n]	34.3	15[c]	0	0
Sorghum	6,750	7.1[s]	4.3	82[e]	37[p]	151.7	14[c]	0	0
Sugar beets	446	0.4[g]	0.2	75[h]	77[m]	25.8	18[c]	10	1.9
Other grain	8,267	2.7[h]	1.3	45[h]	20[g]	74	5[k]	7	26
Alfalfa	10,394	0.1[j]	0.05	1[j]	12[g]	1.2	19[c]	5	0.3
Hay	14,121	0.3[h]	0.15	3[h]	12[g]	5.1	3[k]	5	2.1
Pature	241,564	2.3[g]	1.3	1[j]	12[g]	29	3[k]	2	1.6
TOTALS		207.3	92.6			3,044.1			680.2

[a]USDA, 1986. [b]Pimentel and Levitan, 1986. [c]Chandler et al., 1984. [d]Hanthorn and Duffy, 1983. [e]USDA, 1988. [f]Ferguson, 1984. [g]Estimated. [h]Delvo and Hawthorn, 1983. [i]Bryson, 1988. [j]Duffy, 1982. [k]USDA, 1965. [m]Swenson and Johnson, 1983; Winter and Wiese, 1982. [n]McArthur et al., 1985. [o]Bryson, 1988. [p]Epplin et al., 1984. [q]Perry and Saunders, 1982. [r]Frans, 1985. [s]Szmedra, 1989.

Table 10.4B. Vegetable crop losses from weeds with current herbicide use and estimated costs if herbicides were reduced and several alternatives were substituted.

Crop	Ha × 10[3a]	Total (kg × 10^6) herbicide use		Herbicide treatment			Current crop pest loss, %	Added alternative control cost $/ha[g]	Total added alternative control cost $ × 10^6
		Current	Reduced[g]	Ha treated %	Cost $/ha	Total cost $ × 10^6			
Lettuce	91	0.06[b]	0.04	84[b]	30[g]	2.3	8[c]	10	0.25
Cole	111	0.20[b]	0.10	75[b]	30[g]	2.5	12[c]	12	0.5
Carrots	36	0.03[b]	0.02	63[b]	30[g]	0.68	12[c]	10	0.07
Potatoes	550	0.8[g]	0.4	96[i]	30[h]	15.8	7[c]	0	0
Tomatoes	177	0.1[b]	0.05	86[b]	21[d]	3.2	10[c]	0[f]	0
Sweet corn	241	0.4[b]	0.13	79[b]	50[g]	9.5	11[c]	15	1.9
Onions	47	0.4[b]	0.4	99[e]	24[g]	1.2	13[c]	—	—
Cucumbers	47	0.03[b]	0.015	76[b]	20[g]	0.71	12[c]	0	0
Beans	150	0.8[b]	0.4	95[b]	37[d]	5.3	10[c]	5	0.35
Cantaloupe	45	0.02[b]	0.01	65[b]	20[g]	0.59	10[c]	10	0.15
Peas	143	0.1[b]	0.05	85[b]	20[g]	2.5	13[c]	10	0.61
Peppers	23	0.02[g]	0.01	60[j]	20[g]	0.28	11[c]	10	0.14
Sweet potatoes	43	0.03[g]	0.02	35[j]	20[g]	0.52	9[c]	10	0.15
Watermelons	83	0.01[b]	0.005	45[b]	20[g]	0.85	9[c]	10	0.19
Other vegetables	100	0.04[g]		50[j]	20[g]	1.00	8[k]	0	0
TOTALS		3.04	1.650			46.93			4.31

[a]USDA, 1986. [b]Ferguson, 1984. [c]Chandler et al., 1984. [d]Based on amount applied (Greene et al., 1985) and price of pesticides used (USDA, 1975). [e]Ellerbrock, 1989. [f]Cost can be higher if black plastic is used. [g]Estimated. [h]Fohner and White, 1982; Meyer, 1986b. [i]Parks, 1982. [j]USDA, 1975.

Table 10.4C. Fruit and nut crop losses from weeds with current herbicide use and estimated costs if herbicides were reduced and several alternatives were substituted.

		Total (kg × 10⁶) herbicide use		Herbicide treatment					
Crop	Ha × 10³ [a]	Current	Reduced [g]	Ha treated %	Cost $/ha	Total cost $ × 10⁶	Current crop pest loss, %	Added alternative control cost $/ha [g]	Total added alternative control cost $ × 10⁶
Apples	198	6 [m]	3	45 [b]	30 [e]	2.7	4 [c]	0	0
Cherries	54	0.01 [m]	0.005	28 [b]	35 [g]	0.53	8 [f]	5	0.08
Peaches	94	0.09 [m]	0.045	36 [b]	80 [d]	2.7	7 [c]	5	0.17
Pears	34	0.01 [i]	0.005	16 [b]	30 [g]	0.16	4 [c]	5	0.25
Plums	57	0.1 [g]	0.05	11 [i]	111 [f]	0.69	6 [f]	−10	−0.06
Grapes	308	0.3 [g]	0.2	50 [k]	8 [j]	1.2	15 [c]	5	0.74
Oranges	276	1.7 [n]	0.8	54 [b]	42 [h]	6.3	5 [c]	0	0
Grapefruit	98	0.5 [n]	0.25	53 [b]	42 [h]	2.2	7 [c]	0	0
Lemons	32	0.2 [n]	0.1	30 [b]	42 [h]	0.4	4 [c]	0	0
Other fruit	100	0.1 [g]	0.05	20 [i]	30 [g]	0.6	5 [f]	5	0.10
Pecans	155	0.1 [g]	0.05	31 [i]	30 [g]	1.44	11 [c]	5	0.24
Other nuts	170	0.1 [g]	0.05	31 [i]	111 [f]	5.85	5 [f]	−10	−0.53
TOTALS		9.21	4.605			24.77			0.99

[a] USDA, 1986. [b] Suguiyama and Carlson, 1985. [c] Ferguson, 1984. [d] Folwell et al., 1981. [e] Gerling, 1986. [f] Weakley, 1986. [g] Estimated. [h] Tucker et al., 1980. [i] USDA, 1975. [j] Warner, 1985. [k] Pool, 1989. [m] Shwu and Webb, 1981. [n] Haydu, 1981.

conventional-till for corn were about $24/ha higher than those for no-till. However, the costs for the added fertilizers, pesticides, and seeds in the no-till system were $22/ha higher than conventional-till (Duffy and Hanthorn, 1984). Thus, the costs of inputs and returns for the two systems were quite similar. Ridge-till is a form of no-till that has many more advantages for crop production than no-till (Forcella and Lindstrom, 1988). In addition, ridge-till can be employed without the use of herbicides (R. Thompson, personal communication [PC], 1985).

Also, corn and soybean rotations have been found to provide substantially higher returns than either crop grown separately and continuously (OTA, 1979; Helmers et al., 1986; Cramer,1988).

It might be possible to reduce herbicide use on corn by about 60% if the use of mechanical cultivation and rotations (Forcella and Lindstrom, 1988) were increased (Table 10.4A). If these methods were used more often, however, weed control costs might increase by approximately $15/ha (Table 10.4A).

Cotton

Herbicides are used in approximately 96% of cotton production (Table 10.4A), but all cotton hectarage still receives tillage and mechanical cultivation for weed control (Duffy, 1982). Ropewick application of herbicides can help increase the amount of herbicide reaching the target weeds and at the same time reduce the total amount of herbicide used by about 90% (Dale, 1980; Keeley et al., 1984). New application technologies and selective herbicides may be replacing the ropewick applicator (Frans, 1989, PC). In addition, increased use of mechanical cultivation and rotations might reduce herbicide use by 63% with an added estimated cost of $13/ha (Table 10.4A).

Wheat

The use of herbicides in low-rainfall areas (300 to 350 mm per year) provides an opportunity to leave wheat residues on the surface of the land to conserve soil moisture and protect the land from erosion (Freyman et al., 1982). This technique relies on increased herbicides for weed control but has an economic advantage in reduced labor. In higher-rainfall regions, dense plantings, vigorous plants, tillage, and rotations offer a means of reducing herbicide use (Ayers, 1986; Appleby, 1987). Nearly 98% of the wheat hectarage is tilled prior to planting for weed control (Duffy, 1982). Also, when using effective management practices in a rotation sequence of wheat, soybeans, and spinach in Georgia, amounts of herbicide and other pesticides can be reduced. For example, Tew et al. (1982) reported that pesticide costs were reduced 36-fold and net profits were increased

4.5-fold. Thus, it might be possible to reduce herbicide use in wheat production about 20% (Table 10.4A).

Soybeans

Soybeans are the second-largest user of herbicides, with about 96% of the soybean hectarage treated for weed control (Table 10.4A); 96% of the hectarage also receives some tillage and mechanical cultivation for weed control (Duffy, 1982). The ropewick applicator has been used in soybeans to reduce herbicide use about 90% (Dale, 1980); the applicator was found to increase soybean yields 51% over conventional treatments (Dale, 1978). Also, a new model of recirculating sprayer saves 70–90% of the spray emitted that is not trapped by the weeds (Matthews, 1985).

In addition, alternative techniques are available to reduce herbicide use on soybeans. These include ridge-till, tillage, mechanical cultivation, row spacing, planting date, tolerant varieties, crop rotations, and spot treatments (Wax and Pendleton, 1968; Tew et al., 1982; Walker and Buchanan, 1982; King,1983; Helmers et al., 1986; Wilson et al., 1986; Jordan et al., 1987; Cramer, 1988; Forcella and Lindstrom, 1988; Russnogle and Smith, 1988). Reduced-rate technologies for both preemergence and postemergence herbicide applications have been developed by Arkansas researchers (Baldwin et al., 1988) and allow reductions of from one-half to one-fourth the labeled rates by early applications, timed precisely (postemergence) on just-emerged, specific weeds. Employing several of these alternative techniques in combination might possibly reduce herbicide use in soybeans by about 57% (Table 10.4A). Despite the results of Tew et al. (1982) that indicate no added control costs for the alternatives, we estimated that these techniques would increase weed control costs by $10/ha (Table 10.4A).

Rice

Weeds are a major pest of rice, and about 98% of the hectarage is treated with herbicides (Table 10.4A). An equivalent portion of the crop area is tilled and/or receives mechanical cultivation for weed control (Duffy, 1982). The primary alternatives for weed control include scouting, mechanical cultivation, deep tillage, soil and water management, fertilizer management, certified weed-free seed, transplanting, cultivar type, plant spacing, biological control, and rotations (USDA, 1977; UC, 1983; Moody, 1989; NAS, 1989). Using combinations of these various alternatives, herbicide use might be reduced by one-third at an estimated cost of $10/ha (Table 10.4A).

Peanuts

Weeds are a serious problem in peanuts, and 93% of the hectarage is treated with herbicides (Table 10.4A); 99% is tilled and/or receives mechanical cultivation for weed control (Duffy, 1982). In some cases the most effective means of weed control is a combination of herbicides and cultivation (Wilcut et al., 1987). However, in some cases cultivation alone is most economical (Bridges et al., 1984). During dry years, cultivation is more effective than herbicidal weed control (Wilcut et al., 1987). In addition to mechanical cultivation, crop rotations are highly effective for weed control (Buchanan et al., 1982). Using a combination of alternatives for herbicides, it might be possible to reduce herbicide use by one-third (Table 10.4A) and lower costs by $2/ha.

Sorghum

Weeds are a major problem in sorghum production, and about 82% of the hectarage is treated with herbicides (Table 10.4A), while about 97% is tilled and/or receives mechanical cultivation (Duffy, 1982). Alternatives to herbicides include crop rotations and sweep tillage (Allen et al., 1980; Janzen et al., 1987; Williams et al., 1987). Employing a combination of these techniques, it might be possible to reduce herbicide use about 40% at no added cost (Epplin et al., 1984) (Table 10.4A).

Sugar Beets

About 75% of sugar beet hectarage is treated with herbicides (Table 10.4A). Alternatives for weed control include mechanical cultivation, rotations, and cover crops. Using these alternatives, it might be possible to reduce herbicide use about one-half at an estimated cost of $10/ha (Table 10.4A).

Tobacco

Herbicides are used on 71% of the tobacco hectarage (Table 10.4A), whereas 100% is tilled and/or cultivated for weed control (Duffy, 1982). Again, it might be possible to reduce herbicide use by two-thirds through the substitution of mechanical cultivation (Table 10.4A).

Alfalfa and Hay

Alfalfa and hay land are treated with about 1 million kg of herbicides (Table 10.4A). Through more effective use of the alternatives including rotations, timing of cutting, resistant varieties, and spot treatments, it might be possible to reduce herbicide use about one-half at an estimated cost of $5/ha (Forney et al., 1985; Dawson, 1986) (Table 10.4A).

Cole

Weeds are a serious problem in cole crop production, and about 75% of the crop is treated with herbicides (Table 10.4B). Herbicide use may be reduced by using band applications (Hicks and Rehm, 1986) as well as mechanical cultivation and ropewick application technology (Dale, 1979). Additional methods to reduce weed problems include planting early maturing varieties and using large transplants, thus giving the cole plants a competitive advantage over the weeds (Agamalian, 1984). It is projected that herbicide use might be reduced by one-half with an estimated cost of $12/ha (Table 10.4B).

Potatoes

Losses of potatoes while using herbicides and other weed controls average 7% (Chandler et al., 1984); however, with no weed control, potato losses ranged from 18% to 71% (Tolman et al., 1986). Weed control can be effectively carried out with mechanical cultivation and without herbicides; if conditions in the field are wet, herbicides are more effective (Sieczka, 1984; Meyer, 1986b). Employing mechanical cultivation, it might be possible to reduce herbicide use in potato production about one-half without added costs (Table 10.4B).

Tomatoes

About 86% of the tomato hectarage is treated with herbicides (Table 10.4B). One effective substitute for herbicides is black plastic mulch. The black plastic technique has an added advantage of helping to produce cleaner, higher-quality tomatoes with higher yields (Bhella, 1988). Use of black plastic also results in earlier production (Wien and Minotti, 1988) but costs about $600/ha (Teasdale and Colacicco, 1985; Wolfe and Rutkowski, 1987). Mechanical cultivation is another effective substitute for herbicides, and it costs significantly less (Henne, 1979; Wolfe and Rutkowski, 1987). Without herbicides, Grattan et al. (1988) reported that subsurface-drip irrigation will effectively limit weeds. In addition, it has been demonstrated that using a wiper applicator for glyphosate can reduce the amount of herbicide required for weed control 90% while still obtaining effective control (Dale, 1979; Harrison, 1983). For high-value, fresh-market tomatoes, employing the black plastic can be economically feasible, but for processing tomatoes the use of mechanical cultivation or wiper-applicator technology might be preferred. Thus, it might be possible to reduce herbicide use in tomatoes by about 80% (Table 10.4B). Although the savings in herbicides would offset any added labor and other costs for mechanical cultivation and wiper-applicator technology, the black plastic would be more costly than herbicides.

Sweet Corn

As with field corn, it might be possible to reduce herbicide use for sweet corn production by about two-thirds by substituting mechanical cultivation and rotations (Table 10.4B). These substitute techniques were estimated to cost an added $15/ha.

Onions

Weeds are a major problem in onion production because of competition within the row, and hand weeding appears to be the only alternative (Boldt et al., 1981). Because of the high cost of hand weeding, it was assumed there was no alternative weed control technology for large-scale onion production (Table 10.4B).

Cucumbers

In addition to cultivation, weeds can be controlled by using black plastic mulch (Hemphill and Crabtree, 1988). The mulch, when used with plastic row covers, increased early market fruit yields by 45% and total yield by 16%. This weed control technology raised the estimated net economic return for cucumbers significantly above that with herbicides (Hemphill and Crabtree, 1988). Thus, employing the plastic mulch alternative, it was estimated that herbicide use could be reduced by one-half with increased profits (Table 10.4B).

Beans

Although 95% of the bean crop is treated with herbicides, it is estimated that the amount of herbicide used in beans could be reduced by one-half through the substitution of mechanical cultivation, crop rotations, and more efficient herbicide application techniques (Dale, 1979, 1980; Parker, 1981; Harrison, 1983). As with tomatoes, a hand-held wiper and ropewick applicator to apply glyphosate can achieve fully effective control with only 10% of the usual spray application (Dale, 1979; Harrison, 1983). Employing a combination of these alternatives, it might be possible to reduce the amount of herbicide applied by about one-half. The added cost of these alternatives was calculated to be about $5/ha (Table 10.4B).

Sweet Potatoes

Cultivation can improve weed control in herbicide-treated sweet potatoes (Glaze et al., 1981). We estimated that one-third of the herbicide used in sweet potatoes could be substituted by mechanical cultivation at an estimated cost of $10/ha (Table 10.4B).

Apples

Although herbicides play a role in weed control in orchards, various highly effective alternatives are available that include mowing, tilling, black plastic mulch, and cultivation (Stinchcombe and Stott, 1983; Hogue and

Neilsen, 1987). Employing these alternatives, it might be possible to reduce herbicide use about one-half at no added cost (Weakley, 1986) (Table 10.4C).

Peaches and Plums

Substituting extra cultivation for herbicides reduces the total costs of weed control in peaches and plums (Folwell et al., 1981; Weakley, 1986). Thus, it might be possible to reduce herbicide use about 50% (Table 10.4C).

Oranges, Grapefruit, and Lemons

Mechanical cultivation in citrus groves to remove weeds reduces weed control costs below those of herbicides and is equally effective in maintaining yields (Tucker et al., 1980). Thus, it might be possible to reduce herbicide use in citrus about one-half, at no additional cost (Table 10.4C).

Pecans and Other Nuts

An estimated 31% of pecans and other nut crops are treated with herbicides. Based on walnut treatment costs, treatments for other nuts were estimated to be $111/ha (Weakley, 1986). Employing cultivation for weed control costs only $91/ha. Thus, a saving of $10/ha was possible (Table 10.4C). At the same time herbicide use could be reduced about one-half.

Fungicides

A total of 38 million kg of fungicides is applied to crops in the United States annually (Table 10.1). Because of the extensive use of host-plant resistance for plant pathogen control, the opportunities for reducing fungicide use are not as great as those for insecticides. However, possibilities do exist for reducing the use of fungicides in crop production. These include the use of rotations, scouting, and forecasting.

Cotton and Tobacco

Cotton and tobacco combined use a relatively small amount of fungicide (Table 10.5A). Over a 10-year period in Texas, the adoption of scouting and other integrated pest management (IPM) measures reduced fungicide use on cotton about 82% (OTA, 1979). Management of irrigation water is also reported to reduce cotton diseases (Grimes, 1985). These practices might reduce fungicide use on cotton and tobacco about two-thirds, if adopted for both cotton and tobacco in other regions of the nation (Table 10.5A).

Table 10.5A. Field crop losses from plant pathogens with current fungicide use and estimated costs if fungicides were reduced and several alternatives were substituted.

		Total (kg × 10^6) fungicide use		Fungicide treatment					
Crop	Ha × 10^3 [a]	Current	Reduced[g]	Ha treated %	Cost $/ha	Total cost $ × 10^6	Current crop pest loss, %	Added alternative control cost $/ha[g]	Total added alternative control cost $ × 10^6
Corn	30,419	0	0	0	0	0	10[f]	0	0
Cotton	3,852	0.05[i]	0.02	4[k]	17[j]	2.6	10[r]	10	0.3
Wheat	26,208	0	0	0	0	0	20[d]	0	0
Soybeans	24,933	0.03[d]	0	1[c]	70[n]	17.4	7[m]	0	0
Rice	1,013	0.03[g]	0.02	3[b]	121[p]	3.6	6[f]	3	0.03
Tobacco	279	0.10[g]	0.03	82[c]	30[g]	7.0	11[f]	10	0.20
Peanuts	594	2.5[j]	1.6	93[e]	60[q]	33.1	28[f]	10	5.0
Sorghum	6,750	0	0	0	0	0	9[f]	0	0
Sugar beets	446	0.01[g]	0.006	13[e]	21[h]	1.22	16[f]	5	0.19
Other grain	8,267	0	0	0	0	0	0	0	0
Alfalfa	10,394	0	0	0	0	0	0	0	0
Hay	14,121	0	0	0	0	0	0	0	0
Pasture	241,564	0	0	0	0	0	0	0	0
TOTALS		2.72	1.656			64.92			5.72

[a]USDA, 1986. [b]USDA, 1983. [c]Duffy, 1982. [d]USDA, 1988. [e]USDA, 1975. [f]USDA, 1965. [g]Estimated. [h]Swenson and Johnson, 1983. [i]OTA, 1979. [j]USDA, 1988. [k]El-Zik, 1986. [m]Mulrooney, 1986. [n]Perry and Saunders, 1982. [o]USDA, 1987. [p]USDA, 1982. [q]McArthur et al., 1985. [r]Frans, 1985.

Table 10.5B. Vegetable crop losses from plant pathogens with current fungicide use and estimated costs if fungicides were reduced and several alternatives were substituted.

Crop	Ha × 10[3a]	Total (kg × 10⁶) fungicide use		Fungicide treatment			Current crop pest loss, %	Added alternative control cost $/ha[g]	Total added alternative control cost $ × 10⁶
		Current	Reduced[g]	Ha treated %	Cost $/ha	Total cost $ × 10⁶			
Lettuce	91	0.04[b]	0.032	93[b]	24[g]	2.0	12[c]	5	0.08
Cole	111	0.20[b]	0.06	43[b]	20[g]	0.95	9[h]	5	0.24
Carrots	36	0.06[b]	0.04	41[b]	20[g]	0.30	8[h]	5	0.02
Potatoes	550	2.5[b]	1.7	97[i]	25[e]	13.3	20[f]	5	2.7
Tomatoes	177	2.5[b]	1.25	94[b]	121[d]	21.4	21[h]	10	0.8
Sweet corn	241	0.2[b]	0.1	1[b]	20[g]	0.07	8[h]	6	0.007
Onions	47	0.2[b]	0.13	84[b]	44[g]	1.74	21[h]	5	0.07
Cucumbers	47	0.02[b]	0.01	27[b]	20[g]	0.25	15[h]	5	0.02
Beans	150	0.3[b]	0.24	54[b]	10[d]	0.81	20[h]	5	0.2
Cantaloupe	45	0.05[d]	0.03	83[b]	20[g]	0.75	21[h]	5	0.062
Peas	143	0.06[g]	0.02	54[b]	20[g]	1.54	23[h]	5	0.19
Peppers	23	0.01[c]	0.006	48[i]	20[g]	0.22	14[h]	5	0.06
Sweet potatoes	43	0.01[c]	0.006	1[i]	20[g]	0.01	18[h]	5	0.002
Watermelons	83	0.1[b]	0.006	86[b]	20[g]	1.43	14[c]	5	0.12
Other vegetables	100	0.001[g]	0.001	10[j]	20[g]	0.2	10[h]	0	0
TOTALS		6.251	3.631			44.97			4.571

[a]USDA, 1986. [b]Ferguson, 1984. [c]Tolman et al., 1986. [d]Based on amount applied (Gianessi and Greene, 1988) and price of pesticide used (USDA, 1985). [e]Parks, 1982; Warner, 1985. [f]Love and Tauer, 1987; Teng and Bissonnette, 1985. [g]Estimated. [h]USDA, 1965. [i]USDA, 1975.

Table 10.5C. Fruit and nut crop losses from plant pathogens with current fungicide use and estimated costs if fungicides were reduced and several alternatives were substituted.

Crop	Ha × 10³[a]	Total (kg × 10⁶) fungicide use		Fungicide treatment			Current crop pest loss, %	Added alternative control cost $/ha[g]	Total added alternative control cost $ × 10⁶
		Current	Reduced[g]	Ha treated %	Cost $/ha	Total cost $ × 10⁶			
Apples	198	3.5[m]	2.8	90[b]	170[d]	30.3	8[e]	0	0
Cherries	54	0.5[m]	0.4	85[b]	70[g]	3.2	24[e]	0	0
Peaches	94	2.2[c]	1.2	90[b]	205[c]	17.3	21[e]	0	0
Pears	34	0.14[i]	0.11	86[b]	60[g]	1.8	17[e]	0	0
Plums	57	3[g]	2.4	35[f]	60[g]	1.2	10[e]	0	0
Grapes	308	9[k]	7	95[k]	50[h]	14.6	27[e]	0	0
Oranges	276	1.5[j]	1	84[b]	60[g]	13.9	16[e]	0	0
Grapefruit	98	0.5[j]	0.4	94[b]	60[g]	5.5	2[e]	0	0
Lemons	32	0.3[j]	0.2	48[b]	60[g]	0.92	29[e]	0	0
Other fruit	100	3[g]	2.4	26[f]	50[g]	1.3	20[e]	0	0
Pecans	155	1[g]	0.8	46[f]	50[g]	3.6	21[e]	0	0
Other nuts	170	4[g]	2.6	46[f]	50[g]	3.9	12[e]	0	0
TOTALS		28.64	21.31			97.52			0

[a]USDA, 1986. [b]Suguiyama and Carlson, 1985. [c]Folwell et al., 1981; Warner, 1985. [d]Gerling, 1986; Kovach and Tette, 1988. [e]USDA, 1965. [f]USDA, 1975. [g]Estimated. [h]Whitaker and White, 1982; Warner, 1985. [i]Ferguson, 1985. [j]Haydu, 1981. [k]Pool, 1989. [m]Shwu and Webb, 1981. [n]Haydu, 1981.

Rice

Only 3% of the rice crop is treated with fungicides (Table 10.5A). Employing host-plant resistance, sanitation, early flooding, and fertilizer management may reduce fungicide use about one-third at an estimated cost of $3/ha (Table 10.5A).

Peanuts

A large percentage of the peanut hectarage is treated with fungicides (Table 10.5B). Alternatives for control of the major foliar diseases include tillage, sanitation (burying crop residues), forecasting and the use of appropriate rotations (Porter et al., 1982; Johnson et al., 1985; Grichar and Boswell, 1987). Transparent plastic may be used to increase soil temperatures, which results in reduced soil and stem diseases and increasing peanut yields (Porter et al., 1982). By using a combination of these alternatives it might be possible to reduce fungicide use on peanuts by about one-third at an estimated cost of $10/ha (Table 10.5B).

Cole

About 43% of cole crops are treated with fungicides (Table 10.5B). Cole crop diseases can be reduced by purchasing disease-resistant seeds, using proper crop rotations, improving sanitation, and using appropriate fertilizer (especially lime) (Roberts and Boothroyd, 1972). Employing a combination of several of these technologies, it may be possible to reduce fungicide use on cole crops about two-thirds at an estimated cost of $5/ha (Table 10.5B).

Potatoes

About 97% of the potato hectarage is treated with fungicides (Table 10.5B). Without fungicide treatments, losses from diseases ranged between 5% and 25%, while losses with fungicide treatments were reported to be about 20% (Teng and Bissonnette, 1985; Tolman et al., 1986; Love and Tauer, 1987). Shields et al. (1984) reported that the planting of short-season potatoes in Wisconsin reduced the number of fungicide applications by one-third. Correct storage, handling, and planting of seed tubers and proper management of soil moisture and fertility minimize losses to most diseases (UC, 1986). Forecasting and scouting might also be employed to reduce fungicide use 15% to 25% (Royle and Shaw, 1988; Tette and Koplinka-Loehr, 1989). Fungicides should be applied before infection, and at the same time the crop should be scouted for the appearance of disease symptoms. Employing a combination of these controls, it might be possible to reduce fungicide use on potatoes about one-third at an estimated cost of $5/ha (Table 10.5B).

Tomatoes

A forecasting system employed with tomatoes in Pennsylvania indicated that fungicide use could be reduced 55% while maintaining excellent pathogen control (Madden et al., 1978). Thus, forecasting and scouting methods may enable a 55% reduction in fungicide use in tomato production. This could provide savings of about $65/ha in use of fungicides; however, we estimated that the added alternative control would cost an estimated $10/ha (Table 10.5B).

Sweet Corn

Only about 1% of sweet corn is treated with fungicides, and yield losses to plant pathogens are relatively low (Table 10.5B). Thus, it was assumed that an effective treat-when-necessary program could reduce fungicide use by 50% at an estimated cost of $6/ha (Table 10.5B).

Onions

Diseases are a major limitation in onion production. Losses of onions with fungicide treatments average 21% (USDA, 1965), whereas losses without fungicide treatment average 24% (Tolman et al., 1986). Alternative practices available to reduce the use of fungicide include improved sanitation, rotations, and scouting (Ellerbrock and Lorbeer, 1977; Shoemaker and Lorbeer, 1977). With these methods, fungicide use might be reduced by one-third with an estimated cost of about $5/ha (Table 10.5B).

Cucumbers

About 27% of the cucumber hectarage is treated with fungicides (Table 10.5B). Thompson and Jenkins (1985) reported that improved forecasting and scouting may reduce fungicide use by 50%. Other alternative techniques for reducing diseases in cucumbers include using resistant cucumber varieties and rotations (Lloyd and McCollum, 1940; Sitterly, 1969; Thompson and Jenkins, 1985; Sumner and Phatak, 1987). In addition, the use of photodegradable plastic was found to be significantly more effective than fungicides or other control technologies for control of several diseases (Lewis and Papavizas, 1980). Employing combinations of these alternatives, it might be possible to reduce fungicide use in cucumber production about one-half at an estimated cost of $5/ha (Table 10.5B).

Beans

About half of the bean hectarage is treated with fungicides (Table 10.5B). Bean rust and white mold are the major diseases of beans. Through improved forecasting, scouting, biocontrol, and the use of resistant and mixed varieties (Baker et al., 1985; Schwartz et al., 1987; Mukishi and Trutman,

1988; Stavely, 1988), it is estimated that fungicide use might be reduced 20% with an estimated cost of $5/ha (Table 10.5B).

Sweet Potatoes

Growers normally do not apply pesticides for the control of fungal and viral diseases. However, nematicides are routinely applied for nematodes. An effective method of control for diseases and nematodes is the use of resistant cultivars (Jones et al., 1985, 1989). The use of these cultivars should reduce or eliminate dependency on pesticides.

Apples

About 90% of all fungicides are applied to apples, peaches, citrus, and other fruit crops (Pimentel and Levitan, 1986). IPM data from apples in New York State suggest that fungicide use on apples could be reduced about 10% by scouting and better forecasting of disease, depending on the weather and year (Kovach and Tette, 1988).

In addition, a recent design in spray nozzle and application equipment demonstrated that the amount of fungicide applied for apple scab control could be reduced by 50% (Van der Scheer, 1984). Thus, by employing better weather forecasting and improved application technology combined with scouting, fungicide use could be potentially reduced on apples an estimated 20% (Table 10.5C).

Peaches

About 90% of the peach hectarage requires heavy treatment with fungicides. In some cases as much as $326/ha is spent on fungicides (Kirchner et al., 1987). Hall (1984) reported that fungicide use could be reduced 20% by using the alternate-row spray technique. Although fungicides are important in peach production, the use of scouting, better forecasting, and orchard sanitation can reduce fungicide use 40–50% (Gorsuch et al., 1984; Gorsuch and Miller, 1984) (Table 10.5C).

Grapes

An estimated 95% of the grape hectarage is treated with fungicides (Table 10.5C). Fungicide use can be reduced in both fresh grapes and wine grapes through cultural management, fertilizer management, and other means (Pearson, 1986; NAS, 1989). Employing a combination of these technologies, it might be possible to reduce fungicide use by one-third (Table 10.5C).

Pecans

An estimated 46% of the pecan hectarage is treated with fungicides (Table 10.5C). Farmers in Georgia have demonstrated that the use of a multipest management program can reduce the number of fungicide applications as well as the amount of fungicide applied while at the same time significantly increasing net profit (Gentry et al., 1982; Gottwald and Bertrand, 1988). By employing an effective forecasting and management scheme, it might be possible to reduce fungicide use on pecans by about 20% (Table 10.5C).

Overall Pesticide Reduction Assessment

Substituting nonchemical alternatives for some pesticides used on 40 major crops, we confirm that total agricultural pesticide use can potentially be reduced by approximately 50%. The added costs for implementing these alternatives are estimated to be about $818 million (Tables 10.3, 10.4, and 10.5). This would increase total pest control costs approximately 20% and total food production costs at the farm 0.5%. However, actual retail costs would increase on the average only about 1.5% because farm prices make up only one-third of total retail food prices (USDA, 1986).

It is important to note that if pesticide use were reduced further so that reduced crop yields resulted, the benefits-and-costs relationship would be quite different. For example, each 1% decrease in crop yield in agriculture results in a corresponding 4.5% increase in the farm price of goods (Sisler, 1988). It is also important to note that overproduction is the prime reason that the United States spends $26 billion annually on price supports (USOMB, 1989).

Environmental Costs/Benefits

Balanced against the economic benefits of pesticides are the pesticide control measures that cost about $4.1 billion annually (Tables 10.3, 10.4, and 10.5). This figure does not include the indirect environmental and public health costs, which total about $8 billion annually (see Table 3.6).

Of course, we do recognize that although the nonchemical alternative controls proposed as substitutes for pesticides in this study are significantly safer than pesticides, the alternatives themselves may cause some social and environmental problems (Pimentel et al., 1984). However, if one assumes that reducing pesticide use by 50% might also eventually reduce the environmental and public health risks from presticides from one-quarter to one-half, then the added costs for the nonchemical alternatives ($818 million) would be more than offset by the reduced environmental and public health risks listed in Table 3.6.

Conclusion

From this analysis it is clear that pesticides cause serious public health problems and considerable damage to agricultural as well as natural ecosystems. A conservative estimate suggests that the environmental and social costs of pesticide use in the United States equal approximately $2.2 billion annually, and the actual cost is probably double this amount. In addition to these costs, the nation spends $4.1 billion annually to treat crops with 320 million kg of pesticides.

This study confirms that it might be possible to reduce pesticide use by one-half, at a cost of approximately $818 million. Such a finding supports the projection of the Office of Technology Assessment (1979) and the National Academy of Sciences (1989) and the policy adopted by the Danish and Swedish governments—that pesticide use could be reduced 35% to 50%.

The 50% pesticide reduction in our current assessment would help satisfy the concerns of the majority of the public, who worry about pesticide levels in their food and damage to the environment (Sachs et al., 1987). If pesticide use were reduced by one-half without any decline in crop yield, the total price increase in purchased food is calculated to be only 1.5%. If the public could be assured that pesticides in their food and environment were greatly reduced, they probably would be willing to pay this slight increase in food costs.

In addition, it is clear that the public would accept some reduction in cosmetic standards if this would result in a decrease in pesticide contamination of food (Healy, 1989). This is confirmed by the growing popularity of organic food stores and supermarkets that guarantee pesticide-free foods (Hammit, 1986; Poe, 1988). Furthermore, it has been documented that the processing of soft-bodied insects in catsup and applesauce involves absolutely no risk to public health and even has some nutritional value (Pimentel et al., 1977a). The implementation of higher cosmetic standards today results in greater quantities of pesticides being applied to food crops. This rapidly growing use of pesticides for cosmetic purposes is detrimental to public health and the environment and also contrary to public demand (Pimentel et al., 1977a).

Although some of the data used in this preliminary investigation have limitations, the "best estimates" presented suggest that it is possible to reduce pesticide use by up to one-half. We hope that more complete data will be assembled and that detailed analyses will be made concerning the potential for reducing pesticide use. In particular, more data are needed concerning those agricultural technologies and policies that have contributed to the increase in pesticide use during the past 40 years while simultaneously increasing crop losses to pests.

Implementing a program to reduce pesticide use in agriculture will require the combined education of farmers and public and some new regulations. In addition, it will require that the federal government revise its current policies, like its commodity and price-support program which prevent farmers from employing crop rotations and other sound agricultural practices (NAS, 1989). Several current government policies actually increased pest problems and pesticide use (NAS, 1989).

At the same time, a greater investment is needed in research on alternative pest control practices. Many opportunities exist to reduce pesticides through the implementation of new environmental, cultural, and biological pest controls (See chapters in this book). We strongly support the National Academy of Sciences research recommendations for alternative pest controls (NAS, 1989).

If the public is concerned about pesticides contaminating their food and environment, are the small economic costs necessary to reduce pesticide use worth denying the ecological and public health benefits? Hopefully, the public and state and federal governments will investigate the ecology, economics, and ethics of pesticide reduction in agriculture. This analysis suggests that it is essential that a careful assessment be made to evaluate the benefits and risks of pesticides and the nonchemical alternatives for society.

Acknowledgments

We thank the following people for reading an earlier draft of this article, for their many helpful suggestions, and, in some cases, for providing additional information: B. Barclay and M. El-Ashry, World Resources Institute; R. Frans, University of Arkansas; J. Hatchett, Kansas State University; M. Harris, Texas A&M University; H. Hokkanen, Agricultural Research Centre, Jokioinen, Finland; D. Horn, Ohio State University; C. Huffaker, University of California, Berkeley; H. Janzen, Agriculture Canada; P. Johnson, University of New Hampshire; F. McEwen, University of Guelph; B. Mogensen, National Environmental Research Institute, Denmark; I. Oka, Bogor Research Institute for Field Crops, Indonesia; C. Osteen, USDA, Washington, D.C.; M. Pathak, International Rice Research Institute, Philippines; J. Pierce, Environmental Action, Washington, D.C.; E. Radcliffe, University of Minnesota; J. Schalk, U.S. Veg. Lab, Charleston, N.C.; D. Wen, Chinese Academy of Sciences; K. Stoner, Connecticut Agricultural Experiment Station; G. Surgeoner, University of Guelph; G. Teetes, Texas A&M University.

And at Cornell University, T. Dennehy, E. Glass, R. Roush, J. Tette, and C. Wien.

We gratefully acknowledge the partial support of this study by a William and Flora Hewlett Foundation Grant to the Centre for Environmental Research.

References

Agamalian, H. 1984. Selective weed control in cole crops. Proc. Calif. Weed Conf. **36**:118–120.

Ahrens, C. and H.H. Cramer. 1985. Improvement of agricultural production by pesticides. pp. 151–162 *in* Environment and Chemicals in Agriculture. F.P.W. Winteringham, ed. Elsevier Appl. Sci. Publ., New York.

Akesson, N.B. and W.E. Yates. 1984. Physical parameters affecting aircraft spray application. pp. 95–115 *in* Chemical and Biological Controls in Forestry. W.Y. Garner and J. Harvey, eds. Am. Chem. Soc. Ser. 238. Washington, D.C.

Akins, M.B., L.S. Jeffery, J.R. Overton, and T.H. Morgan, Jr. 1976. Soybean response to preemergence herbicides. Proc. South. Weed Sci. Soc. **29**:50.

Allen, R.R., J.T. Musick, P.W. Unger, and A.F. Wiese. 1980. Soil, water, and energy conserving tillage—Southern Plains. pp. 94–101 *in* Proc. Am. Soc. Agr. Eng. Conf., Crop Production in the 80's, Palmer House, Chicago.

Altieri, M.A. and M. Liebmann. 1988. Weed Management in Agroecosystems: Ecological Approaches. CRC Press, Boca Raton, Fla. 354 pp.

Antle, J.M. and S.K. Park. 1986. The economics of IPM in processing tomatoes. Calif. Agr. **40(3/4)**:31–32.

Appleby, A.P. 1987. Weed control in wheat. pp. 396–415 *in* Wheat and Wheat Improvement. E.G. Heyne, ed. Am. Soc. Agron., Madison, Wisc.

Armbrust, E.J., B.C. Pass, D.W. Davis, R.G. Helgesen, G.R. Manglitz, R.L. Pienkowski, and C.G. Summers. 1980. General accomplishments toward better insect control in alfalfa. pp. 187–216 *in* New Technology of Pest Control. C.B. Huffaker, ed. John Wiley and Sons, New York.

Arrington, L.G. 1956. World survey of pest control products. U.S. Dept. of Commerce, U.S. Govt. Printing Office. 213 pp.

Asquith, D., B.A. Croft, S.C. Hoyt, E.H. Glass, and R.E. Rice. 1980. The systems approach and general accomplishments toward better insect control in pome and stone fruits. pp. 249–317 *in* New Technology of Pest Control. C.B. Huffaker, ed. John Wiley and Sons, New York.

Ayers, L.J. 1986. Low-cost weed control that works. The New Farm **8(3)**:28–33.

Baker, C.J., J.R. Stavely, and N. Mock. 1985. Biocontrol of bean rust by *Bacillus subtilis* under field conditions. Plant Dis. **69**:770–772.

Baldwin, F.L., L. Oliver, and T. Tripp. 1988. Arkansas' experience with reduced-rate herbicide recommendations. WSSA Abstracts, No. 126. p. 45 (1988 Meeting of the Weed Science Society of America, Las Vegas, Nevada, Feb. 2–4).

Barber, G.W. 1942. Mineral-oil treatment of sweet corn for earworm control. U.S. Dept. of Agr. Circ. No 657, Washington, D.C. 16 pp.

Bhella, H.S. 1988. Tomato response to trickle irrigation and black polyethylene mulch. J. Am. Soc. Hort. Sci. **113**:543–546.

Bieber, J.L., Y.N. Lin, and D.W. Parvin. 1981. Short season cotton production systems as an alternative to heavy application of insecticides in the Mississippi Delta. AEC Res. Rep. No. 125, Miss. Agr. For. Exp. Sta., Miss. State Univ.

Boldt, P., A. Putnam, and L. Binning. 1981. Economic analysis of nitrogen use on onions grown in Minnesota, Michigan, and Wisconsin. Abst., Proc. North Central Weed Control conf. **36**:57–58.

Bridges, D.C., R.H. Walker, J.A. McGuire, and N.R. Martin. 1984. Efficiency of chemical and mechanical methods for controlling weeds in peanuts (*Arachis hypogea*). Weed Sci. **32**:584–591.

Bryson, C. 1988. Economic losses due to weeds in southern states. Proc. Annu. Mtg. South. Weed Sci. Soc. **41**:390.

Buchanan, G.A., D.S. Murray, and E.W. Hauser. 1982. Weeds and their control in peanuts. pp. 206–249 *in* Peanut Science and Technology. H.E. Pattee and C.T. Young, eds. American Peanut Research Education Society, Yoakum, Tex.

Burrows, T.M. 1981. The demand for pesticides and the adoption of integrated pest management. Diss. Abst. Int. A **42(5)**:2223A.

Byers, R.E. and C.G. Lyons. 1985. Effect of chemical deposits from spraying adjacent rows on efficacy on peach bloom thinners. HortScience **20(60)**:1076–1078.

Cadoux, M. 1984. Chlorpyrifos dissipation in muck soil and maggot resistance in onions: Implications for management of the onion maggot. MPS Project Report, Department of Vegetable Crops, Cornell University, Ithaca, N.Y. 16 pp.

Calkin, J. 1984. Filbert pest management. pp. 86–90 *in* Proc. 69th Annu. Mtg. Nut Growers Soc. of Oregon, Washington, and British Columbia, Portland, Ore.

Calkin, J. 1985. Integrated pest management. pp. 101–107 *in* Proc. 70th Annu. Mtg. Nut Growers Soc. of Oregon, Washington, and British Columbia, Portland, Ore.

Calkin, J. 1987. IPM program review. pp. 81–86 *in* Proc. 72nd Annu. Mtg. Nut Growers Soc. of Oregon, Washington, and British Columbia, Portland, Ore.

Cantwell, G.E. and W.W. Cantelo. 1984. Control of the Colorado potato beetle with *Bacillus thuringiensis* variety *thuringiensis*. Am. Pot. J. **61**:451–459.

Chandler, J.M. 1985. Economics of weed control in crops. pp. 2–20 *in* The Chemistry of Allelopathy. A.C. Thompson, ed. Am. Chem. Soc., Washington, D.C. ACS Symposium Series 268.

Chandler, J.M., A.S. Hamill, and A.G. Thomas. 1984. Crop Losses Due to Weeds in Canada and the United States. Weed Science Society of America, Champaign, Ill.

Chang, W.L. 1965. Comparative study of weed control methods in rice. J. Taiwan Agr. Res. **14(1)**:1–14.

Cochran, M.J. 1985. Economic methods and implications of IPM strategies for cotton. Integrated Pest Management on Major Agricultural Systems, Washington, D.C.

COPR (Center for Overseas Pest Research). 1976. Pest control in rice. PANS Manual No. 3. London, UK. 295 pp.

CR. 1987. Cornell Recommendations for Commercial Vegetable Production. NYS Coll. of Agr. and Life Sci., Cornell University, Ithaca, N.Y. 99 pp.

Cramer, C. 1988. 222-bushel corn—without chemicals. The New Farm **10(1)**:12–15.

Cranshaw, W.S. and E.B. Radcliffe. 1984. Insect contaminants of Minnesota processed peas. Tech. Bull. AD-T-2211, Univ. Miss. Agr. Exp. Sta.

Culver, D.J. 1978. Insect pest management makes more profit. Diamond Walnut News **60(2)**:5–7.

Dale, J.E. 1978. The rope-wick applicator—A new method of applying glyphosate. Proc. South. Weed Sci. Soc. **31**:332.

Dale, J. 1979. A non-mechanical system of herbicide application with a rope wick. PANS **25**:431–436.

Dale, J. 1980. Rope wick applicator—tool with a future. Weeds Today **11(2)**: 3–4.

Davis, D.W. 1985. Integrated pest management: New ways to manage aphids and other alfalfa pests. Utah Sci. **46(1)**:24–27.

Dawson, J.H. 1986. Dodder control in alfalfa. pp. 149–153 *in* Proc. 38th Annu. Calif. Weed Conf.

Delvo, H. and M. Hawthorn. 1983. Inputs Outlook and Situation. USDA Econ. Res. Serv., Washington, D.C.

Dennehy, T.J. 1989. Personal communication. Department of Entomology, NYS Agricultural Experiment Station, Geneva, N.Y.

Dennehy, T.J., C.J. Hoffman, J.P. Nyrop, and M.C. Saunders. 1989. Development of low spray, biological and pheromone approaches for control of grape berry moth, *Endopiza viteana* Clemens, in the eastern United States. *in* Monitoring and Integrated Management of Arthropod Pests of Small Fruit Crops. Intercept, London, pp. 261–282.

Douce, G.K. and E.F. Suber. 1985. Summary of losses from insect damage and costs of control in Georgia, 1985. Spec. Publ. 40, Ga. Agr. Exp. Sta., College of Agr., Univ. Ga.

Douce, G.K. and R.M. McPherson. 1988. Summary of losses from insect damage and costs of control in Georgia, 1987. Spec. Publ. 54, Ga. Agr. Exp. Sta., College of Agr., Univ. Ga. December 1988.

Duffy, M. 1982. Pesticide use and practices, 1982. Econ. Res. Serv. Agr. Info. Bull. No. 462, USDA.

Duffy, M. and M. Hanthorn. 1984. Returns to corn and soybean tillage practices. USDA Econ. Res. Serv., Agr. Econ. Rep. No. 508.

Edelson, J.V., B. Cartwright, and T.A. Royer. 1986. Distribution and impact of *Thrips tabaci* on onion. J. Econ. Entomol. **79**:502–505.

Eichers, T.R., P.A. Andrilenas, and T.W. Anderson. 1978. Farmers' use of pesticides. USDA, Econ. Stat. Coop. Serv., Agr. Econ. Rep. No. 418.

Ellerbrock, L.A. 1989. Personal communication. Department of Vegetable Crops, Cornell University, Ithaca, N.Y.

Ellerbrock, L.A. and J.W. Lorbeer. 1977. Sources of primary inoculum of *Botrytis squamosa*. Phytopathology **67**:363–372.

Elliot, B.R., J.M. Lumb, T.G. Reeves, and T.E. Telford. 1975. Yield losses in weedfree wheat and barley due to post-emergence herbicides. Weed Res. **15**:107–111.

El-Zik, K.M. 1986. Half a century dynamics and control of cotton disease: dynamics of cotton diseases and their control. pp. 29–33 *in* Proc. Beltwide Cotton Prod. Res. Conf. Natl. Cotton Council and the Cotton Foundation.

EPA. 1975. A study of the efficiency of use of pesticides in agriculture. Office of Pesticide Programs, U.S. Environmental Protection Agency, Washington, D.C. 240 pp.

Epplin, F.M., E.G. Kreuzer Jr., and T.F. Peeper. 1984. Economics of conservation tillage versus conventional tillage for selected crops in Oklahoma. Publ. 110, Great Plains Agr. Council, Bozeman, Mont.

Falcon, L.A. 1976. Problems associated with the use of arthropod virus pest control. Annu. Rev. Entomol. **21**:305–324.

FAO. 1979. Elements of integrated control of sorghum pests. FAO Plant Protection Paper No. 19. FAO, Rome.

Farrar, C.A., T.M. Perring, and N.C. Toscano. 1986. A midge predator of potato aphids on tomatoes. Calif. Agr. **40(11)**:9–10.

Ferguson, W.L. 1984. 1979 pesticide use on vegetables in five regions. U.S. Dept. Agr., Econ. Res. Serv., Washington, D.C. ERS Staff Report No. AGES 830920.

Ferguson, W.L. 1985. Pesticide use on stored crops, aggregated data, 1977–80. Agr. Info. Bull. No. 494. Econ. Res. Serv., USDA. 25 pp.

Flanders, R.V. 1985. Biological control of the Mexican bean beetle: potentials for and problems of inoculative releases of *Pediobius foveolatus*. pp. 686–694 *in* Proc. World Soybean Research Conference III: Proceeding. R. Shibles, ed. Westview Press, Boulder, Colo.

Fohner, G.R. and G.B. White. 1982. Cost of pesticides for potatoes in upstate New York, 1981. A.E. Res. 82–30. Dept. of Agr. Econ., Cornell University.

Folwell, R.J., D.L. Fagerlie, G. Tamaki, A.G. Ogg, R. Comes, and J.L. Baritelle. 1981. Economic evaluation of selected cultural methods for suppressing the green peach aphid as a vector of virus diseases of potatoes and sugarbeets. Bull. 0900, Coll. Agr. Res. Cent., Wash. State University.

Forcella, F. and M.J. Lindstrom. 1988. Movement and germination of weed seeds in ridge-till crop production systems. Weed Sci. **36**:56–59.

Ford, R.J. 1986. Field trials of a method for reducing drift from agricultural sprayers. Can. Agr. Eng. **28(2)**:81–83.

Forney, D.R., C.L. Foy, and D.D. Wolf. 1985. Weed suppression in no-till alfalfa (*Medicago sativa*) by prior cropping of summer annual forage grasses. Weed Sci. **33**:490–497.

Frans, R. 1985. A summary of research achievements in cotton. pp. 53–61 *in* Integrated Pest Management on Major Agricultural Systems. R.E. Frisbie and P.L. Adkisson, eds. Texas Agricultural Experiment Station, College Station, Tex.

Freyman, S., C.J. Palmer, E.H. Hobbs, J.F. Dormaar, G.B. Schaalje, and J.R. Moyer. 1982. Yield trends on long-term dryland wheat rotations at Lethbridge. Can. J. Plant Sci. **62**:609–619.

Frisbie, R. 1985. Regional implementation of cotton IPM. Integrated Pest Management on Major Agricultural Systems, Washington, D.C.

Gentry, C.R., J.S. Smith, and K.H. Reichelderfer. 1982. Evaluation of a multipest management program in Georgia (U.S.A.) pecan orchards. Prot. Ecol. **4**:339–351.

Gerling, W.D. 1986. Grower production costs. New England Fruit Meetings **82**:28–35.

Gianessi, L.P. and C.R. Greene. 1988. The use of pesticides in the production of vegetables: benefits, risks, alternatives and regulatory policies. Vegetable and specialties situation and outlook report. USDA, Econ. Res. Serv. Rep. No. 245, pp. 27–42.

Glass, E.H. and S.E. Lienk. 1971. Apple insect and mite populations developing after discontinuance of insecticides: 10 year record. J. Econ. Entomol. **64**:23–26.

Glaze, N.C., S.A. Harman, and S.C. Phatak. 1981. Enhancement of herbicidal weed control in sweet potatoes (*Ipomoea batatas*) with cultivation. Weed Sci. **29**:275–281.

Gorsuch, C.S. and R.W. Miller. 1984. Reduced pesticide program for peaches. pp. 57–61 *in* Proc. Annu. Mtg. Ark. State Hort. Soc., Myrtle Beach, S.C.

Gorsuch, C.S., E.I. Zehr, and R.W. Miller. 1984. Reduced use of fungicides and insecticides on peaches. pp. 87–95 *in* Proc. Joint National Peach Council and Southeastern Peach Convention.

Gottwald, T.R. and P.F. Bertrand. 1988. Effects of an abbreviated pecan disease control program on pecan scab disease increase and crop yield. Plant Dis. **72**:27–32.

Grattan, S.R., L.J. Schwankl, and W.T. Lanini. 1988. Weed control by subsurface drip irrigation. Calif. Agr. **42(3)**:22–24.

Greene, C.R., R.A. Kramer, G.W. Norton, E.G. Rajotte, and R.M. McPherson. 1985. An economic analysis of soybean integrated pest management. Am. J. Agr. Econ. **67**:566–572.

Grichar, W.J. and T.E. Boswell. 1987. Comparison of no-tillage, minimum, and full tillage cultural practices on peanuts. Peanut Sci. **14**:101–103.

Grimes, D.W. 1985. Cultural techniques for management of pests in cotton. pp. 365–382 *in* Integrated Pest Management on Major Agricultural Systems. R.E. Frisbie and P.L. Adkisson, eds. Texas Agricultural Experiment Station, College Station, Tex.

Hall, F.R. 1984. Evaluation of alternate row middle (ARM) spraying for apple orchards. Res. Circ. 283, Ohio Agr. Res. Dev. Cent., Wooster, Ohio.

Hall, F.R. 1985. The opportunities for improved pesticide application. pp. 69–76 *in* Proc. National Peach Council, 44th Annu. Convention, Nashville, Tenn.

Hammit, J.K. 1986. Estimating consumer willingness to pay to reduce food-borne risk. Rand Corporation, Santa Monica, Calif. 77 pp.

Hanthorne, M. and M. Duffy. 1983. Returns to corn pest management practices. USDA Agr. Econ. Rep. 501.

Harris, M.K., H.W. Van Cleave, and G.M. McWhorter. 1978. Minimum pecan insect management. Department of Entomology, Texas Agr. Exp. Sta. paper TA-14168.

Harrison, H.F. 1983. Hoeing and hand-held wiper application of glyphosate for weed control in vegetables. HortScience **18**:333–334.

Hatcher, J.E., M.E. Wetzstein, and G.K. Doucc. 1984. An economic evaluation of integrated pest management for cotton, peanuts, and soybeans in Georgia. Univ. Ga. Coll. Agr. Exp. Sta., Res. Bull. 318.

Hatchett, J.H., K.J. Starks, and J.A. Webster. 1987. Insect and mite pests of wheat. pp. 625–675 *in* Wheat and Wheat Improvement. E.G. Heyne, ed. American Society of Agronomy, Madison, Wisc.

Haydu, J.J. 1981. Pesticide use in United States citrus production. 1977. USDA Econ. Stat. Serv., Nat. Res. Econ. Div.

Headley, J.C. 1968. Estimating the production of agricultural pesticides. Am. J. Agr. Econ. **50**:13–23.

Headley, J.C. and M.A. Hoy. 1986. The economics of integrated mite management in almonds. Calif. Agr. **40**:28–30.

Headley, J.C. and M.A. Hoy. 1987. Benefit/cost analysis of an integrated mite management program for almonds. J. Econ. Entomol. **80**:555–559.

Healy, M. 1989. Buyers prefer organic food. USA Today. March 20, 1989.

Heath, R.R., J.A. Coffelt, P.E. Sonnett, F.I. Proshold, B. Dueben, and J.H. Tumlinson. 1986. Identification of sex pheromone produced by female sweet potato weevil, *Cylas formicarius elegantulus* (Summers). J. Chem. Ecol. **12**:1489–1503.

Helmers, G.A., M.R. Langemeir, and J. Atwood. 1986. An economic analysis of alternative cropping systems for east-central Nebraska. Am. J. Alt. Agr. **4**:153–158.

Hemphill, D.D., Jr. and G.D. Crabtree. 1988. Growth response and weed control in slicing cucumbers under row covers. J. Am. Soc. Hort. Sci. **113**:41–45.

Henne, R.C. 1979. Weed control systems for transplanted tomatoes. Northeast Weed Sci. Soc. Proc. **33**:166–169.

Hicks, D.R. and G.W. Rehm. 1986. Corn production costs. Crops Soils Mag. **38**:17–19.

Hoffmann, M.P., L.T. Wilson, F.G. Zalom, and L. McDonough. 1986. Lures and traps for monitoring tomato fruitworm. Calif. Agr. **4(9/10)**:17–18.

Hogue, E.J. and G.H. Neilsen. 1987. Orchard floor vegetation management. Hort. Rev. **9**:377–430.

Horn, D.J. 1988. Ecological Approach to Pest Management. The Guilford Press, New York.

Huffaker, C.B. 1980. New Technology of Pest Control. John Wiley and Sons, New York.

Hyslop, J.A. 1938. Losses occasioned by insects, mites, and ticks in the United States. E-444, USDA, Washington, D.C. 57 pp.

ICAITI. 1977. An Environmental and Economic Study of the Consequences of Pesticide Use in Central American Cotton Production. Final Report. Central American Research Institute for Industry, United Nations Environment Programme.

Jaques, R.P. 1988. Field tests on control of the imported cabbageworm (*Lepidoptera: Pieridae*) and the cabbage looper (*Lepidoptera: Noctuidae*) by mixtures of microbial and chemical insecticides. Can. Entomol. **120**:575–580.

Jaques, R.P. and D.R. Laing. 1989. Effectiveness of microbial and chemical insecticides in control of the Colorado potato beetle (*Coleoptera: Chrysomelidae*) on potatoes and tomatoes. Can. Entomol. **121**:1123–1131.

Janzen, H.H., D.J. Major, and C.W. Lindwall. 1987. Comparison of crop rotation for sorghum production in southern Alberta. Can. J. Plant Sci. **67**:385–393.

Jimenez, M.J., N.C. Toscano, D.L. Flaherty, P. Ilic, F.G. Zalom, and K. Kido. 1988. Controlling tomato pinworm by mating disruption. Calif. Agr. **42(6)**:10–12.

Johns, G.F., ed. 1966. On the Way to Plant Protection. Rodale Press, Emmaus, Pa. 355 pp.

Johnson, A.W. 1978. Effects of tobacco budworm control at different treatment levels with several insecticides on flue-cured tobacco. J. Econ. Entomol. **71**:183–185.

Johnson, C.S., P.M. Phipps, and M.K. Beute. 1985. Cercospora leafspot management decisions: an economic analysis of a weather-based strategy for timing fungicide applications. Am. Peanut Sci. **12**:82–85.

Johnston, R.L. and G.W. Bishop. 1987. Economic injury levels and economic thresholds for cereal aphids (*Homoptera: Aphididae*) on spring-planted wheat. J. Econ. Entomol. **80**:478–482.

Jones, A., P.D. Dukes, J.M. Schalk, M.G. Hamilton, M.A. Mullen, R.A. Baumgardner, D.R. Paterson, and T.E. Boswell. 1985. 'Regal' sweet potato. HortScience **20**:781–782.

Jones, A., P.D. Dukes, J.M. Schalk, and M.G. Hamilton. 1989. 'Excel' sweet potato. HortScience **24**:171–172.

Jones, A., J.M. Schalk, and P.D. Dukes. 1987. Control of soil insect injury by resistance in sweet potato. J. Am. Soc. Hort. Sci. **112**:195–197.

Jones, D.C., G.A. Herzog, R.M. McPherson, and H. Womack. 1988. Tobacco insects. *In* 1986 Summary of Losses from Insect Damage and Costs of Control in Georgia. Spec. Publ. No. 46. Agr. Exp. Sta., Coll. Agr., Univ. Ga. 28 pp.

Jordan, T.N., H.D. Coble, and L.M. Wax. 1987. Weed Control. pp. 429–457 *in* Soybeans: Improvement, Production, and Uses. J.R. Wilcox, ed. Am. Soc. Agron., Madison, Wisc.

Karel, A.K. and C.L. Rweyemamu. 1985. Resistance to foliar beetle, *Ootheca bennigseni* (Coleoptera: Chrysomelidae) in common beans. Environ. Entomol. **14**:662–664.

Karel, A.K. and A.V. Schoonhoven. 1986. Use of chemical and microbial insecticides against pests of common beans. J. Econ. Entomol. **79**:1692–1696.

Keeley, P.E., R.J. Thullen, C.H. Carter, and J.H. Miller. 1984. Control of johnsongrass (*Sorghum halepense*) in cotton (*Gossypium hirsutum*) with glyphosate. Weed Sci. **32**:306–309.

King, A.D. 1983. Progress in no-till. J. Soil Water Cons. **38**:160–161.

King, E.G., J.R. Phillips, and R.B. Head. 1986. Thirty-ninth annual conference report on cotton insect research and control. pp. 126–135 *in* Proc. Beltwide Cotton Prod. Res. Conf., Memphis, Tenn.

Kirby, R.D. and J.E. Slosser. 1984. Composite economic threshold for three lepidopterous pests of cabbage. J. Econ. Entomol. **77**:725–733.

Kirchner, D., C. Price, R. Rom, and C. Garner. 1987. Economic analysis of commercial, fresh market, irrigated peach production in Arkansas. Ark. Agr. Exp. Sta., Spec. Rep. No. 129.

Kovach, J. and J.P. Tette. 1988. A survey of the use of IPM by New York apple producers. Agr. Ecosyst. Environ. **20**:101–108.

Kramer, N.W. 1987. Grain sorghum production and breeding—historical perspectives to future prospects. pp. 1–9 *in* Proc. 42nd Annu. Corn and Sorghum Industry Research Conf., Publ. No. 42, American Seed Trade Association, Washington, D.C.

Krishnaiah K., N.J. Mohan, and V.G. Prasad. 1981. Efficacy of *Bacillus thuringiensis* Ber. for the control of lepidopterous pests of vegetable crops. Entomon **6**:87–93.

Lewis, J.A. and G.C. Papavizas. 1980. Integrated control of Rhizoctonia fruit rot of cucumber. Phytopathology **70**:85–89.

Liapis, P.S. 1983. Economic evolution of alternative tobacco insect control technologies. ERS Staff Report No. AGES 830329, USDA ERS.

Lloyd, J.W. and J.P. McCollum. 1940. Fertilizing onion sets, sweet corn, cabbage and cucumbers in a four-year rotation. Bull. Univ. Ill. Agr. Exp. Sta. **464**:217–235.

Lockeretz, W., G. Shearer, and D.H. Kohl. 1981. Organic farming in the corn belt. Science **211**:540–547.

Love, J. and L.W. Tauer. 1987. Crop biotechnology research: the case of viruses. Agr. Econ. Res. 87-15, Dept. of Agr. Econ., Cornell University.

Madden, L., S.P. Pennypacker, and A.A. MacNab. 1978. FAST, a forecast system for *Alternaria solani* on tomato. Phytopathology **68**:1354–1358.

Mahrt, G.G., R.L. Stoltz, C.C. Blickenstaff, and T.O. Holtzer. 1987. Comparisons between blacklight and pheromone traps for monitoring the western bean cutworm (Lepidoptera: Noctuidae) in south central Idaho. J. Econ. Entomol. **80**:242–247.

Maiteki, G.A. and R.J. Lamb. 1985. Spray timing and economic threshold for the pea aphid, *Acyrthosiphon pisum* (Homoptera: Aphididae), on field peas in Manitoba. J. Econ. Entomol. **78**:1449–1454.

Marlatt, C.L. 1904. The annual loss occasioned by destructive insects in the United States. pp. 461–474 *in* Yearbook of the Department of Agriculture. U.S. Government Printing Office, Washington, D.C.

Matthews, G.A. 1985. Application from the ground. pp. 93–117 *in* Pesticide Application: Principles and Practice. P.T. Haskell, ed. Clarendon Press, Oxford.

Mayer, D.F., J.D. Lunden, and L. Rathbone. 1987. Evaluation of insecticides for *Thrips tabaci* and effects of thrips on bulb onions. J. Econ. Entomol. **80**:930–932.

Mazariegos, F. 1985. The use of pesticides in the cultivation of cotton in Central America. UNEP Industry and Environment. July/August/September. pp. 5–8.

McArthur, W.C., R.D. Krenz, and G.D. Garst. 1985. U.S. peanut production practices and costs. Staff Report 850108, USDA, Econ. Res. Serv. 50 pp.

McLeod, D. 1986. Economic effectiveness: an alternate approach to insecticide evaluation in sweet corn. J. Agr. Entomol. **3**:272–279.

McPherson, R.M. 1983. Soybean insect management guidelines, number 8—trap crops. Va. Polytech., Ext. Div. Publ. 444-048. 3 pp.

Metcalf, C.L., W.P. Flint, and R.L. Metcalf. 1962. Destructive and Useful Insects. 4th ed. McGraw-Hill, New York. 1087 pp.

Meyer, J.R. 1986a. What's bugging your peaches? pp. 31–38 *in* Proc. National Peach Council.

Meyer, R.J. 1986b. Weed control systems in potatoes. Proc. Calif. Weed Control Conf. **38**:98–101.

Mogensen, B.B. 1989. Personal communication. National Environmental Research Institute, Copenhagen, Denmark.

Moldenhauer, W.C. and N.W. Hudson, eds. 1988. Conservation Farming on Steep Lands. Soil and Water Conservation Society, Ankeny, Iowa.

Moody, K. 1991. Weed management in rice. *In* Handbook on Pest Management in Agriculture. D. Pimentel, ed. 2nd ed. CRC Press, Boca Raton, Fla., pp. 301–328.

Mukishi, P. and P. Trutman. 1988. Can diseases be effectively controlled in traditional varietal mixtures using resistant varieties? Annu. Rep. Bean Improv. Coop. **31**:104–105.

Mullen, M.A. and K.A. Sorensen. 1984. Proceedings Sweet Potato Weevil Workshop. Jan. 24, New Orleans, La. 73 pp.

Mulrooney, R.P. 1986. Soybean disease loss estimate for southern United States in 1984. Plant Dis. **70(9)**:893.

NAS. 1975. Contemporary Pest Control Practices and Prospects. Vol. I. Pest control: an assessment of present and alternative technologies. National Academy of Sciences, Washington, D.C. 506 pp.

NAS. 1989. Alternative Agriculture. National Academy of Sciences, Washington, D.C.

NBA. 1988. Action programme to reduce the risks to health and the environment in the use of pesticides in the agriculture. General Crop Production Division, The National Board of Agriculture, Stockholm, Sweden.

Nissen, S.J. and M.E. Juhnke. 1984. Integrated crop management for small grain production in Montana. Plant Dis. **68(9)**:748–752.

Oatman, E.R., I.M. Hall, K.Y. Arakawa, G.R. Plantner, L.A. Bascom, and L.L. Beagle. 1970. The corn earworm on sweet corn in southern California with a nuclear polyhedrosis virus and *Bacillus thuringiensis*. J. Econ. Entomol. **63**:415–421.

Oka, I.N. and D. Pimentel. 1976. Herbicide (2,4-D) increases insect and pathogen pests on corn. Science **193**:239–240.

OTA. 1979. Pest Management Strategies. Vol. II. Working papers. Office of Technology Assessment, Washington, D.C. 169 pp.

Palladino, P.S.A. 1989. Entomology and ecology: the ecology of entomology. The "insecticide crisis" and the entomological research in the United States in the 1960s and 1970s: Political, institutional, and conceptual dimensions. Ph.D. Thesis, University of Minnesota. 316 pp.

Parker, R. 1981. Weed control in field bean. Coop. Ext. Serv. Rep. No. EB0765. Washington State University, Pullman, Wash.

Parks, J.R. 1982. Pesticide use on fall potatoes in the Northeast region, 1979. Natural Resource Economics Division. Econ. Res. Serv., USDA, Washington, D.C. 20 pp.

Paul, J. 1989. Getting tricky with rootworms. Agrichemical Age **33(3)**:6–7.

Pearson, R.C. 1986. Fungicides for disease control in grapes—advances in development. pp. 145–155 *in* Fungicide Chemistry Advances and Practical Applications. M.B. Green and D.A. Spilker, eds. American Chemical Society, Washington, D.C.

Perry, C.E. and F.B. Saunders. 1982. Cost and returns for selected crop enterprises at the Southeast Georgia Branch Station, 1978–80 with comparisons for the 18-year period, 1963–80. Res. Rep. 397. Ga. Coll. Agr. Exp. Sta. 15 pp.

Pimentel, D. 1961. Species diversity and insect population outbreaks. Ann. Entomol. Soc. Am. **54**:76–86.

Pimentel, D. 1971. Ecological Effects of Pesticides on Non-target Species. U.S. Govt. Print. Off., Washington, D.C. 220 pp.

Pimentel, D. 1976. World food crisis: energy and pests. Bull. Entomol. Soc. Am. **22**:20–26.

Pimentel, D. 1986. Agroecology and economics. pp. 299–319 *in* Ecological Theory and Integrated Pest Management Practice. M. Kogan, ed. John Wiley and Sons, New York.

Pimentel, D. and C.A. Shoemaker. 1974. An economic and land use model for reducing insecticides on cotton and corn. Environ. Entomol. **3**:10–20.

Pimentel, D. and L. Levitan. 1986. Pesticides: amounts applied and amounts reaching pests. BioScience **36**:86–91.

Pimentel, D. and D. Wen. 1990. Technological changes in energy use in U.S. agricultural production. Chapter *in* Agroecology. C.R. Carroll, J.H. Vandermeer, and P.M. Rosset, eds. McGraw Hill, New York. pp. 147–164.

Pimentel, D., J. Krummel, D. Gallahan, J. Hough, A. Merrill, I. Schreiner, P. Vittum, F. Koziol, E. Back, D. Yen, and S. Fiance. 1978. Benefits and costs of pesticide use in United States food production. BioScience **28**:772, 778–784.

Pimentel, D., E.C. Terhune, W. Dritschilo, D. Gallahan, N. Kinner, D. Nafus, R. Peterson, N. Zareh, J. Misiti, and O. Haber-Schaim. 1977a. Pesticides, insects in foods, and cosmetic standards. BioScience **27**:178–185.

Pimentel, D., C. Shoemaker, E.L. LaDue, R.B. Rovinsky, and N.P. Russell. 1977b. Alternatives for reducing insecticides on cotton and corn; economic and environmental impact. Report on Grant No. R802518-02, EPA, Washington, D.C. 147 pp.

Pimentel, D., D. Andow, R. Dyson-Hudson, D. Gallahan, S. Jacobson, M. Irish, S. Kroop, A. Moss, I. Schreiner, M. Shepard, T. Thompson, and B. Vinzant. 1980a. Environmental and social costs of pesticides: a preliminary assessment. Oikos **34**:127–140.

Pimentel, D., C. Glenister, S. Fast, and D. Gallahan. 1984. Environmental risks of biological pest controls. Oikos **42**:283–290.

Poe, C.A. 1988. Where cleanliness means profits. Time. September 5, p. 51.

Pool, R.M. 1989. Personal communication. Dept. of Horticultural Sciences, NYS Agricultural Experiment Station, Geneva, N.Y.

Porter, D.M., D.H. Smith, and R. Rodriguez-Kabana. 1982. Peanut plant diseases. pp. 326–410 *in* Peanut Science and Technology. H.E. Pattee and C.T. Young, eds. American Peanut Research and Education Society, Yoakum, Tex.

Prokopy, R.J. 1988. Beyond the first stage of apple IPM in Massachusetts. Proc. Annu. Mtg. Mass. Fruit Growers Assoc. **94**:78–81.

Prokopy, R.J., R.G. Hislop, R.G. Adams, K.I. Hauschild, E.O. Owens, C.A. Ackes, and A.W. Ross. 1978. Towards integrated management of apple insects and mites. Proc. Annu. Mtg. Mass. Fruit Growers Assoc. **84**:38–44.

Prokopy, R.J., W.M. Coli, R.G. Hislop, and K.I. Hauschild. 1980. Integrated management of insect and mite pests in commercial apple orchards in Massachusetts. J. Econ. Entomol. **73**:529–535.

PSAC. 1965. Restoring the Quality of Our Environment. Report of the Environmental Pollution Panel, President's Science Advisory Committee, The White House, Washington, D.C. November 1965.

Radcliffe, E.B., K.L. Flanders, D.W. Ragsdale, and D.M. Noetzel. 1991. Potato insects—pest management systems for potato insects. *In* Handbook on Pest Management in Agriculture. D. Pimentel, ed. 2nd ed., CRC Press, Boca Raton, Fla. pp. 587–622.

Ridgway, R. 1980. Assessing agricultural crop losses caused by insects. pp. 229–233 *in* Crop Loss Assessment: Proc. of E.C. Stakman Commemorative Symp. Univ. of Minnesota, St. Paul.

Ritcey, G. and F.L. McEwen. 1984. Control of the onion maggot with furrow treatments. J. Econ. Entomol. **77**:1580–1584.

Roberts, D.A. and C.W. Boothroyd. 1972. Fundamentals of Plant Pathology. W.H. Freeman, San Francisco. 402 pp.

Roberts, D.W., R.A. Lebrun, and M. Semel. 1981. Control of the Colorado potato beetle with fungi. pp. 119–137 *in* Advances in Potato Pest Management. J.H. Lashomb and R. Casagrande, eds. Hutchinson Ross, Stroudsburg, Pa. 288 pp.

Roush, R.T. and J.A. McKenzie. 1987. Ecological genetics of insecticide and acaracide resistance. Annu. Rev. Entomol. **32**:361–80.

Royle, D.J. and M.W. Shaw. 1988. The costs and benefits of disease forecasting in farming practice. pp. 231–246 *in* Control of Plant Diseases: Costs and Benefits. B.C. Clifford and E. Lester, eds. Blackwell Scientific, Palo Alto, Calif.

Ruesink, W.G., C.A. Shoemaker, A.P. Gutierrez, and G.W. Fick. 1980. The systems approach to research and decision making for alfalfa pest control. pp. 217–247 *in* New Technology of Pest Control. C.B. Huffaker, ed. John Wiley and Sons, New York.

Russnogle, J. and D. Smith. 1988. More dead weeds for your dollar. Farm J. **112(2)**:9–11.

Sachs, C., D. Blair, and C. Richter. 1987. Consumer pesticide concerns: a 1965 and 1984 comparison. J. Consum. Aff. **21**:96–107.

Schalk, J.M. and A. Jones. 1985. Major insect pests. *In* Sweet Potato Products: A Natural Resource for the Tropics. J.C. Boww Kamp, ed. CRC Press, Inc., Boca Raton, Fla., pp. 59–78.

Schalk, J.M. and R.H. Radcliffe. 1977. Evaluation of the United States Department of Agriculture program on alternative methods of insect control: Host plant resistance to insects. FAO Plant Prot. Bull. **25**:9–14.

Schalk, J.M., C.S. Creighton, R.L. Fery, W.R. Sitterly, B.W. Davis, T.L. McFadden, and A. Day. 1979. Reflective film mulches influence insect control and yield in vegetables. J. Am. Soc. Hort. Sci. **104(6)**:759–762.

Schwartz, H.F., D.H. Casciano, J.A. Asenga, and D.R. Wood. 1987. Field measurement of white mold effects upon dry beans with genetic resistance or upright plant architecture. Crop Sci. **27**:699–702.

Schweizer, E.E. 1989. Weed free fields not key to highest profits. Agricultural Research, USDA, May 1989. 14–15.

Seiber, J.N. 1988. California's initiatives in support of improved application efficiency. pp. 59–66 *in* Improving On-target Placement of Pesticides. Agr. Res. Inst., 9650 Rockville Pike, Bethesda, Md. 220 pp.

Shaunak, R.K., R.D. Lacewell, and J. Norman. 1982. Economic implications of alternative cotton production strategies in the lower Rio Grande Valley of Texas, 1923–1978. Tex. Agr. Exp. Sta. B-1420. 25 pp.

Shields, E.J., J.R. Hygnstrom, D. Curwen, W.R. Stevenson, J.A. Wyman, and L.K. Binning. 1984. Pest management for potatoes in Wisconsin—a pilot program. Am. Pot. J. **61**:508–516.

Shoemaker, P.B. and J.W. Lorbeer. 1977. Timing initial fungicide application to control botrytis leaf blight epidemics on onion. Phytopathology **67**:412–413.

Shwu, E. and H. Webb. 1981. Staff Report. Preliminary data. Pesticide use on selected deciduous fruits in U.S. 1978. USDA Staff Report No. AGE55810626.

Sieczka, J.B. 1984. Results of cultivation herbicide experiment in potatoes. *In* 1983 Results of Weed Control Experiments on Long Island. J.B. Sieczka, J.F. Creighton, D.D. Moyer, W.J. Sanok, and M. Soto, eds. Dept. of Veg. Crops, L.I. Hort. Res. Lab., Riverhead, N.Y. 13 pp.

Sisler, D.G. 1988. Personal communication. Dept. of Agricultural Economics, Cornell University, Ithaca, N.Y.

Sitterly, W.R. 1969. Effect of rotation on cucumber gummy stem blight. Plant Dis. Rep. **53**:417–449.

Smith, J.W. and C.S. Barfield. 1982. Management of preharvest insects. pp. 250–325 *in* Peanut Science and Technology. H.E. Pattee and C.T. Young, eds. American Peanut Research and Education Society, Yoakum, Tex. 825 pp.

Stavely, J.R. 1988. Bean rust resistance in the United States in 1987. pp. 130–131 *in* Annual Report of the Bean Improvement Cooperative. H.F. Schwartz, ed. Vol. 31. Fort Collins, Colo.

Stemeroff, M. and J.A. George. 1983. The benefits and costs of controlling destructive insects on onions, apples, and potatoes in Canada 1960–1980: Summary. Bull. Entomol. Soc. Can. **15**:91–97.

Stinchcombe, G.R. and K.G. Stott. 1983. A comparison of herbicide-controlled orchard ground cover management systems on the vigour and yield of apples. J. Hort. Sci. **58**:477–489.

Stockwin, W. 1988. Sweeping away pests with BugVac. Am. Veg. Grow. **36(11)**:34–38.

Straub, R.W. 1988. Suppression of cabbage root maggot (Diptera: Anthomyiidae) damage to cruciferous transplants by incorporation of granular insecticide into potting soil. J. Econ. Entomol. **81**:578–581.

Straub, R.W. and J.C. Heath. 1983. Patterns of pesticide use on New York State produced sweet corn. New York Food Life Sci. Bull. **102**:1–6.

Street, R.S. 1989. The bug sucker. Agrichemical Age **33(3)**:38–39.

Süddeutsche Zeitung. 1989. Hollands Bauern sollen Milliarden zahlen.

Suguiyama, L.F. and G.A. Carlson. 1985. Fruit crop pests: growers report the severity and intensity. USDA Econ. Res. Serv., Agr. Info. Bull. No. 488. 27 pp.

Summers, M. and C.Y. Kawanishi. 1978. Viral pesticides: present knowledge and potential effects on public and environmental health, symposium proceedings. Health Effects Research Laboratory, Office of Health and Ecological Effects, U.S. EPA, Research Triangle Park, N.C. 311 pp.

Sumner, D.R. and S.C. Phatak. 1987. Control of foliar diseases of cucumber with resistant cultivars and fungicides. Appl. Agr. Res. **2**:324–329.

Swenson, A.L. and R.G. Johnson. 1983. Sugarbeet production in the Red River Valley and Southern Minnesota. Dept. Agr. Econ., Agr. Exp. Sta., N. Dakota State Univ. Report No. 68.

Szmedra, P.I. 1991. Pesticide use in agriculture. Chapter *in* Handbook of Pest Management in Agriculture, 2nd ed. D. Pimentel, ed. CRC Press, Boca Raton, Fla., pp. 649–678.

Taylor, F., G.S.V. Raghaven, S.C. Negi, E. McKyes, B. Vigier, and A.K. Watson. 1984. Corn grown in a Ste. Rosalie clay under zero and traditional tillage. Can. Agr. Eng. **26(2)**:91–95.

Teasdale, J.R. and D. Colacicco. 1985. Weed control systems for fresh market tomato production on small farms. J. Am. Soc. Hort. Sci. **110**:533–537.

Teetes, G.L. 1982. Sorghum insect pest management—I. *In* Proc. Internat. Symp. Sorghum, 2–7 Nov. 81, Patancheru, A.P. India. ICRISAT.

Teetes, G.L. 1985. Insect resistant sorghums in pest management. Insect Sci. Applic. **6(3)**:433–451.

Teetes, G.L. and J.W. Johnson. 1978. Insect resistance in sorghum. pp. 167–189 *in* Proc. Annu. Corn Sorghum Res. Conf., Chicago, Ill.

Teetes, G.L., M.I. Becarra, and G.C. Peterson. 1986. Sorghum midge management with resistant sorghum and insecticide. J. Econ. Entomol. **79(4)**:1091–1095.

Teng, P.S. and H.L. Bissonnette. 1985. Potato yield losses due to early blight in Minnesota fields, 1981 and 1982. Am. Pot. J. **62**:619–628.

Tette, J.P. and C. Koplinka-Loehr. 1989. New York State Integrated Pest Management Program: 1988 Annual Report. IPM House, N.Y. State Agr. Exp. Sta., Geneva, N.Y. 66 pp.

Tette, J.P., J. Kovach, M. Schwarz, and D. Bruno. 1987. IPM in New York apple orchards—development, demonstration and adoption. New York's Food and Life Sci. Bull. No. 119.

Tew, B.V., M.E. Wetzstein, J.E. Epperson, and J.D. Robertson. 1982. Economics of selected integrated pest management production systems in Georgia. Res. Rep. 395, Univ. Ga. Coll. Agr. Exp. Sta., Athens, Ga. 12 pp.

Thompson, D.C. and S.F. Jenkins. 1985. Influence of cultivar resistance, initial disease, environment, and fungicide concentration and timing on anthracnose development and yield loss in pickling cucumbers. Phytopathology **75**:1422–1427.

Thompson, R. 1985. Personal communication. Boone, Iowa.

Tolman, J.H., D.G.R. McLeod, and C.R. Harris. 1986. Yield losses in potatoes, onions and rutabagas in southwestern Ontario, Canada—the case for pest control. Crop. Prot. **5**:227–237.

Tucker, D.P.H., R. Muraro, and B. Abbitt. 1980. Two weed control systems for Florida citrus. Proc. Fla. State Hort. Soc. **93**:30–33.

UC. 1983. Integrated Pest Management for Rice. Div. Agr. Sci. Publ. 3280, University of California, Berkeley. 94 pp.

USBC. 1971. Statistical Abstract of the United States. 1970. 92nd Edition. U.S. Dept. of Commerce, U.S. Bureau of the Census, Washington, D.C.

USBC. 1988. Statistical Abstract of the United States. 1987. 108th Edition. U.S. Dept. of Commerce, U.S. Bureau of the Census, Washington, D.C.

USDA. 1936. Agricultural Statistics 1936. U.S. Department of Agriculture, U.S. Government Printing Office, Washington, D.C.

USDA. 1954. Losses in Agriculture. Agr. Res. Serv. 20-1. 190 pp.

USDA. 1961. Agricultural Statistics 1961. U.S. Government Printing Office, Washington, D.C.

USDA. 1965. Losses in Agriculture. Agr. Handbook No. 291. Agr. Res. Serv., U.S. Government Printing Office, Washington D.C.

USDA. 1973. Rice in the United States: varieties and production. Agriculture Handbook No. 289. Agr. Res. Serv., U.S. Government Printing Office, Washington D.C.

USDA. 1975. Farmers' use of pesticides in 1971 . . . extent of crop use. Econ. Res. Serv., Agr. Econ. Rep. No. 268. 25 pp.

USDA. 1977. Weed control in U.S. rice production. Agriculture Handbook No. 497. Agr. Res. Serv., U.S. Government Printing Office, Washington D.C.

USDA. 1982a. Guidelines for the control of insect and mite pests of foods, fibers, feeds, ornamentals, livestock and households. pp. 329–343 *in* Agr. Res. Serv. Agr. Handbook No. 584.

USDA. 1982b. Farm pesticide supply-demand trends. Econ. Res. Serv. Bull. No. 485. 23 pp.

USDA. 1983. Agricultural Handbook 589. Agr. Res. Serv. Chap. 20. Page 534.

USDA. 1985. Agricultural Prices 1984 Summary. Crop Rep. Board, Stat. Res. Serv., U.S. Dept. of Agr., Washington, D.C.

USDA. 1986. Agricultural Statistics 1986. United States Government Printing Office. [1985 data used.]

USDA. 1987. ERS (1987) Inputs situation and outlook report. Agricultural Resources No. 5 (AR-5). Econ. Res. Serv.

USDA. 1988. Agricultural Resources. Inputs situation and outlook report. Econ. Res. Serv, AR-9.

USOMB. 1989. Budget of the United States Government. Office of Management and Budget. U.S. Government Printing Office, Washington D.C.

Van den Bosch, R. and P.S. Messenger. 1973. Biological Control. Intext Educational Publishers, New York.

Van der Scheer, H.A. Th. 1984. Testing of crop protection chemicals in fruit growing. pp. 70–77 *in* Annu. Rep., Research Station for Fruit Growing, Wilhelminadorp, The Netherlands.

Van Doren, D.M., G.B. Triplett, Jr., and J.E. Henry. 1977. Influence of long-term tillage and crop rotation combinations on crop yields and selected soil parameters for an Aeric Ochraqualf soil [Maize]. Res. Bull. Ohio Agr. Res. Dev. Cent. 1091.

Walker, R.H. and G.A. Buchanan. 1982. Crop manipulation in integrated weed management systems. Weed Sci. Suppl. 1 **30**:17–24.

Ware, G.W. 1983. Reducing pesticide application drift-losses. Coop. Ext. Serv., Coll. Agr., Univ. Arizona, Tucson. 27 pp.

Ware, G.W., W.P. Cahill, P.D. Gerhardt, and J.M. Witt. 1970. Pesticide drift. IV. On target deposits from aerial application of insecticides. J. Econ. Entomol. **63**:1982–1983.

Warner, M.E. 1985. Enterprise budgets for potatoes, wheat, cauliflower, peaches, and table grapes on Long Island, New York: A comparison of costs, returns and labor requirements. A.E. Res. 85-12. Dept. Agr. Econ., Cornell Univ., pp. 50–73.

Wax, L.M. and J.W. Pendleton. 1968. Effect of row spacing on weed control in soybeans. Weed Sci. **16**:462–464.

Weakley, C.V. 1986. Making cost comparisons between weed management systems. Proc. Calif. Weed Conf. **38**:174–176.

Whitaker, D.B. and G.B. White. 1982. Economic profiles for apple orchards and vineyards, 1981 and five year average, 1977–81. A.E. Res. 82-48, Dept. Agr. Econ., Cornell Univ. Agr. Exp. Sta.

Wien, H.C. and P.L. Minotti. 1988. Response of fresh market tomatoes to nitrogen fertilizer and plastic mulch in short growing season. J. Am. Soc. Hort. Sci. **113**:61–65.

Wilcox, J.R., ed. 1987. Soybeans: Improvement, Production and Uses. American Society of Agronomy, Madison, Wisc. 888 pp.

Wilcut, J.W., G.R. Wehtje, and M.G. Patterson. 1987. Economic assessment of weed control systems for peanuts (*Arachis hypogaea*). Weed Sci. **35**:433–437.

Williams, J.R., O.S. Johnson, and R. Guin. 1987. Tillage systems for wheat and sorghum: an economic and risk analysis. J. Soil Water Conserv. **43**:120–123.

Wilson, H.P., M.P. Mascianica, T.E. Hines, and R.F. Walden. 1986. Influence of tillage and herbicides on weed control in a wheat (*Triticum aestivum*)—soybean (*Glycine max*) rotation. Weed Sci. **34**:590–594.

Wilson, M.C., F.C. Chen, and M.C. Shaw. 1984. Susceptibility of the alfalfa weevil to a *Bacillus thuringiensis* exotoxin. J. Ga. Entomol. Soc. **19**:366–371.

Winter, S.R. and A.F. Wiese. 1982. Economical control of weeds in sugarbeets (*Beta vulgaris*). Weed Sci. **30**:620–623.

Wolfe, D.W. and E. Rutkowski. 1987. Use of plastic mulch and row covers for early season vegetable production. Veg. Crops Rep. No. 355. Dept. Veg. Crops, Cornell Univ., Ithaca, N.Y.

Wright, R.J. 1984. Evaluation of crop rotation for control of Colorado potato beetles in commercial potato fields on Long Island. J. Econ. Entomol. **77**:1254–1259.

Wright, R.J., D.P. Kain, R. Loria, J.B. Sieczka, and D.D. Mayer. 1986. Final report of the 1985 Long Island potato integrated pest management pilot program. Cornell Univ. L.I. Hort. Res. Lab, Riverhead, N.Y. 22 pp.

Zehnder, G.W. and J.J. Linduska. 1987. Influence of conservation tillage practices on population of Colorado potato beetle (Coleoptera: Chrysomelidae) in rotated and non-rotated tomato field. Environ. Entomol. **16**:135–139.

Part III

Government Policy and Pesticide Use

11

Government Policies That Encourage Pesticide Use in the United States

Kenneth A. Dahlberg

Introduction

While primarily associated in most people's minds with agriculture, pesticides, insecticides, herbicides, rodenticides, fumigants, and other synthetic chemicals designed to kill undersirable insects, plants, and animals are used widely in society.[1] Among the many nonagricultural uses are those in mosquito abatement programs, forestry, road and railroad rights-of-way management, urban pest management, lawn care, and household pest control.

This chapter seeks to place these pesticide uses in a larger value, social, and conceptual framework. After reviewing the underlying cultural values and institutions which facilitated the development and use of pesticides, emerging challenges to their continued use are discussed—first generally and then in terms of the critiques made of the conventional production paradigm by those seeking a more sustainable agriculture. The major policies, interest, and institutions involved at the state, national, and international level are reviewed—with current controversies placed in historical context. By taking a systems perspective based upon goals of maintaining the health, sustainability, and regenerative capacities of the larger food system, the need for major reforms becomes much clearer.

The Larger Value and Institutional Context

The widespread use of pesticides throughout industrial society reflects certain Western cultural values as well as a number of specific industrial values and assumptions. Some of the deepest Judeo-Christian beliefs and values relate to humankind's separation from and dominance over nature and the hierarchical and patriarchal nature of society and the family. With

[1]For stylistic simplicity, the term "pesticides" will be used throughout as shorthand for the various synthetic chemicals used in these different ways.

the Renaissance strong beliefs in reason, science, and technology as the path toward social progress emerged. During the Agricultural Revolution and then the Industrial Revolution which it made possible (Grigg, 1984), new concepts such as individualism, nationalism, and representative democracy emerged and became the basis of powerful social movements. The rapid growth and expansion of capitalist economies flowed from and also strengthened these social forces. In pursuing its vision of the future, industrial society has from the beginning stressed industrial, urban, formal, and technological systems while it has correspondingly neglected agricultural, rural, informal, and natural systems (Dahlberg, 1990c).

The discovery and exploration of the New World in many ways complemented the growth of these values, beliefs, and institutions—both through geographic expansion and by relieving population and political pressures in Europe. This expansion reflected beliefs that Western culture and technologies were superior to indigenous ones and should supplant them. These beliefs also provided a rationale for many types of exploitation, ranging from outright plunder to a great strategic search for exotic plants and animals which might be exported or transplanted to the Old World. There was also a tendency to export Old World agricultural practices, tools, animals, and crops to the New World, at least in the "white settler" colonies (Crosby, 1986). More generally, non-European habitats, social and agricultural practices, as well as institutions were all seen as subject to restructuring or replacement along "modern" lines. Indeed, as machines became increasingly important as the Industrial Revolution progressed, they became the measure of men (Adas, 1989).

The development of modern chemistry grew out of, and was compatible with the mechanical view of the world embodied in 18th- and 19th-century scientific theories and the functional specialization of industry. The development of pesticides clearly fits in with and reflects these broad values and assumptions. In addition, the extensive use of DDT during and immediately after World War II tended to shape the larger context within which pesticides have been used since. The language, logic, and actual conduct of a "war campaign" against malaria (Farvar and Milton, 1972) provided imagery of "enemies" who are "attacked" and "defeated" through the use of the "miraculous" technological fixes (or "silver bullets") provided by modern science. The shift of DDT to agricultural uses carried with it much of this imagery, albeit with a different rationale. While the malaria programs were largely preventative in approach and aimed at improving public health (something later complemented on the curative side with the equally "miraculous" powers of penicillin), the "wars" against "attacking" agricultural pests were economic and were designed to improve productivity and profitability by reducing crop losses.

The institutional patterns accompanying the industrialization of agriculture through mechanical cum chemical approaches are very complex. For our purposes here, it is sufficient to note that, in addition to the broad cultural and social values discussed above, they are based on (1) functional specialization (whether in agri-industry, farming, food processing, agricultural research, or education) and (2) the so-called "production paradigm" which measures "success" in terms of a narrowly defined set of economic and productivity criteria.[2]

These values and patterns are now combined with (and reinforce) an increasingly concentrated structure of farmland ownership (OTA, 1986) and increasing oligopoly among farm input suppliers and output purchasers (Constance and Heffernan, 1989). These institutional patterns and structures underlie many of the policies that encourage pesticide use. (See below.) However, in the broad societal terms we are dealing with here, it is the power and rigidity of these institutions which must be noted because they make adaptive change to more sustainable and regenerative systems difficult—even though the need for such change is becoming increasingly apparent.

Emerging Challenges

A number of global trends and uncertainties threaten all sectors of modern industrial society. These include deep uncertainties about the stability and viability of the international economic system, about the price and availability of fossil fuels over the longer term, about the climatic impacts of continued use of fossil fuels, and about the loss of genetic and biological diversity—the very *source* of the regenerative capacity of our various renewable resource systems. (See Dahlberg, 1987.) These trends and uncertainties have increased the calls for fundamental changes in modern industrial society.

---

[2]A paradigm is a widely accepted model within a given field that is based upon shared assumptions about the nature of the problems that field should address and the approaches for dealing with those problems. The term gained widespread use as a result of Thomas Kuhn's use of it in analyzing *The Structure of Scientific Revolutions* (1962). The "production paradigm" in agriculture assumes that the top-priority problem to be addressed is increasing production—rather than what has been seen as the top-priority problem in the eyes of most peasants throughout history: having dependable production each year.

Within the production paradigm, the accepted approaches and criteria are those of scientific experimentation (based on quantifiable measurements) and neoclassical economics. The experiment station model (now being challenged by farming systems and on-farm approaches) stresses replicable experiments dealing with only a few variables at a time. The neoclassical economic model fails to address many important "externalities," such as the health, environmental, and social consequences of production systems. Also, it typically employs short time-horizons and when it does seek to address the future (through "discount rates"), it uses very dubious assumptions. (See Howarth and Norgaard, 1990.)

In terms of agriculture, these calls have included those seeking more sustainable and regenerative approaches. I prefer the latter term because it more clearly points to the basic reproductive and generational questions that are crucial to the health of individuals, populations, and societies. Also, the term is less subject to being co-opted by those who would define "sustainability" in terms of maintaining current systems and privileges (Dahlberg, 1991).

Regenerative (or sustainable) systems also need to be evaluated with a broader set of criteria than the narrow economic ones associated with the "production paradigm." Besides broadening economic criteria to include all relevant "externalities" (Dahlberg, 1986), three other "e's" need to be included as criteria for evaluating the sustainability of systems. (See Douglass, 1984.) Regenerative (or sustainable) food and agricultural systems must be (1) *ecologically* sound and have the capacity to provide long-term food sufficiency; (2) *equitable* in terms of providing social justice; and (3) *ethical* in terms of respecting both future generations and other species—each of which requires a respect for, and preservation of, its present diversity, whether social, cultural, genetic, and/or biological.

The above conceptualization involves another significant expansion: from the narrow sectoral and production focus of *agriculture* (whether conventional or alternative) to a comprehensive systems approach. One aspect of this means that in addition to agriculture, forestry, and fisheries need to be included. More importantly, such a *food and fiber systems* approach needs to include all phases from production, processing, distribution, and use (preparation, cooking, preservation, and storage) to recycling, composting, and disposal. Finally, these food-system phases exist and need to be analyzed at different levels—from the household level on up to the global. (See Dahlberg, 1992, for a detailed discussion.)

Such an approach necessarily involves not only systems analysis, but what ecologists call "hierarchy" theories or what I prefer to call "contextual analysis," a term that carries fewer historical connotations of the superiority of higher-level systems (and thus a rationale for their dominance). (See Dahlberg, 1990b.) In terms of the policy focus of this chapter, this means that while the emphasis will be on national-level systems, there will be some discussion of the relevance and interactions of policies at the next lower (state) and higher (international) levels. It will be noted that as we now shift to short time horizons and away from the above broad-gauge cultural/historical overview, the language and analysis will also shift and become closer to that used in policy and regulatory studies.

State and Local Policies That Encourage Pesticide Use

Most state and local policies relating to the use of pesticides in commercial agricultural and forestry production follow federal legislation closely.

Nonproduction uses of pesticides are often the provenance of individual states and localities. These nonproduction uses are often significant, if not well documented. Different types of figures suggest their importance. Urban lawns, gardens, trees, and forests represent considerable acreages and often high levels of pesticide use.

For example, there are 31 million acres of lawn in the United States (The Lawn Institute, 1991). The levels of pesticide use in these nonproduction areas are often higher on a per-acre basis than those in commercial agricultural production. In terms of active ingredients (in thousands of pounds), a 1979 study (NRC, 1980a:109) gave the following breakdown: Farm uses—840,500 (73%); Commercial/Government uses—222,500 (19%); and Home and Garden uses—87,000 (8%). These figures were gathered prior to the period of rapid growth of the urban lawn-care industry. They also include household uses of pesticides (to control mice, ants, cockroaches, termites, etc.), some of which involve nonfood and fiber uses.

There are few figures either on the extent of urban/suburban forests or on the pesticides used in their management. In part this is due to changing urban and suburban boundaries. More significantly, no federal agency nor trade association collects national figures, although data are available for some cities (NAS, 1975). Nor is it clear how extensively pesticides are used on the many rural and urban gardens in the United States, gardens which annually produce some $18 billion worth of produce (NGA, 1989), a figure that rivals the value of the U.S. corn crop, even though these gardens are largely ignored by the USDA.

Another major area where pesticides are used is to manage vegetation on rights-of-way. There are an estimated 60 million acres of vegetation-covered roadside, railroad, and electric, telephone, and gas-line rights-of-way in the United States (NAS, 1975). For many years, blanket sprays of 2,4-D and 2,4,5-T were used to try to prevent tree and large shrub growth in these areas. Different private and public agencies have jurisdiction for the maintenance of these rights-of-way, although their choices and uses of pesticides as a management technique can be limited by local, state, or federal regulations.

Management of urban pests is another area where few detailed figures on pesticide use are available. General categories of urban pests include structural pests; stored-product pests; fabric and paper pests; nuisances; public health pests; and yard, garden, tree, and forest pests. These various urban pests are mentioned because pesticides used to attack them are more likely to generate conflicts between goals of economic well-being, environmental health, and public health. A classic example of these conflicts was seen in the outbreaks of the Medfly in California—where ultimately, concerns over the economic well being of the agricultural sector overrode concerns about environmental and public health. However, it is in the area of public health that the states and localities retain the greatest legislative and policy authority—a point clarified by a recent Supreme Court decision (Wisconsin Public Intervenor v. Mortier) which ruled that the Federal Insecticide, Fungicide, and Rodenticide Act (FIFRA) does not preempt the power of states and localities to regulate more stringently than the federal standards. Many states choose to delegate mosquito control to local governments or mosquito abatement districts, while the control of rodents and insects is often delegated to local health agencies or those in charge of housing (NRC, 1980a). Sanitary codes are sometimes enacted on the state level, but more often at the local level. However, local waste management policies have been increasingly subject to state and/or federal review.

Because rural agricultural and forestry uses of pesticides are usually kept separate (both conceptually and in terms of data) from urban uses, this is where a regenerative food and fiber systems approach (which would include urban food and fiber systems) offers a more integrated analysis of these factors. This approach also involves a basic shift in evaluative models from an economic to a health model that stresses the long-term health and regenerative capacity of both the natural and social systems. Thus, it can bring public health and food systems concerns more closely together—whether in urban or rural settings.

While state and local governments can restrict the use of pesticides for purposes of public health, such restrictions are a limited exception to the more general pattern where most states encourage the use of pesticides in agriculture and forestry in the name of economic growth. This pattern reflects the predominant power in most states of corporate and special interests which encourage pesticide use. This is reinforced by the more general reluctance of states to regulate businesses for environmental reasons. "Right-to-know" legislation, which enables workers to know what chemicals they are working with, has been fought in many states. In many states, labor/management legislation is geared toward industrial workers and makes organization of farm workers difficult. And when it comes to farmers themselves, state legislatures are typically reluctant to regulate them in any way. Add to this the general exemption of farmers and farm workers from both federal and state occupational safety and health programs and one has a setting favorable to widespread use of pesticides in agriculture and forestry.

There are, however, a number of areas where states have passed legislation regulating pesticide use. Some of these policy areas are delegated to the states under federal legislation, while others are related to areas of legal controversy where the state is forced to act. The latter include a range of issues, such as those involving the drift of pesticide applications to adjacent areas, the legal liability of pesticide companies and consultants, physician reporting requirements for observed pesticide exposures, etc.

Political controversies have also emerged over the public health threats of pesticides. California Proposition 65, which was passed by voters in 1986, seeks to limit the use of a list of chemicals that the state has determined can cause cancer or birth defects. All products, including food, which contain even trace elements of these chemicals, must be so labeled. Also, industry is prohibited from releasing these chemicals in any way whereby they might enter drinking water. Some states, such as Iowa, which have high groundwater contamination, have started taxing fertilizer and/or pesticide sales, earmarking the taxes to support alternative agricultural approaches.

Interwoven throughout these various policy arenas are the extension, research, and educational policies of each state's land-grant institutions. Farm organizations, input suppliers, chemical and seed companies, and commodity groups have all sought to influence the direction and content of the programs of these institutions through such techniques as lobbying state legislatures, serving on advisory committees in state departments of agriculture, bringing court cases, and funding land-grant research. As at other levels of government, the basic question becomes—Who defines what is in the "public interest"?

National Policies That Encourage Pesticide Use

As indicated above, the basic institutional framework of agriculture (and other economic sectors) is one of functional specialization. The dominant evaluative criteria is economic productivity. Other important context-setting conditions are (1) long-term trends which have changed urban/rural population and political balances; (2) the long-term increase in the number of large and very small farms, with a decline in the number of middle-sized farms; and (3) the increasing power of other sectors of society vis-à-vis agriculture and of agribusiness interests vis-à-vis farmers. All of these factors have led to what many call the "cheap food policy" of the United States. In fact, it is not a policy, but rather the result—by default—of the lack of a U.S. food policy (Heffernan, 1986).

General Policies and Practices

In addition to the above context-setting conditions, there are a number of general policies and practices (which apply to all economic sectors) which make possible and/or provide important support for the various specific policies which encourage pesticide use. Four areas will be discussed.

The first involves general corporate and labor law. Several aspects are of particular relevance. Antitrust laws, policy and enforcement have offered little more than a minor hindrance to growing concentration and oligopoly throughout the agribusiness, banking, and food-processing industries. This has weakened the relative power of farmers in economic, political, and informational terms and has made it more difficult for Congress to regulate corporate abuses.

The extension of patent and intellectual property rights to living organisms promises to have a profound effect on the structure of agriculture both domestically and internationally. It has already had a significant effect upon research priorities within the land-grant system—whereby basic ecological, taxonomic, and plant breeding and maintenance work has languished while large amounts have been invested in biotechnology research. In addition, many conflicts of interest have emerged, such as university researchers consulting or working part time for private firms. In addition biotechnology research sponsored by large chemical companies tends to dictate the direction of "public" research (Busch et al., 1991; Kloppenburg, 1988).

As indicated in the discussion on state policies, general labor/management policies and practices influence the ability of farm workers to seek protection from pesticide exposures. The exemption of both farmers and farm workers from Occupation Safety and Health Administration (OSHA) coverage has already been noted. A final area relates to general corporate liability and insurance laws. The limited liability corporation was one of the key social innovations facilitating the Industrial Revolution. In the United States, this was complemented by the legal fiction that corporations are "persons" entitled to the same First Amendment rights and protections as individuals—a fiction which, given their corporate power and resources, provides them with a significant advantage over individuals in any legal proceeding. By extension, this has also meant that each chemical or pesticide is "innocent until proven guilty." Long and difficult political battles have been required to shift the burden of proof on the use certain potentially dangerous chemicals (pharmaceuticals and various toxics) from a demonstration by the regulatory agency that the proposed or current use is unsafe to a demonstration by the manufacturer that its use will not be harmful to individual or public health.

General economic policies and practices have a bearing upon pesticide use in various ways. At the broadest level, the dominance of economic measures, indicators, and priorities and their institutionalization over the past several centuries has meant that major social, health, safety, and environmental concerns have had a hard time gaining recognition and inclusion in decision making. Preferences for economic growth are also reflected in the tax code. A host of tax provisions relating to the oil industry (for example, the oil-depletion allowance) have encouraged low fossil-fuel prices which in turn have encouraged the mechanization of farming and the use of fertilizers and pesticides.[3] Other tax provisions—such as rapid depreciation schedules—have reinforced capital-intensive approaches to agriculture. Also, although reduced in the recent tax reform, there are still a number of specific provisions or "loopholes" designed to benefit particular types of producers.

The rules and practices of political parties and the three branches of government, especially the Congress, influence the direction of agricultural policy. The historic overrepresentation of the "farm block" in Congress and on key committees was not broken until the 1960s. And it was not until the last decade that the power of the agriculture and appropriations subcommittees to monopolize agricultural policy was overcome. The already-noted lack of a national food and fiber policy has been a result in large part of the fragmented structure of Congress and the entrenched power of agricultural interests. A range of other policies, such as the very strict rules on lobbying efforts by tax-exempt groups which existed until the 1970s, discouraged many environmental groups from taking an active political role in pesticide and other environmental controversies. Other factors that influence policy in favor of agricultural production (and thus chemical intensive approaches) include the increasing power of political action committees (PACs) and the corresponding decline in the power of political parties. Even so, budget pressures and smaller rural constituencies have led agricultural interests to seek new allies.

The increasing incorporation of agriculture and agricultural policy into U.S. international trade and aid policies has also tended to encourage pesticide use. Because U.S. farm exports have been a major hedge against increasing balance-of-trade deficits, international trade and finance officials have sought to increase them. Efforts begun in the 1950s to find new markets for U.S. grain surpluses led to the "Food for Peace" program (PL 480), which offered these surpluses to needy countries like India at low cost (often hurting local farmers and agricultural markets in the process). PL 480 shipments have declined in recent years, in part due to the national-security abuses during the Vietnam war and more because they now must

[3]In effect these various tax breaks mean that the general taxpayer subsidizes exploration for and production of oil rather than each user paying for these costs in proportion to their use of oil. Prices are thus kept lower than would be the case with a "user-pays" approach. This discourages energy conservation. To create a "level playing field" either these subsidies should be removed or equal subsidies should be provided for energy-conservation efforts. The latter course would mean that taxpayers would be subsidizing both oil production and conservation. Greater energy conservation and more rational market allocation would be achieved by having users pay the full cost of their energy, although important equity questions would remain regarding the ability of the poor to have access to sufficient energy for basic needs. Another emerging issue is how to try to incorporate one of the main "externalities" of fossil-fuel use—the impacts of greenhouse gasses—into the price of these fuels. Some have suggested a "carbon tax."

be paid for with cash appropriations. On the trade front, we have recently seen the latest in a long-running series of battles between the United States and Europe, this time in the General Agreement on Tariffs and Trade (GATT) negotiations, where the United States again sought to open the European Community's markets to U.S. farm products. These negotiations also have significant implications for U.S. pesticide policy. (See the later section, International Policies That Encourage Pesticide Use.)

The Legislative and Regulatory Framework for Pesticides

Overall, then, the above general policies and practices have provided a setting which has encouraged and supported the dominant economic and productivity conceptions of agriculture. Another source of encouragement has been the significant support which the federal government provided to agriculture for over 100 years. This support began with the creation of the land-grand colleges (Morrill Act, 1862) and was followed by the creation of the state experiment station system (Hatch Act, 1887). This system, which later became a major engine for increasing agricultural productivity, grew in large part out of earlier attempts by states to provide consumer protection to farmers in their fertilizer purchases. This was done by establishing state chemists who checked for adulteration (Marcus, 1985). Consumer protection for farmers was also the major source of the Federal Insecticide Act (FIA) of 1910, a kind of "truth-in-packaging" law which required a listing of ingredients and uses which USDA then monitored. Major insecticide producers did not oppose the legislation since it standardized labeling requirements nation-wide and provided a gatekeeping mechanism to keep out "fly-by-night" operators, thereby helping to stabilize markets (Bosso, 1987).

A parallel, but stronger, type of consumer protection emerged at about the same time, but only as a result of a major public outcry. While numerous earlier attempts to pass pure food legislation had failed in a rural- and farm-dominated Congress, Upton Sinclair's expose of the meat packing industry in *The Jungle* plus Teddy Roosevelt's strong political support led to passage of the 1906 Pure Food and Drug Act (PFDA). This regulatory law forbade the manufacture, sale, or transport of poisonous, adulterated, or misbranded foods, drugs, liquors, or medicines. Federal authorities were given the power to set standards and enforce them and to seize products deemed contaminated or a public health hazard. Administrative responsibility for enforcement of both FIA and PFDA was given to USDA—and by the 1920s inevitable conflicts of interest arose between agricultural interests and urban food consumers over the permitted levels of pesticide residues on food. USDA consistently sided with agricultural interests. Several internal reorganizations failed to deal with these problems and after a long and difficult political battle, what had by then become the Food and Drug Administration (FDA) was moved out of USDA in 1940. The FDA, however, was still subject for many more years to the power of the farm-block-dominated House Subcommittee on Agricultural Appropriations (Bosso, 1987).

These debates regarding the protection of the economic interests of agriculture vs. the public health of urban consumers were updated and transformed in the immediate post–World War II period by the appearance of DDT and other "miraculous" pesticides. In 1947, the Federal Insecticide, Fungicide, and Rodenticide Act (FIFRA) was passed with little legislative fanfare. The act updated the 1910 legislation and added a provision for premarket registration and clearance. However, since the legislation allowed industry to "protest" any administrative decision to deny an application, and since a further denial could be taken to court, the burden of proof for banning a product as unsafe still lay with the USDA rather than with the industry to prove its safety.

Concern about the impact of new chemicals upon urban consumers became the highly visible focus of the Delaney Committee hearings in 1950–1951. The main points to emerge were that "(1) [the] food and chemical industries did not have consumer health as their primary orientation, and (2) the FDA had no mechanism for knowing beforehand which chemicals reached the consumer, and with what effects" (Bosso, 1987:75). The Pesticides Control Amendment (PCA) of 1954 gave the FDA a formal role in pesticide regulation by allowing it to set maximum tolerances for spray residues found in or on raw agricultural products. This role was strengthened by the Delaney clause in the 1958 Food Additives Amendment to the PCA. This clause, which prohibited cancer-causing food additives, was the one pesticide-related piece of legislation to pass over the opposition of the agricultural subsystem until the mid-60s. With this one exception, the long dominance of economic and production interests and orientations continued.

With the 1960s, a new consideration—the environment—emerged onto the Congressional scene and has become increasingly intertwined with the earlier concerns over economics and public health. The great controversies caused by the 1962 appearance of Rachel Carson's *Silent Spring* challenged not only the wonders of DDT, but the objectivity of industry-sponsored scientists. In a deeper sense, it challenged the whole pesticide/production paradigm and its links to conventional beliefs in science and technology as the paths to progress.

As the policy process has become more open to environmental and public health concerns, legislative and regulatory measures dealing with pesticides have become much more complex. In the 1970s, the Environmental Protection Agency (EPA) became a new player. The Federal Environmental

Pesticides Control Act of 1972 sought to set up a new and more comprehensive regulatory framework for pesticides. Complex registration and testing procedures were established, often with unrealistic deadlines. The 1978 Amendments to FIFRA satisfied no one. The redirection of EPA and FDA toward reduced regulation under the Reagan Administration led to increased frustrations on the part of public health and environmental advocates.

Efforts to reform FIFRA and the Federal Food, Drug and Cosmetic Act have dragged on until the present.[4] Debate and controversy have revolved around the 1987 recommendations from the Board on Agriculture of the National Academy (NAS, 1987). They recommended (1) the creation of a consistent standard for pesticide residues in both raw and processed foods; (2) consistent regulatory treatment for all types of pesticides as well as doing away with the distinction between "new" and "old" pesticides (the age divide being 1972); (3) dropping the Delaney clause and applying a consistent "negligible-risk" standard (typically identified as risking less than a one-in-a-million chance of death); (4) having EPA focus on the most threatening pesticides used on the most consumed crops; and (5) having EPA develop the data and tools to systematically assess dietary risks from pesticides. Another issue (discussed below) is the "circle of poison," which involves demands for greater regulation or restrictions on the export of pesticides banned in the United States since they often reappear as residues on imported foods.

While constantly evolving, the above legislative and regulatory framework for pesticides also interacts with other more specific production, processing, distribution, food use, recycling, and waste-management policies. These will be examined in turn.

Production Policies

Domestic Farm Programs

A recent overview (Knutson et al., 1986) analyzes 20 domestic farm programs and groups them under three categories: (1) supply control,

---

[4]The recent Supreme Court decision (Wisconson Public Intervenor v. Mortier, No. 89-1905) that FIFRA does not preempt more stringent state or local regulations of pesticides will add a new dimension to the current battles. The chemical industry has indicated that it will seek amendments to FIFRA that would preempt state or local regulation. Environmentalists, farm workers, and public health groups will resist these efforts at the national level. If it cannot amend FIFRA to preempt more stringent state and local regulations, the industry is then likely to adopt the same strategy as the National Rifle Association has done with gun control—that is, to focus its efforts at the state level, seeking to prevent more stringent state standards and trying to have the state preempt any local powers to set more stringent standards. As noted later, the proposed pesticide regime under GATT would represent an even lower-common-denominator approach and would preempt not only stricter state and local regulations but any national regulations stricter than GATT standards.

(2) price supports, and (3) income supports. This patchwork of programs has evolved to meet changing economic, political, and farm conditions. Each program has its own objectives, which often are not achieved and which may conflict with or compromise other programs. Only a couple of examples will be discussed here. Among different approaches to supply control, acreage reductions, set-asides, and diversions have been used extensively, but with only modest success. The reason for this is that farmers tend to take their poorest land out of production while seeking to increase production on the remaining acres available for that crop—often through increased fertilizer and pesticide applications. Long-term land retirement and conservation reserve programs are similar, but operate over longer time periods. The 1985 and 1990 Farm Bills have sought to take highly erosive land out of production voluntarily by giving farmers an annual rental fee plus a portion of the cost of establishing a required cover crop. While this program encourages soil and land conservation, there is the same tendency to farm the remaining land more intensively. Complementing these measures have been "swampbuster" and "sodbuster" provisions which are designed to prevent wetlands from being drained and new marginal lands from being brought into cultivation.

Price- and income-support programs have been identified as one of the major barriers to moving toward more diversified farming systems since they encourage crop specialization at both the farm and regional levels (GAO, 1990). Since only certain crops are supported (and pasture was excluded from a farmer's crop base until the 1990 Farm Bill), farmers have tended to grow only two or three supported crops and have avoided pasture-based rotations which significantly reduce fertilization and pesticide requirements.

Direct farmer supports and subsidies are also reinforced by a variety of marketing and credit programs. Marketing programs include such things as demand expansion and food assistance (commodity distributions programs; school lunch programs; food stamps; and the women's, infants, and children [WIC] program), market organization and control (cooperatives, marketing boards, and marketing orders), and market facilitators (market news price reporting, crop and livestock production estimates and reports, and grading and standards) (Knutson, et al., 1986). While their main goal is to maintain or increase farmer income through market expansion, control of supply, and/or price stabilization, several of these programs have some impact on the use of pesticides. This is primarily because quality grades or standards are used to sort out "surpluses," placing a premium on avoiding pest damage which would affect either the size or appearance of the product.

Credit programs include a wide range of debt restructuring, loan, and guarantee programs. While these programs would appear to be neutral in terms of their influence upon pesticide use, many are still linked to the narrow range of supported crops and the specialization and intensification that result therefrom. Also, it is only with the beginning of national organic certification (contained in the 1990 Farm Bill) that marketing programs may come to apply to organic produce.

The types of distortion caused by these domestic programs are magnified by a number of international trade programs designed either to increase crop exports or to protect domestic producers. Knutson et al. (1986) have identified 20 such programs, broken down as follows: (1) trade barrier reduction (U.S. participation in GATT and the Generalized Systems of Preferences [GSP] which gives certain developing countries duty-free entry for certain products); (2) export subsidies; (3) domestic industry protection; (4) embargoes; and (5) long-term bilateral and international commodity agreements. Often, these programs encourage greater exports of domestically supported crops, thereby intensifying production and further reinforcing crop specialization. They also are clearly linked to the economic interest of the great grain-trading companies in expanding their shipments of grain overseas.

Direct Sectoral Supports

Beyond these specific government programs there are a range of traditional sectoral supports for agriculture. These include the various supports provided by the land-grant system: agricultural education, research and development, and extension. USDA also conducts extensive research. Again, while theoretically neutral regarding pesticide use, these institutional support systems became strong advocates of pesticides early on as the production paradigm became dominant after World War II. This was part of a larger trend whereby reductionist scientific models became predominant throughout academia, while government policy became increasingly subject to the narrow evaluative criteria of neoclassical economics.[5] Reductionism also reinforced the role and power of specialized disciplinary approaches—which have become entrenched in powerful departmental and professional fortresses. The difficulties of integrated pest management (IPM) in developing more interdisciplinary approaches illustrate these structural (and ideological) difficulties.

In addition, the political structures surrounding the land-grant system also shape its agendas. From their founding, the extension services fostered

---

[5]Reductionist approaches—generally and in neonclassical economics—are based on several unprovable assumptions: (1) that the essential characteristics of any phenomenon are captured best by analyzing its parts; (2) that there is a sharp distinction between facts and values; (3) that only those facts that are measurable are indeed facts; and (4) that these measurable facts are more valid than other types of information or knowledge.

the development of a nation-wide network of county-level Farm Bureaus which then organized at the state and national levels (the American Farm Bureau Federation)—something that has had a profound impact on the politics of agriculture at all governmental levels. (See Bosso, 1987:35–38, for a discussion.) Local and state Farm Bureaus offered a common meeting ground for large farmers, local business men, bankers, and commodity dealers. Also, many Farm Bureaus run farm-supply cooperatives which sell fertilizers, pesticides, and other inputs. The Farm Bureau has opposed state and federal regulation of farming, whether for health, safety, or environmental reasons. In addition to the various federal support programs which the Farm Bureau has sought, state bureaus seek special support from state legislatures, departments of agriculture, and land-grant universities to fund or conduct research and extension for the major crops and livestock of that state (Bosso, 1987). The net result is a complicated institutional system—a "farmer–agribusiness–land grant–USDA complex"—which is committed to the production paradigm and understands pesticides to be an integral part of that.

The legislative battles described above that occurred in an effort to try to obtain some public health and environmental protection from the consequences of current agricultural production practices have clearly gone against both the political and ideological grain of this complex. Also, efforts to encourage alternative approaches have had similar difficulties—beginning with IPM and continuing on with farming-systems approaches to the efforts now to encourage sustainable agriculture. While the label "sustainable" has become faddish, the larger content of sustainable agriculture—which includes a range of social justice, equity, and ethical concerns as well as environmental ones (Allen and Van Dusen, 1990)—risks being lost or diluted as agencies seek to co-op this term (Dahlberg, 1991).

Policies Affecting the Larger Food System

Similar efforts to protect public health and the environment can be seen throughout the rest of the food system. In terms of processing and distribution, one of the earliest battles, as mentioned, involved the meat-packing industry and efforts to ensure pure foods for urban consumers. Federal inspection of meat and especially poultry facilities continues to be a political battleground, especially now that the large confinement facilities used in production routinely use growth hormones and/or drugs which may harm consumers. The increased use of mechanized and batch poultry processing has raised questions about increased levels of *Salmonella* contamination, especially in a period of reduced federal inspection. Many beef and poultry processing plants are also involved in controversies over their lack of adequate health and safety standards and practices. Budget pressures and a hostility toward regulation, especially in the Reagan Administration, required consumer groups to once again seek redress in Congress.

Inspection and quarantine procedures for imported foods are another area of concern. Here again, maintenance of consumer protection has required investigative journalism (the publication of the book *Circle of Poison*, Weir and Schapiro, 1981) and consumer activism (coordinated by groups like the Pesticides Action Network) to try to alert Congress to the dangers of pesticide residues on imported foods. These residues are often from chemicals banned for use in the United States but exported by U.S. chemical companies. Another area of concern, but one that has involved less consumer activism, is the whole question of "cosmetic standards," especially for fruits and vegetables. As discussed in detail in chapter 4, the FDA has set "defect action levels" (DALs) for the presence of insects and mites without evaluating the health impacts of the increased pesticide use and residues required to meet the DALs—which seem to have been set more in terms of consumer repugnance than health threats. The emphasis on "perfect" fruits and vegetables embodied in the private grading standards of processors, wholesalers, and retailers has also led to higher pesticide use and residue levels.

Food labeling regulations have also had a rocky history. Questions of food purity, safety, additives, and nutrition are all involved, and this means that the often-conflicting interests of farmers, chemical companies, food processors, marketers, nutritionists, and consumer groups must be reconciled. (See Clancy, 1988, for a discussion.) An additional controversy which is associated with food labeling involves the "recommended daily allowances" (RDA) for a healthy diet. The battles over these RDAs—which have huge implications for the dairy and beef industries—have gone on for years. They also have implications for agriculture more generally (to the degree that people heed them) because a reduction in meat consumption by Americans would greatly reduce the domestic market for grain production. While one response might be to seek more exports, another would be to pursue less-intensive and more-sustainable production systems at home. The former would continue current levels of pesticide use while the latter would significantly reduce them.

Food recycling and food waste composting and disposal has received little attention among agriculturalists—whether of conventional or sustainable orientation. Clearly those promoting sustainable approaches are conceptually more open to such ideas since they already encourage on-farm composting, the use of crop residues, and the recycling of animal and green manures. Each stage of the food system involves important and complex waste issues. Food processors generate large amounts of organic waste. Distributors (wholesale and retail) also generate significant levels of organic wastes which are often simply landfilled. Food users rarely think about their wastes (other than getting them out of sight). While still used in a few places, there was much greater recycling of urban food wastes to local pig farms until the 1950s, when grain surpluses led growers' associations to seek to replace these systems.

All of these organic "wastes" are potentially valuable resources, and this is starting to be recognized once again. One New Zealand tomato and vegetable processor has taken advantage of these by building a biogas converter which uses them to power his plant. Some distributors are exploring composting systems for their "wastes." Local food banks—often coordinating their efforts through Second Harvest—seek to recycle mislabeled or slightly damaged products as well as locally gleaned foods to food pantries and cupboards run by local hunger and poverty groups. Many states are now beginning to ban yard wastes from landfills—something that should encourage household and neighborhood composting, which in turn could expand interest in household and community gardens. Such composting and recycling systems encourage users to examine more closely any dangerous chemicals and additives they use on their lawns or in their gardens. Finally, the idea of source separation of wastes in recycling programs might offer a prelude to the idea of separating organic from industrial wastes in sewage systems—something that would make their recycling more feasible and less dangerous.

Generally increasing the energy and resource efficiency of the food system would help to reduce the estimated $150 billion of waste found there (Pimentel, 1990). Such increases in energy and resource efficiency would also reduce the overall levels of production needed. This would create both opportunities and problems. By freeing up land, more-extensive, but less-energy- and pesticide-intensive cultivation practices involving crop rotations would become more feasible. These practices could also be combined with efforts to better maintain rural landscapes and biodiversity. The problem of sustaining farm families and rural communities economically (a necessary condition for their regeneration) would remain. Current efforts to support family farmers indirectly through crop-support programs have largely failed since large agribusiness firms have captured most of the dollars. It would be better to support farm families directly through some sort of negative income tax which could have as its justification not only the traditional American support for this way of life but support for a new role in helping to maintain rural lands, biodiversity, and landscapes. The needed for a general rural development policy for the United States would remain.

International Policies That Encourage Pesticide Use

The international context for policy is one that has been changing steadily since World War II. It is only in the last couple of years—with the dismantling of the Berlin Wall and the end of the Cold War—that there has been a recognition that the main issues are increasingly those associated with the fundamental economic disparities between rich and poor countries and less with ideological conflicts between the capitalist and socialist forms of industrial society.

However, other deep myths and ideologies remain, especially those that suggest that it is through the "transfer" of modern technologies to the Third World that these disparities in wealth and development will be overcome. The myth of technological neutrality has deep cultural roots (Dahlberg, 1990a).[6] It also serves the interests of the technologically and organizationally powerful—nationally and internationally—and is a particularly useful myth for both multinational corporations and those running international development programs. Besides reinforcing general beliefs in Western cultural superiority, it provides strong support for the doctrines of international free trade—doctrines which have served the dominant international economic powers since at least the 19th century.

The idea of "sustainable development," which is a cornerstone of the United Nation's World Commission on Environment and Development report, *Our Common Future* (WCED, 1987), implicitly challenges many of these myths and doctrines. It calls for major, but different, changes in the energy, industrial, agricultural, urban, and rural policies of both North and South in order to meet the common global challenges of population, resources, food security, and protection of biodiversity in ways which will be ecologically sound and socially and intergenerationally just. As the latest focal point in the ongoing debate about the limits to growth, the report highlights the very different paradigms that underlie traditional international economic and strategic doctrines and those linked to emerging concepts of sustainability and the Gaia hypothesis.

General Policies and Practices

In spite of international agencies like the World Bank adopting (and co-opting?) the term "sustainability," the fundamental emphasis of development programs upon economic growth remains. Economic development

[6]This myth posits that technologies are themselves ethically and socially neutral and that negative consequences of their use are the responsibility of the users/operators. This view is nicely captured by the bumper sticker stating that "Guns don't kill people, people kill people." This ignores the fact that the basic design principle of guns is to deliver a deadly projectile with high accuracy and speed. In addition to their design principles, technologies are not neutral because they reflect the physical and social environments in which they were developed. It is for this reason that the Japanese seeking modernization in the late 19th century were very careful not to simply adopt "transferred" Western technologies; rather they sought either to modify and upgrade existing Japanese technologies or to make major changes in Western technologies to adapt them to their environment and culture.

programs are also strongly affected by the fact that the two largest multinational funders, the World Bank and the International Monetary Fund, are politically controlled by the rich industrial countries through a system of weighted voting. This means that rich-country conceptions, priorities, and technologies dominate the development agenda. While there is now a somewhat greater recognition of the importance of rural and agricultural development, the "production paradigm" is still dominant in specialized agencies like the Food and Agriculture Organization (FAO). The South has struggled with little success to deal with the distortions in basic North–South terms of trade and "terms of technology" (i.e., the control of the North over access to the sources of technological innovation).[7] The difficulties of effecting institutional change are even greater at the international level than at the national level because it is much harder for citizens' groups and nongovernmental organizations to organize effective opposition to the power of national governments and multinational corporations.

International trade controversies have clustered around the very different conceptions of the General Agreement on Tariffs and Trade (GATT) and the UN Conference on Trade and Development (UNCTAD), which developed its demands for a "New International Economic Order" in the late 1970s in an effort—largely unsuccessful—to redress long-standing imbalances in international terms of trade.[8] The dominant trade forum of both the great powers and the multinational corporations (MNCs) is GATT. GATT is based on five principles which are designed to make trade nondiscriminatory, to reduce nontariff barriers to trade (quotas, foreign exchange restrictions, health, safety, and sanitation regulations, technological standards, packaging and labeling regulations, etc.), and to provide a forum for settling disputes. It now appears that the MNCs are seeking to use GATT (as well as the free-trade negotiations between the United States and Canada and Mexico) to try to get rid of what they see to be bothersome health, safety, labor, and environmental restrictions and regulations found at the national level by preempting them with lowest-common-denominator

[7]For a detailed discussion of this, giving examples of how the developed countries have sought to maintain their current dominance in biotechnology research and development, see Goldstein (1989).

[8]There were several reasons for this lack of success. While UNCTAD called for major reforms in the terms of trade, these would have had to have been carried out by the rich countries themselves and/or by other organizations in which the Third World countries have much less power and influence. UNCTAD became the forum for these Third World demands because it is based on the principle of one country, one vote, so the poor majority controls it. However, the forums of international economic and monetary power—like the World Bank and the International Monetary Fund—are dominated by the great powers through weighted voting systems. Since the UNCTAD reforms have been largely unsuccessful, Third World countries have increasingly sought to pursue their individual national trade interests through the GATT rounds.

standards at the international level. A similar process has often occurred within the United States when particular states have sought more stringent standards than those found at the national level. Particularly relevant are the proposals for the treatment of pesticide residues on food. (See below.)

In addition, it is primarily through GATT that the industrial countries have been seeking to open up new markets for genetically engineered products by establishing an international legal regime which would extend the patenting of genes and living organisms to all signatories. Each signatory would have to recognize such patent rights in general and recognize in their own country the patents of any other signatory or be subject to trade sanctions and retaliation. While many developing countries have resisted the idea of intellectual property rights, they will be tempted to sign any eventual GATT agreement because of the greater access to First World commodity markets that current proposals offer.

If they do sign, they will weaken other Third World efforts to protect the Vavilof centers of crop germ-plasm diversity (primarily located in the Third World) and to guarantee to the countries of origin some of the value added to germ plasm through plant breeding and/or genetic engineering. The forum for these efforts has been the FAO, where several conventions seeking these goals are under negotiation. The Third World position argues that germ plasm is a "natural resource" subject to national jurisdiction and protection. The First World argues that it is a "natural heritage of mankind," which means that it is freely available to all. Neither side seems willing to consider germ plasm to be the "common heritage of humankind," which would place it under international jurisdiction and establish procedures for its equitable treatment—such as has occurred under the Law of the Sea Convention in regard to those waters falling outside the exclusive economic zones of member states.

International Production Policies

International support for agricultural production comes in various forms. One is the research support provided by the World Bank's Consultative Group on International Agricultural Research (CGIAR), which funds 13 research centers around the world. The most famous of these are CIMMYT (the International Maize and Wheat Improvement Center, Mexico) and IRRI (the International Rice Research Institute, the Philippines), which were instrumental in developing the high-yielding varieties (HYVs) of maize and rice of the Green Revolution (Dahlberg, 1979). While there has been some movement away from the highly capital- and chemical-intensive emphasis of the early years and serious discussion of sustainability (FAO, 1988), the still-dominant "production paradigm" limits the discussion to how to reduce environmental impacts through lower chemical inputs. The larger social, health, equity, and rural development dimensions of sustainability continue to be neglected.

Broad-gauge agricultural development programs have been supported over the years by the World Bank, the FAO, and the International Fund for Agricultural Development (IFAD). These programs, as well as national programs, have also been complemented by a wide range of educational and technical assistance programs. However, the dominance of the "production paradigm" has made it difficult to include specific systems-oriented approaches such as IPM—something that illustrates the organizational rigidity of international agencies.

Policies Affecting the Larger International Food System

The policies and controversies here involve the same groups and similar issues to those found at the national level. Producers (here the large chemical and pharmaceutical companies, rather than farm groups), distributors (large commodity trading companies), and food multinationals seek to maintain and/or increase their access to world markets through a variety of trade and regulatory mechanisms. Consumers seek guarantees that their food is safe and does not contain dangerous hormones, additives, or residues. Environmentalists and public health advocates are concerned with the impacts of pesticides and other dangerous chemicals upon the environment and farm workers.

Basic differences between the First and Third World affect the controversies and policies as well as the types of alliances that develop. There appears to be much more of an alliance between First and Third World environmentalists on the issue of seeking to prevent industrial countries from dumping banned or discontinued pesticides and pharmaceuticals in the Third World than there is on protecting consumers through food inspection and quarantine measures. Equally, Third World governments are much more in unison regarding efforts to require multinational companies to disclose the legal status, safety, etc., of products they are exporting (Norris, 1982) than in supporting the efforts of First World consumers to protect their foods from the other half of the "circle of poison," whereby residues of pesticides banned in the industrial countries return on fruit and vegetables from the Third World (Weir and Schapiro, 1981).

There are also divisions within the First World on food safety and quarantine issues—particularly between the United States and the European Communities (EC). The EC ban on U.S. beef produced with growth hormones is only the most recent example of a long series of battles. However, many of these battles reflect not only a concern to protect European consumers, but also an effort to protect European farmers and their markets.

Controversies over the dangers of radioactive contaminants in European foods following the Chernobyl disaster showed other divisions.

While these high-visibility issues give the impression that consumers are well protected, the proposals for treatment of pesticide residues in the latest GATT negotiations illustrate in detail the general threat to national health, safety, labor, and environmental standards mentioned above. Under these proposals (strongly supported by the U.S. administration and the major multinationals), countries would not be able to set pesticide residue standards higher than those adopted by the Codex Alimentarius Commission (CAC) without being subject to trade sanctions and retaliation. Thus, the most powerful "stick" which GATT and its members can employ will be aimed at countries which try to protect their citizens above these lowest-common-denominator standards.

While it is important to develop international standards and codes, it must be recognized that all international standards tend toward such lowest common denominators. However, in most areas, such as in pollution or ocean dumping standards or maritime and air safety standards, states are not prevented or penalized if they choose to set more stringent standards. Thus, the primary evaluative criteria involved in the GATT proposals for pesticide and food safety regulation are clearly economic ones aimed at protecting economic sectors, companies, and jobs rather than health and/or safety criteria aimed at protecting individuals. It is also important to note that while this particular standard-setting body operates under the auspices of the Joint FAO/WHO Food Standards Program, it and its committees are strongly influenced by agricultural production perspectives and by heavy involvement of the pesticide industry in its meetings (Boardman, 1986).

What implementation of the proposed GATT/GAC standards would mean is illustrated by the distinct possibility that the permitted DDT pesticide residue levels for food products imported into the United States could increase 10–50 times (depending on the product) over current Food and Drug Administration levels.[9] It would also mean that more stringent

[9]There is some controversy and uncertainty regarding what would or would not be permitted. Recently negotiated additions to the proposed GATT treaty would permit countries to claim more stringent "scientifically based health standards" which could supersede the GATT standards. These additions appear to be aimed in part at the EC limitations on beef produced with growth hormones, which the United States claims have no scientific basis. The real question (beyond whether the GATT treaty will be agreed to and ratified) is who will decide what are or are not "scientifically based health standards." As noted earlier, the Codex Alimentarius Commission is dominated by groups with strong production and pesticide perspectives and has no representation of farm workers, environmentalists, or unions. In any case, any country that seeks an exemption will have to pursue a complicated and lengthy process to try to prove that its standards are scientifically based and needed to protect public health. Besides potentially discriminating against poor Third World countries, such a procedure means that with an occasional exception, the lowest common denominator will still predominate.

measures which a state might choose to adopt on food sold within its borders would become illegal. Ratification of these GATT proposals would thus preempt any national or state legislation to the contrary.

Similar risks are involved in the free-trade agreements and negotiations between the United States, Canada, and Mexico. The "harmonization" of technical and agricultural standards has already led to Canada's more stringent pesticide regulation policies being replaced by the U.S. "risk-benefit" model. Significantly, there is no inclusion of environmentalists on the advisory committees set up to implement such harmonization procedures.

There are relatively few other international policies affecting other aspects of the food system, although there is increasing concern about the dumping of contaminated foods and food wastes. After Chernobyl, there were several incidents of European exporters seeking to ship radioactively contaminated foods to the Third World. Then there was the famous "garbage ship" which sailed around the world seeking someplace to dump its wastes. The problem is, of course, a much broader one that involves all types of toxic and hazardous wastes which multinationals seek to dispose of cheaply.

Conclusions

The ways in which society understands and treats pesticides illustrates a number of conflicting beliefs, values, and interests. While only one component and example in a larger debate about the future of industrial societies, pesticides are particularly significant because they represent the one area where society consciously applies massive amounts of toxic chemicals to the environment. Pesticides are not a "by-product" of production processes (the still-dominant view of industrial pollution); they are a crucial and integral part of industrial agriculture as currently structured. Thus, challenges to pesticide use become a direct challenge to many basic Western cultural and industrial beliefs.

Deep cultural beliefs about humankind's separation from and dominance over nature are challenged by an increasing awareness that we are destroying our life-support systems. A number of industrial beliefs and values which emphasize that the continued elaboration of science and technology lead us to social progress are also challenged by the increasing exploitation and degradation of individuals, peoples, and environments. Such general challenges are made concrete and pointed in the case of pesticides because society is consciously permitting massive applications of chemicals which threaten the health, welfare, and safety of farmers, farm workers, food processors, and food consumers.

While safety measures always cost money and there are always choices involved in determining what is an "acceptable" level of risk, pesticides differ from other areas such as auto or air safety. This is because if one chooses to drive or fly, there are necessarily certain risks (which can to some extent be mitigated through safety measures). However, in agriculture, there are a number of traditional and modern production techniques which do not require the use of toxic chemicals. These techniques, while not always quite as productive in the short term, are much more sustainable over the long term and have the added benefit of being less fossil-fuel intensive, thus producing fewer greenhouse gasses than current techniques.

As the tangled regulatory, legislative, and political battles reviewed above demonstrate, the road to trying to achieve more healthful, safe, and sustainable food systems is a rocky one. While most analysts and activists focus on the power of the various vested interests involved, it is also important to recognize the ways in which they seek to invoke increasingly obsolescent beliefs and myths to try to protect their power and privilege. This is why volumes like this one which seek to examine the deeper cultural, value, and ethical beliefs and issues which surround and undergird the agri-pesticide complex are crucial.

References

Adas, M. (1989) *Machines and the Measure of Men: Science, Technology, and Ideologies of Western Dominance*, Cornell University Press, Ithaca, NY.

Allen, P. and Van Dusen, D. (1990) Sustainability in the balance: raising fundamental issues. Agroecology Program. University of California, Santa Cruz.

Bosso, C.J. (1987) *Pesticides and Politics: The Life Cycle of a Public Issue*, University of Pittsburgh Press, Pittsburgh.

Boardman, R. (1986) *Pesticides in World Agriculture*, St. Martin's Press, New York.

Busch, L., Lacy, W.B., Burkhardt, J. and Lacy, L.R. (1991) *Plants, Power, and Profit: Social, Economic and Ethical Consequences of the New Biotechnologies*, Basil Blackwell, Cambridge, MA.

Clancy, K.L. (ed) (1988) *Consumer Demands in the Market Place: Public Policies Related to Food Safety, Quality, and Human Health*, Resources for the Future, Washington, DC.

Constance, D. and Heffernan, W. (1989) The demise of the family farm: the rise of oligopoly in agricultural markets." Paper presented at the 1989 Meeting of the Agriculture, Food, and Human Values Society, Little Rock, Arkansas.

Crosby, A.W. (1986) *Ecological Imperialism: the Biological Expansion of Europe, 900–1900*, Cambridge University Press, Cambridge, UK.

Dahlberg, K.A. (1979) *Beyond the Green Revolution: The Ecology and Politics of Global Agricultural Development*, Plenum, New York.

______. (ed) (1986) *New Directions for Agriculture and Agricultural Research: Neglected Dimensions and Emerging Alternatives*, Rowman & Allanheld, Totowa, NJ.

______. (1987) Redefining development priorities: genetic diversity and agroecodevelopment. *Conservation Biology*, **1**(4),311–322.

______. (1990a) The value content of agricultural technologies. *Agricultural Ethics*, **2**(2),87–96.

______. (1990b) Towards a theory of regenerative food systems: an overview. Paper prepared for the Conference on Redefining Agricultural Sustainability, University of California, Santa Cruz, June 28–30, 1990.

______. (1990c) The industrial model and its impacts on small farmers: the green revolution as a case, in *Agroecology and Small Farm Development*, (eds M.A. Altieri and S.B. Hecht), CRC Press, Boca Raton, FL, pp. 83–90.

______. (1991) Sustainable agriculture: fad or harbinger? *BioScience*, **41**(5),337–340.

______. (1992) Renewable resource systems and regimes: key missing links in global change studies. *Global Environmental Change*, **2**(2), 128–152.

Douglass, G.A. (1984) *Agricultural Sustainability in a Changing World Order*, Westview Press, Boulder, CO.

Farvar, M.T. and Milton, J.P. (eds) (1972) *The Careless Technology: Ecology and International Development*, Natural History Press, New York.

Food and Agriculture Organization of the United Nations (FAO) (1988) *Sustainable Agriculture: Implications for International Agricultural Research*, TAC Secretariat, Food and Agriculture Organization, Rome.

General Accounting Office (GAO) (1990) *Alternative Agriculture: Federal Incentives and Farmers' Options*, BAO/PEMD-90-12, U.S. General Accounting Office, Washington, DC.

Goldstein, D.J. (1989) Ethical and political problems in Third World biotechnology. *J. Agricultural Ethics*, **2**(1):5–36.

Grigg, D.B. (1984) The agricultural revolution in Western Europe, in *Understanding Green Revolutions* (eds T.P. Bayless-Smith and S. Wanmali), Cambridge University Press, Cambridge, UK, Ch. 1.

Heffernan, W. (1986) Review and evaluation of social externalities, in *New Directions for Agriculture and Agriculture Research* (ed K.A. Dahlberg), Rowman & Allanheld, Totowa, NJ, pp. 199–220.

Howarth, R.B. and Norgaard, R.B. (1990) Intergenerational resource rights, efficiency, and social optimality. *Land Economics*, **66**:1–11.

Kloppenburg, J., Jr. (1988) *First the Seed: The Political Economy of Plant Biotechnology, 1492–2000*, Cambridge University Press, New York.

Knutson, R.D., Richardson, J.W., Klinefelter, D.A., Paggi, M.S., and Smith, E.G. (1986) *Policy Tools for U.S. Agriculture*, Agricultural and Food Policy Center, B-1548, Texas A&M University, College Station.

Kuhn, T. (1962) *The Structure of Scientific Revolutions*, University of Chicago Press, Chicago.

Lawn Institute, The. (1991) Personal Communication. The Lawn Institute, Pleasant Hill, TN.

Marcus, A.I. (1985) *Agricultural Science and the Quest for Legitimacy*, Iowa State University Press, Ames.

National Research Council (NRC) (1975) *Forest Pest Control*, National Academy Press, Washington, DC.

National Research Council (NRC) (1980a) *Urban Pest Management*, National Academy Press, Washington, DC.

National Gardening Association (NGA) (1989) National gardening fact sheet, National Gardening Association, Burlington, VT.

Norris, R. (ed) (1982) *Pills, Pesticides & Profits*, North River Press, Croton-on-Hudson, NY.

Office of Technology Assessment (OTA) (1986) *Technology, Public Policy, and the Changing Structure of American Agriculture*, OTA-F-285, U.S. Government Printing Office, Washington, DC.

Pimentel, D. (1990) Environmental and social implications of waste in U.S. agriculture and food sectors. *J. Agricultural Ethics*, **3**(1):5–20.

Weir, D. and Shapiro, M. (1981) *Circle of Poison: Pesticides and People in a Hungry World*, Institute for Food and Development Policy, San Francisco.

World Commission on Environment and Development (WCED) (1987) *Our Common Future*, Oxford University Press, New York.

12

Pesticide Use Trends and Issues in the United States

Craig Osteen[1]

Pesticides have been used in U.S. agriculture since the late 1800s, but their use grew dramatically from the late 1940s to the early 1980s and then stabilized. The development and growing use of synthetic organic pesticides have been an integral part of a technological revolution in U.S. agriculture that increased productivity by 2.2 times between 1947 and 1988 (USDA, 1990). Growth in pesticide use has created many controversies about potential effects of pesticides on food safety, groundwater quality, worker safety, and wildlife mortality. The controversies reflect two major ideas: (1) using more pesticides is not necessarily a panacea for pest control, and (2) undesirable health or environmental effects of using some pesticides may outweigh their production benefits. Today, many people fear the risks of unknown or poorly understood hazards and are impatient with the U.S. Environmental Protection Agency's (USEPA) slow and deliberate resolution of pesticide controversies. There are also people arguing for a policy of limiting or reducing the overall level of pesticide use, which is a different approach than restricting or banning individual pesticides.

This paper discusses major pesticide use trends in the United States, pesticide regulatory policy with a focus on balancing risks and benefits, and several current policy issues. Important topics include (1) the effects of such factors as pesticide productivity, farm programs, and pesticide regulations on use; (2) the effects of increased pesticide use on productivity and pest losses; (3) the effects of changing attitudes toward pesticides on regulatory policy; and (4) a major shortcoming in the regulatory process for balancing risks and benefits.

---

[1]The author is an agricultural economist with the Resource and Technology Division, ERS, USDA, Washington, D.C. The views presented are those of the author and do not represent the official views of any agency or organization.

Pesticide Use Trends

Effective chemical control of agricultural pests originated in the late 1800s (Klassen and Schwartz, 1985). Paris green (copper acetoarsenite) was developed in the United States in the 1870s to combat the potato beetle, and Bordeaux mixture (quicklime and copper sulfate) was developed in France in the 1880s to control disease in grape culture. Prior to World War II, arsenicals, sulfur compounds, and oils were commonly used. However, the development of synthetic organic materials, such as 2,4-D and DDT, during World War II heralded the modern age of chemical pesticides.

Aggregate Trends

Synthetic organic pesticide use grew rapidly from the late 1940s to the early 1980s before stabilizing. USEPA (1990) estimates that agricultural pesticide use grew from 320 million pounds active ingredient (a.i.) in 1964 to 880 million pounds a.i. in 1982 but fell to 845 million pounds a.i. in 1988 (Fig. 12.1). There was rapid growth during the 1960s and 1970s, but by the late 1970s markets for pesticides became saturated, and growth slowed. Pesticide use since 1980 has been heavily influenced by crop acreage, and reduced crop acreage helped to stabilize pesticide use after 1982.

U.S. Department of Agriculture (USDA) pesticide surveys show that use on major field crops (corn, cotton, soybeans, sorghum, rice, tobacco,

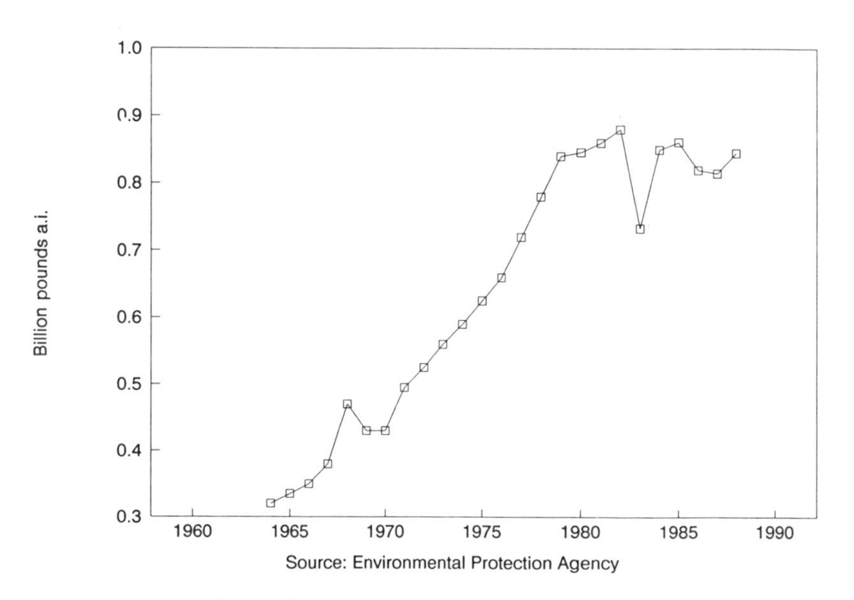

Figure 12.1. Quantity of agricultural pesticides.

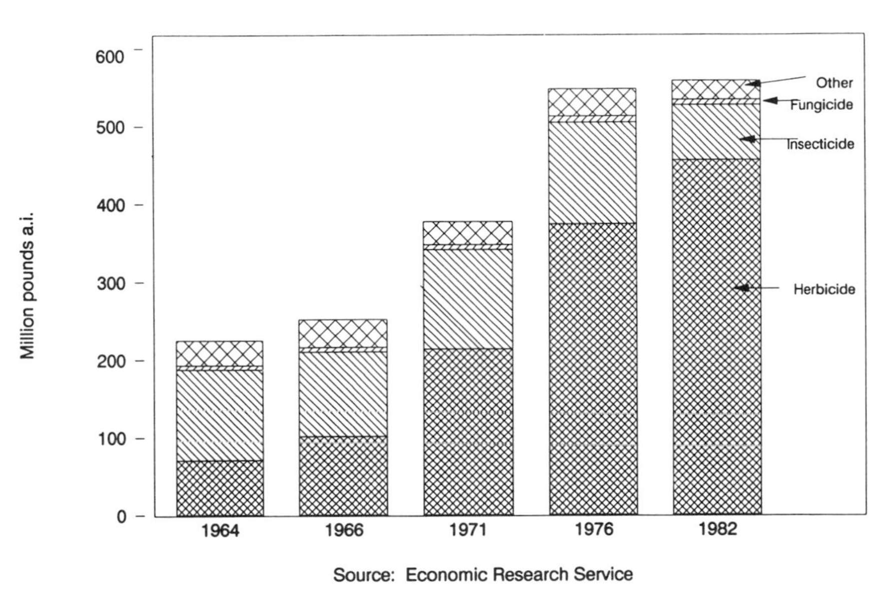

Figure 12.2. Pesticide use on major crops.

peanuts, wheat, other small grains, alfalfa, other hay, and pasture) grew from 225 million pounds a.i. in 1964 to 548 million pounds a.i. in 1976 to 558 million pounds a.i. in 1982 (Fig. 12.2, Table 12.1).[2] Major components in that trend were (1) a rise in pesticide use on corn and soybeans from 50 million pounds a.i. (22% of major field crop use) in 1964 to 412 million pounds a.i. (74%) in 1982, (2) a rise in herbicide use on major field crops from 71 million pounds a.i. (32%) in 1964 to 456 million pounds a.i. (82%) in 1982, and (3) a rise in insecticide use from 117 million pounds a.i. in 1964 to 130 million pounds a.i. in 1976, and then a dramatic fall to 71 million pounds a.i. in 1982.[3] The use of fungicides and other pesticides on major crops was relatively stable between 1964 and 1982.

Insecticides

By the 1950s, insecticides were being widely used on a variety of high-value crops including cotton, tobacco, peanuts, potatoes, fruits, and veg-

---

[2]The 1982 survey was restricted to 33 states, while the earlier surveys included 48 states. California, the second-biggest cotton-producing state, was among those excluded. The result is that use in 1982 is underestimated. Examination of cotton pesticide use data for 1976 and 1979 indicates that the estimates of total use could be 4–20 million pounds a.i. low, which has little effect on the overall trend.

[3]Examination of 1976 and 1979 cotton data indicate that 1–8 million pounds a.i. of insecticides could be missed in 1982. However, the general trend from 1976 to 1982 does not change.

Table 12.1. Pesticide use on major crops.[a]

Year	Herbicides	Insecticides	Fungicides	Other	Total
		Million pounds a.i.			
1964	70.5	116.7	5.8	31.7	224.7
1966	101.2	108.3	6.0	35.7	251.1
1971	213.1	127.9	6.4	29.8	377.2
1976	373.9	130.3	8.1	35.3	547.6
1982	455.6	71.2	6.6	24.3	557.7

Sources: Andrilenas, 1974; Eichers et al., 1968, 1970, 1978; and USDA, 1983.

[a]Active ingredients excluding sulfur and petroleum. Major crops are cotton, corn, soybeans, sorghum, rice, tobacco, peanuts, wheat, other small grains, alfalfa, other hay, and pasture.

etables (Table 12.2). Somewhat later, insecticide use on major field crops, particularly corn, grew rapidly. For example, less than 10% of corn acreage was treated with insecticides during the mid-1950s, but 35–45% was treated by 1975 (Fig. 12.3). The primary use was for soil insects in continuous corn rotations. Since the early 1980s, the proportion of corn acres treated has fallen from 45% to 32%.

Cotton, corn, and soybeans accounted for 82% of total quantity of insecticide use on major field crops in 1982 (Fig. 12.4, Table 12.3). The decline in major crop insecticide use between 1976 and 1982 occurred primarily on cotton, where quantity fell from 73 million pounds a.i. in 1971 to 64 million pounds a.i. in 1976 to 17 million pounds a.i. in 1982. In 1989, cotton insecticide use was approximately 18 million pounds a.i. (Crutchfield, 1990). Corn and soybean insecticide quantity increased from 21 million pounds a.i. in 1964 to 41 million pounds a.i. in 1982.

The decline in insecticide use between 1976 and 1982 largely reflects the changing composition of compounds used. Organochlorine use fell steadily from about 70% of insecticide quantity in 1966 to only 6% in 1982 because of pest resistance and regulatory actions (Table 12.4). Organophosphate quantity grew from about 20% of the total in 1966 to almost 70% in 1982. The growth of carbamates was less dramatic. Pyrethroids, introduced in the late 1970s, accounted for about 4% of insecticide quantity in 1982. Due to their low rates of application, pyrethroids account for a much greater percentage of total insecticide treatments than quantity. The use of pyrethroids on cotton to control cotton bollworms, tobacco budworms, and other pests has contributed to the major reduction in insecticide quantity applied to cotton. Cotton insecticide application rates fell from 5–6 pounds a.i. per crop-acre before 1977 to about 1.6 pounds after 1977, even though the percentage of acreage treated varied between 50 and 70%, with no obvious trend.

Table 12.2. Share of crop acres treated with insecticides.[a]

Year	Corn	Cotton	Soybeans	Wheat	Sorghum	Fruits and nuts	Potatoes	Vegetables	Tobacco	Peanuts	Rice	Other grains	Alfalfa	Other hay
	Percent													
1952	1	48	NA[b]	NA	NA	82	75	61	47	NA	NA	NA	NA	NA
1958	6	66	NA	NA	NA	81	80	74	58	NA	NA	NA	NA	NA
1966	33	54	NA	NA	NA	87	91	58	82	NA	NA	NA	NA	NA
1971	35	61	8	7	39	90	84	58	77	87	35	3	8	2
1976	38	60	7	14	27	NA	NA	NA	76	55	11	5	13	2
1979	NA	48	NA	NA	NA	NA	94	74	NA	NA	NA	NA	NA	NA
1980	43	NA	11	NA	4	NA	NA	NA	NA	NA	NA	NA	NA	NA
1982	37	36	12	3	26	NA	NA	NA	85	48	16	1	7	1
1984	42	63	8	NA	NA	NA	NA	NA	NA	NA	NA	NA	NA	NA
1985	45	65	7	5	NA	NA	NA	NA	NA	NA	NA	NA	NA	NA
1986	41	NA	4	7	NA	NA	NA	NA	NA	NA	NA	NA	NA	NA
1987	41	61	3	7	17	NA	NA	NA	NA	NA	NA	NA	NA	NA
1988	35	61	8	4	NA	NA	89	NA	NA	NA	18	NA	NA	NA
1989	32	68	3	11	NA	NA	91	NA	NA	NA	22	NA	NA	NA
1990	31	NA	NA	5	NA	NA	88	NA	NA	NA	10	NA	NA	NA

[a]Some of the year-to-year variation in estimates may be due to differences in area of crop acreage covered in the survey.

[b]NA = not available.

Sources: Eichers et al., 1978; USDA, 1983, 1984, 1987, 1988, 1989, 1990, 1991.

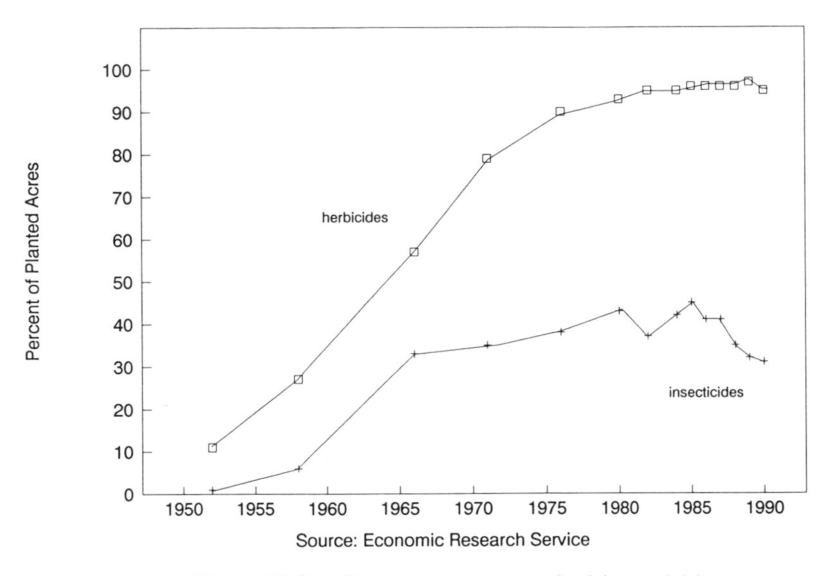

Figure 12.3. Corn acreage treated with pesticides.

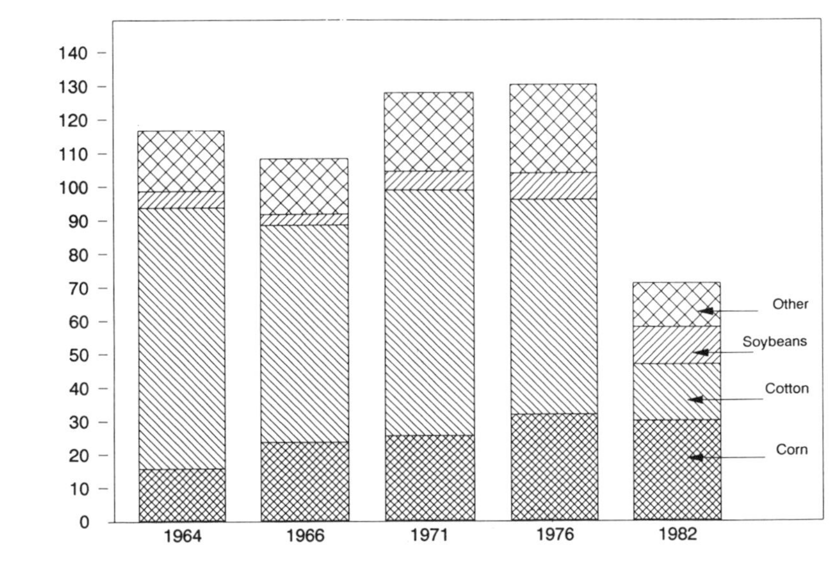

Figure 12.4. Insecticide use on major crops.

Table 12.3. Insecticide use on major crops.[a]

Crop	1964	1966	1971	1976	1982
	Million pounds a.i.				
Cotton	78.0	64.9	73.4	64.1	16.9
Corn	15.7	23.6	25.5	32.0	30.1
Soybeans	5.0	3.2	5.6	7.9	11.1
Other crops	18.0	16.5	23.4	26.3	13.1
TOTAL	116.7	108.2	127.9	130.3	71.2

Sources: Andrilenas, 1974; Eichers et al., 1968, 1970, 1978; and USDA, 1983.

[a]Active ingredients excluding petroleum. Major crops are cotton, corn, soybeans, sorghum, rice, tobacco, peanuts, wheat, other small grains, alfalfa, other hay, and pasture.

Table 12.4. Shares of insecticide classes on major crops.[a]

Year	Organo-chlorines	Organo-phosphates	Carbamates	Pyrethroids	Other
	Percent				
1964	70	20	8	0	2
1966	70	22	7	0	1
1971	45	39	14	0	2
1976	29	49	19	0	3
1982	6	67	18	4	5

Sources: Andrilenas, 1974; Eichers et al., 1968, 1970, 1978; and USDA, 1983.

[a]Active ingredients excluding petroleum. Major crops are cotton, corn, soybeans, sorghum, rice, tobacco, peanuts, wheat, other small grains, alfalfa, other hay, and pasture.

Herbicides

Rapid herbicide growth began in the late 1950s. Herbicide use grew more and stabilized later than insecticide use. Approximately 10% of the acreage of corn, cotton, and wheat was treated with herbicides in 1952 (Fig. 12.3, Table 12.5). Herbicide use on corn, cotton, and soybeans (for which there are no data before 1971) appears to have stablized at 90–97% of acres planted since 1980. Wheat herbicide use may be stabilizing in the range of 50–60% of planted acreage.

Corn and soybeans account for the major portion of herbicide use on major field crops. The quantity of herbicides applied to these crops grew from 30 million pounds a.i. in 1964 (42% of all herbicides used) to 370 million pounds a.i. in 1982 (81%) (Fig. 12.5, Table 12.6). The quantity of herbicides used on other crops has also grown, but not as dramatically.

Table 12.5. Share of crop acres treated with herbicides.[a]

Year	Corn	Cotton	Soybeans	Wheat	Sorghum	Potatoes	Vegetables
	Percent						
1952	11	5	NA[b]	12	NA	NA	NA
1958	27	7	NA	20	NA	NA	NA
1966	57	52	NA	29	NA	NA	NA
1971	79	82	68	41	46	NA	NA
1976	90	84	88	38	51	NA	NA
1979	NA	91	NA	NA	NA	73	84
1980	93	NA	92	NA	61	NA	NA
1982	95	97	93	42	59	NA	NA
1984	95	93	94	NA	NA	NA	NA
1985	96	94	95	44	NA	NA	NA
1986	96	NA	96	53	NA	NA	NA
1987	96	94	95	61	82	NA	NA
1988	96	95	96	53	NA	NA	NA
1989	97	92	96	61	NA	77	NA
1990	95	95	95	51	NA	79	NA

	Tobacco	Rice	Peanuts	Other grains	Alfalfa	Other hay
	Percent					
1971	7	95	92	31	1	2
1976	55	83	93	35	3	2
1982	71	98	93	45	1	3
1988	NA	98	NA	NA	NA	NA
1989	NA	97	NA	NA	NA	NA
1990	NA	98	NA	NA	NA	NA

[a]Some of the year-to-year variation in estimates may be due to differences in area of crop acreage covered in the survey.

[b]NA = not available.

Sources: (Eichers et al., 1978; USDA, 1983, 1984, 1987, 1988, 1989, 1990, 1991).

There is some evidence that herbicide use has decreased since 1982. For example, corn herbicide use declined about 10% from 243 million to about 217 million pounds a.i. between 1982 and 1990 (USDA, 1991). A major portion of the decline is due to reduced crop acreage; there was 7% less corn acreage covered by the 1990 survey than by the 1982 survey. Herbicide rates declined from 3.1 to 2.9 pounds a.i. per treated corn acre. Interestingly, the application rate of atrazine, a widely used corn herbicide, fell from 1.5 pounds a.i. per treated acre in 1982 to 1.2 pounds in 1990, while

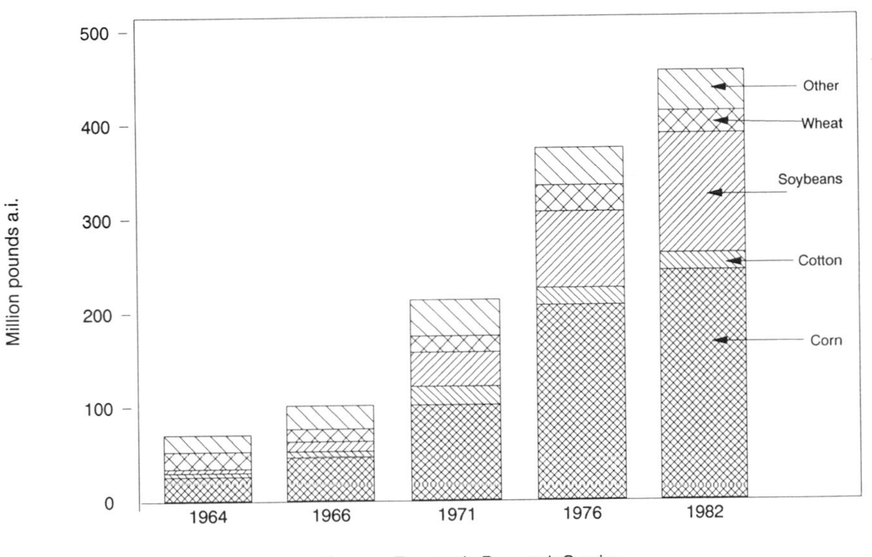

Figure 12.5. Herbicide use on major crops.

corn acreage treated with atrazine remained virtually unchanged. The result was a 17% decline in total atrazine use on corn.

The patterns of herbicide compounds that farmers use have also changed (Table 12.7). Phenoxy use fell from about 45% of total quantity used in 1964 to about 5% in 1982. That percentage decline is somewhat deceptive. Annual phenoxy use was 30–42 million pounds a.i. during 1964–1976 with

Table 12.6. Herbicide use on major crops.[a]

Crop	1964	1966	1971	1976	1982
	Million pounds a.i.				
Cotton	4.6	6.5	19.6	18.3	18.3
Corn	25.5	46.0	101.1	207.1	243.4
Soybeans	4.2	10.4	36.5	81.1	127.0
Wheat and small grains	18.3	13.2	17.0	27.4	24.0
Other crops	17.9	25.1	38.9	40.0	42.9
TOTAL	70.5	101.1	213.1	373.9	455.6

Sources: Andrilenas, 1974; Eichers et al., 1968, 1970, 1978; and USDA, 1983.

[a]Active ingredients excluding petroleum. Major crops are cotton, corn, soybeans, sorghum, rice, tobacco, peanuts, wheat, other small grains, alfalfa, other hay, and pasture.

Table 12.7. Proportions of herbicide classes used on major crops.[a]

Year	Phenoxys	Amides	Triazines	Nitro-phenols	Carbamates	Anilines	Other
				Percent			
1964	46	6	15	1	7	1	24
1966	38	5	21	2	7	4	23
1971	16	22	28	21	5	6	2
1976	11	30	31	8	10	8	2
1982	5	31	27	8	17	8	4

Sources: Andrilenas, 1974; Eichers et al., 1968, 1970, 1978; and USDA, 1983.

[a]Active ingredients excluding petroleum. Major crops are cotton, corn, soybeans, sorghum, rice, tobacco, peanuts, wheat, other small grains, alfalfa, other hay, and pasture.

no particular trend, before falling to 26 million pounds a.i. in 1982. During this time, the quantities of amides, triazines, nitrophenols, carbamates, and dinitroanilines all grew significantly. An important new trend is the increasing use of low-application-rate imidazolinone and sulfonylurea herbicides on wheat and soybeans.

Economic Factors Affecting Pesticide Use

According to economic efficiency criteria, producers should choose the combination of pest control methods that maximizes the difference between pest damage reductions and control costs. They should increase the use of a pest control input until the marginal value of damage reduction of the last input equals the marginal cost. As a result, pesticide use should be influenced by crop prices and the costs of pesticides and alternative control methods.

However, financial risk (variability of returns) and uncertainty (incomplete information about outcomes) are also important considerations. Farmers do not know precisely what level of pest damage there would be without control, the reduction in damage from using a control, or the value of the reductions. They must develop expectations of crop value and potential yield savings from control. Rational decisions will retrospectively appear suboptimal if pest infestations or crop values were different than expected. Because reducing the risk of large financial losses is important to many producers, some may find it rational to apply pesticides or other inputs in excess of profit-maximizing levels.

Pesticide Cost Efficiency

One argument is that pesticide use increased because pesticides often cost less and contributed to higher, less-variable yields than previously used nonchemical methods. In addition, indices show that pesticide prices have generally fallen relative to machinery and fuel prices since 1965 (Fig. 12.6). Pesticide prices generally have fallen relative to crop prices and wages since 1950 (Fig. 12.7). These trends would reduce the costs of pesticides relative to nonchemical control methods and encourage the substitution of pesticides for labor, fuel, and machinery used in pest control (Daberkow and Reichelderfer, 1988). These trends may have induced technological change to take advantage of cheap pesticides (Capalbo and Vo, 1988). During the early and mid-1980s, pesticide prices rose or stabilized relative to crop and other input prices, which may have contributed to the stabilizing of pesticide use. During the late 1980s, pesticide prices again declined relative to the other input prices.

Several studies have shown pesticides to be cost-efficient inputs from the farmer's perspective. Headley (1968) used 1963 data to estimate that $1 spent on pesticides had a $4 return. Campbell (1976) estimated a return of $5 to $13 per insecticide dollar for apple production. However, these two studies have been criticized for using a model specification that could overestimate productivity (Lichtenberg and Zilberman, 1986). Hawkins et al. (1977) estimated an average return of $3.30 to 4.90 per herbicide dollar relative to cultivation in corn production.

Other studies show lower returns. Carlson (1977) showed that the productivity of cotton insecticides declined from the period 1964–1966 to

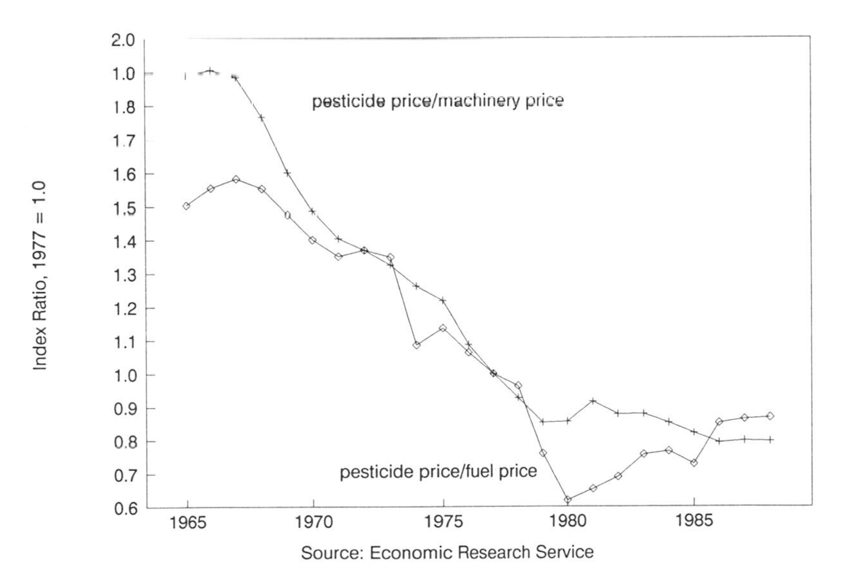

Figure 12.6. Relative price of pesticides compared with machinery and fuel prices.

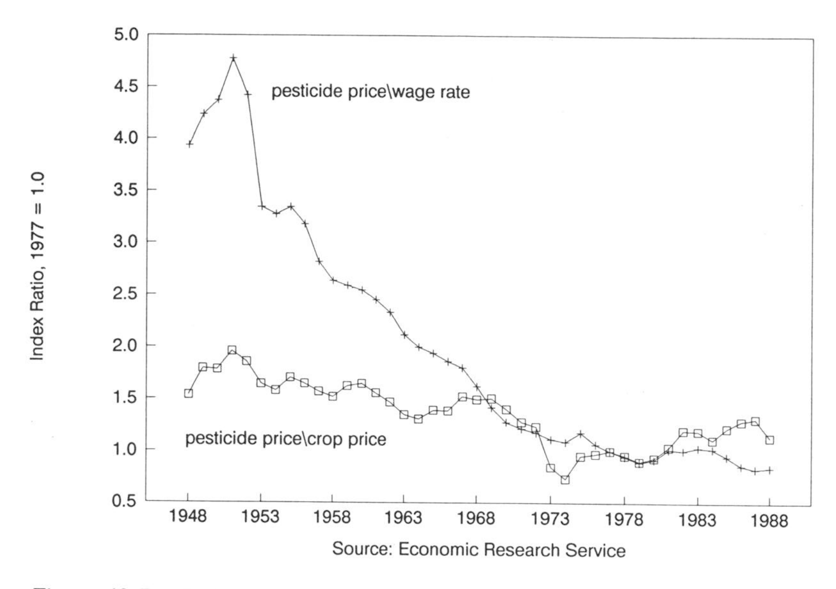

Figure 12.7. Relative price of pesticides compared with crop price and wage rate.

1966–1969 due to increased insect resistance and that demand shifted to organophosphates. Lee and Langham (1973) used 1964–1968 data to estimate that the marginal returns in citrus production were less than marginal cost, which implied overuse of pesticides.

Miranowski (1975) estimated returns of $2.02 for insecticides and $1.23 for herbicides per dollar spent on corn production in 1966. During that same year, the returns for cotton were $0.09 for insecticides and $1.82 for herbicides. While it appears that cotton insecticide use did not justify expenses that year, it is not clear whether farmers' expectations justified that expense. Duffy and Hanthorn (1984) showed average returns in 1980 to be $1.03 for corn insecticides and $1.05 for herbicides. For 1980 soybeans, their estimates were $0.57 for insecticides and $1.13 for herbicides. The lower returns to pesticides in later years may be a result of the equation specifications used in estimation, but such a decline in marginal productivity as use increases is expected.

Effect of Farm Programs

Many economists argue that the combination of price supports and acreage restrictions in commodity programs encourages more pesticide use than would be optimal under free markets (Headley, 1971; Miranowski, 1975). However, it is not clear to what extent the current level of pesticide use is due to the cost-effectiveness of pesticides or to incentives created by commodity programs.

Farm programs could encourage more pesticide use in a variety of ways (Miranowski, 1975): (1) Target prices and acreage restrictions could increase returns to program crops, which would encourage greater per-acre use of pesticides and other inputs to increase crop yields. (2) Acreage restrictions could encourage farmers to substitute pesticides and other inputs for land (and increase per-acre use) to maintain higher production. (3) Higher per-acre returns and lower financial risk could encourage farmers to plant more acres of high-pesticide-use program crops and plant less acreage of lower-pesticide-use nonprogram crops. (4) Inflexible base-acreage rules and reduced financial risk could encourage continuous cropping and discourage pesticide-reducing crop rotations. Alternatively, the acreage restrictions of commodity programs can reduce total pesticide use in comparison to previous years by reducing the acreage planted to high-pesticide-use program crops.

Pesticide use grew rapidly during the 1960s when farm programs restricted crop acreage. This period is often used to argue that farm programs encouraged more pesticide use than free markets would have encouraged. Richardson (1973) argued that the programs hastened agriculture's adjustment to an optimal mix, but by 1965–1969 had not encouraged overuse of pesticides. From the mid-1970s to the early 1980s, acreage restrictions were relaxed, and export demand for U.S. commodities was high. Rising prices and crop acreages were associated with pesticide use growing to market saturation.

During the 1980s, low crop prices, acreage diversion, and land retirement contributed to reductions in pesticide use. However, Carlson (1990) found that acreage diversion programs during the 1980s caused small but statistically significant increases in the number of herbicide treatments per acre for corn and soybeans and insecticide treatments per acre for cotton.

Acreage allotments, which were eliminated in 1977, related program benefits to historical production patterns and discouraged shifts away from those patterns (Eriksen and Collins, 1985). Those programs may have encouraged farmers to grow crops, such as corn or cotton, in areas of high pest infestation and to use more pesticides (Pimentel and Shoemaker, 1974). The elimination of allotments encouraged adjustments in cropping patterns. For example, cotton production shifted to the West and Southwest, where insecticide use is less than in the Southeast and mid-South (Stults and others, 1989).

Recent changes in farm legislation should reduce the link between farm programs and pesticide use. Under the 1985 Food Security Act (FSA), USDA administratively froze farm program yields, which should reduce or eliminate the effect of target prices on per-acre use of pesticides. However, there were still incentives to plant program crops and maintain acreage base to receive future payments. Various approaches, with such names as 50/92 and 0/92, were used under the 1985 FSA to give farmers some flexibility to grow nonprogram crops without reducing the crop acreage base for future payments. Under the 1990 FSA, the triple base concept increases planting flexibility without reducing the crop acreage base for future payments, which should reduce incentives to plant some high-pesticide-use crops and increase incentives to use pesticide-reducing crop rotations. Changes in farm programs may be contributing to the reduced percentage of corn acreage treated with insecticides (Fig. 12.3). It is often alleged that commodity programs encourage farmers to plant continuous corn, and insecticide use is much higher on continuous than on rotated corn (Daberkow et al., 1988). Preliminary analysis of 1990 pesticide survey data show no significant difference in continuous corn planted on farms participating in commodity programs and those not participating, which implies that current commodity programs do not affect insecticide use on corn (Gill and Daberkow, 1991).

Pesticide Use and Productivity

Many of the studies cited above show that pesticides were cost-effective and thus contributed to productivity by reducing cost per unit of output. Indeed, much of the evidence shows use to be close to economic optimality from the farmer's viewpoint. Pesticide use increased dramatically while agricultural productivity was also increasing. The USDA (1990) index of productivity increased by a factor of 2.2 between 1947 and 1988, while the index of agricultural chemical use (pesticides and fertilizers) increased by a factor of 21 (Fig. 12.8). (These indices do not measure the contribution of any factor to changes in productivity.) However, one of the major concerns about pesticides is that increasing their use is not a panacea for pest control.

Effect of Pesticide Use on Yield Losses

What effect has increased pesticide use had on yield losses to pests, and what does that effect mean for productivity? Pimentel and others (1991) claim that "The share of crop yields lost to insects has nearly doubled during the past 40 years despite a more than 10-fold increase in the amount of toxicity of synthetic insecticides used." They report percentage crop losses from a variety of sources and show USDA estimates of insect losses rising from 7% for 1942–1951 to 13% losses for 1951–1960 (USDA 1954, 1965) (Table 12.8). Schwartz and Klassen (1981, 1990) have an alternative interpretation of information in the same USDA reports. They say that

Figure 12.8. Productivity and agrichemical use.

while the pest species causing losses have changed, yield losses to pests have not decreased appreciably over the same time period. All of these authors identified pest resistance and changes to monoculture, uniform varieties, reduced tillage, and other practices as reasons for pest losses not decreasing. Additionally, Pimentel and others identify reduced FDA tolerances for insects and insect parts in foods; more stringent "cosmetic standards"; and higher-yielding, less-pest-resistant varieties as reasons for rising insect losses.

Table 12.8. Percentage annual pest losses reported by Pimentel.

Period	Insects	Diseases	Weeds	Total	Source
	Percent				
1904	9.8	NA[a]	NA	NA	Marlatt, 1904
1910–1935	10.5	NA	NA	NA	Hyslop, 1938
1942–1951	7.1	10.5	13.8	31.4	USDA, 1965
1951–1960	12.9	12.2	8.5	33.6	USDA, 1954
1974	13.0	12.0	8.0	33.0	Pimentel, 1976
1986	13.0	12.0	12.0	37.0	Pimentel, 1986

Source: Pimentel et al., 1991.

[a]NA = not available.

Table 12.9. Pest losses in crop, range, and pasture production, USDA reports.[a]

Period	Insects	Diseases	Weeds	Nematodes	Total
1951–60					
Value					
(billion dollars)	2.2	3.3	2.5	0.4	8.3
Percent	7.5	11.0	8.3	1.3	28.1
1942–1951					
Value					
(billion dollars)	2.0	3.3	2.3	*	7.6
Percent	7.5	12.0	8.3	*	27.8

* = Included with diseases.

[a]Sales of crops and livestock are used to compute percent losses: $30 billion for 1951–1960 and $27 billion for 1942–1951, as reported in Table 2, p. 682, of Pimentel et al. (1991). Crops include field crops, forage seed crops, hay, range and pasture, fruits and nuts, and vegetables.

Sources: USDA, 1954 and 1965; Pimentel et al., 1991.

The percentages in Table 12.8 were computed by Pimentel and others (1991) and do not appear in the two USDA publications cited. My view is that that the value of losses used in the computations were not comparable for the two time periods. Specifically, insect losses in turf production; the maintenance of lawns, golf courses, and other turf areas; and flowers grown in private, public, and commercial gardens, which were not for sale, were included in the 1951–1960 estimates but not the 1942–1951 estimates. These insect losses accounted for 42% of the insect losses reported for 1951–1960. Also, losses on range and pasture were included in the computations for 1951–1960, but not for 1942–1950. Recomputing the percentages by excluding insect losses to turf and flowers not grown for sale for 1951–1960 and adding range and pasture losses for 1942–1951 shows that losses, as a percent of crop and livestock sales, were virtually identical for the two time periods (Table 12.9). (Please note that the computations presented in Tables 12.9 and 12.10 do not account for the economic effects of price changes caused by production losses.)

I also estimated pest damages to crops as a percentage of value of production and of potential production.[4] The estimates include losses for field crops, forage seed crops, hay, range and pasture, fruits, nuts, and vegetables but exclude losses for livestock, forestry, ornamentals, turf, and

[4]Crop value includes the value of crops sold and of those produced and consumed on-farm—for example, feed grain produced and fed to cattle on the same operation. These estimates were obtained from *Agricultural Statistics* for the years 1944–1962. Potential production for each time period adds all production losses shown in the appropriate USDA report.

Table 12.10. Pest losses in crop production.[a]

Period	Insects	Diseases	Weeds	Nematodes	Total
1951–1960					
Value of loss (billion dollars)	2.1	3.1	1.8	0.4	7.4
Percent of production	11.2	16.0	9.6	1.9	38.7
potential production	7.2	10.4	6.2	1.3	25.1
1942–1951					
Value of loss (billion dollars)	1.9	2.8	1.8	*	6.6
Percent of production	11.7	17.1	10.8	*	39.6
potential production	7.3	10.7	6.8	*	24.8

* = Included with diseases.

[a]Included are field crops, alfalfa and hay, forage seed crops, fruits and nuts, vegetables, and horticultural specialties grown for sale. For 1951–1960, the value of production is $19.1 billion, and the potential value (value of production plus production losses) is $29.5 billion. For 1942–1951, the value of production is $16.6 billion and the potential value is $26.5 billion.

Sources: USDA, 1954 and 1965; USDA, *Agricultural Statistics*, 1944–1962.

other commodities. My examination shows that the percentage pest losses for these crops were virtually identical for the two time periods (Table 12.10). (Range and pasture are included in the computations for Table 12.9 but excluded for those in Table 12.10).

All these estimates should be viewed as having a significant degree of uncertainty, because, for most crops and pests, experts made subjective estimates of highly variable pest losses. Different experts at different times had different perceptions of infestations and damages and used different methods and assumptions for estimating losses. The Foreword to the 1965 USDA report said the experts may have had "a better basis for estimating losses" than used for the 1954 report. As an indication, the 1954 report allocated only half of estimated insect losses and none of the weed losses to specific crops, but the 1965 report allocated all estimated losses to specific crops and pests. As a result, it is difficult to make meaningful crop-by-crop comparisons between the two reports.

While percentage pest losses might be rising, the 1954 and 1965 USDA reports on pest losses do not provide evidence to support that argument. The interpretation that percentage pest losses are not falling is consistent with my examination of the reports. My view is that pesticides have contributed to a more efficient agricultural technology from the farmer's perspective, while weak evidence shows pesticides helping to maintain relatively stable pest losses, as a proportion of production.

Counterproductive Pesticide Applications

Despite the apparent contribution to production efficiency, increased pesticide use is not a panacea for all pest problems. While pest losses may be relatively stable over time, some growers might make counterproductive pesticide applications because they did not account for all effects of pesticides on pest damage and pest control costs. (1) Scheduled or prophylactic treatments to control low pest infestations may have little effect on yield, and the value of damage reduction might not exceed cost. (2) Some applications destroy beneficial organisms and natural enemies to pests. As a result, secondary outbreaks could require additional treatments, while species that were adequately controlled by natural enemies become pests. (3) Continued exposure of pest populations to a chemical often leaves the most resistant individuals, which reduces the effectiveness of the chemical, creates the potential for pest outbreaks, and encourages further counterproductive pesticide use. (4) Continuous plantings of some crops, such as corn, can encourage pest population growth and greater use of pesticides than rotating several crops. A monoculture of genetically uniform, high-yielding varieties and overuse of pesticides without regard for beneficial species or pest resistance can create the potential for damaging pest outbreaks. As a result, reducing pesticide use could lower pest damage and control costs in some circumstances.

Economic Thresholds

The economic threshold, which entomologists began discussing around 1960, defines an alternative to scheduled or prophylactic treatments. The theory of the economic threshold is based on the notion that pests should be controlled only when the value of damage reduction exceeds the cost of control (Stern et al., 1959; Hillebrandt, 1960; and Headley, 1972). Information, including pest monitoring and damage projection, is used to eliminate uneconomic pesticide applications. Higher crop prices or lower control costs increase optimal dosages or lower thresholds, according to economic theory. By eliminating uneconomic applications, thresholds can reduce pest control costs; destruction of beneficial species and natural enemies of pests; pest-resistance development; and adverse health, safety, and environmental effects. The costs of using thresholds include pest monitoring and additional management skills.

Risk and uncertainty encourage more pesticide use through higher dosages or lower thresholds. Risk encourages risk-averse farmers to increase

pesticide use to reduce probabilities of large losses. As a result, Turpin (1977) suggested that crop insurance might reduce the risk of infrequent, but severe, damage at less cost than pesticide use. Improved monitoring information about pest damage can reduce uncertainty and thus reduce dosages or increase thresholds.

Integrated Pest Management

Integrated pest management (IPM) is an approach used to reduce counterproductive pesticide applications. IPM focuses on optimizing the use of chemical, biological, and cultural controls, including varietal resistance to pests, trap crops, augmentation of natural enemies, and crop rotation, to manage pest problems rather than rely solely on chemical use (Smith et al., 1976). IPM generally includes pest monitoring and economic thresholds. Closely related are large-area control programs that attempt to coordinate grower actions to control mobile pests (Graebner et al., 1984; Good et al., 1977). Several studies show that risk-averse farmers may choose nonchemical practices to reduce pest damage and reduce variability of returns (Greene et al., 1985; Lazarus and Swanson, 1983; Liapis and Moffitt, 1983). However, one study showed that premature insecticide applications to soybeans in Georgia had little effect on net returns when compared with strict threshold compliance, allowing farmers to maintain a high level of crop protection without incurring the costs of an IPM program (Szmedra et al., 1988).

The most successful adoption of IPM has come with such crops as cotton, fruits, and vegetables, where per-acre use and costs of insecticides are high. Gianessi and Greene (1985) cited several studies where IPM reduced pesticide applications and costs for vegetables and stated that adoption by growers varies considerably by crop. Ferguson (1990) reported that 56% of cotton acreage was professionally scouted in 1989.

Pesticide Regulatory Policy

Pesticide use has grown within the context of regulatory policy. One of the most important issues shaping U.S. pesticide policy has been the balance of production benefits against the health and environmental hazards of pesticide use (Conner et al., 1987). There have been major public reactions to the alleged health and environmental hazards of increased pesticide use. One major idea that changed regulatory policy is that the hazards of using some pesticides might outweigh their benefits. Currently, regulatory policy recognizes a role for pesticides in crop production but emphasizes protection from hazards. The focus is on removing "unsafe" pesticides from the market—those pesticides where risks outweigh benefits—not on restricting the extent of pesticide use.

Pesticide Registration

Before a pesticide can be used in the United States, it must be registered under the Federal Insecticide, Fungicide, and Rodenticide Act (FIFRA), currently administered by USEPA. Pesticide registrations specify sites (such as specific crops or livestock) where pesticides can be applied, methods of use, or locations of use for pesticide products. Currently, registration decisions consider potential health, safety, and environmental hazards, as well as economic benefits of use.

Before a pesticide can be registered for use on a food crop, the Federal Food, Drug, and Cosmetic Act (FDCA) requires residue tolerances or exemptions from tolerance for the raw commodity and all processed commodities, rotational crops, and livestock where residues can be found. Under the Delaney clause, a tolerance cannot be established for a carcinogenic pesticide on a processed food if the pesticide residue concentrates in that food. The Delaney clause does not apply if the carcinogenic pesticide does not concentrate in the processed food. The establishment of residue tolerances considers both benefits and risks to public health, except that benefits are not considered in cases subject to the Delaney clause. Currently, USEPA establishes residue tolerances, while FDA monitors residues and enforces the tolerances.

An important part of the regulatory process, and the most publicized, is the procedure for modifying or cancelling registrations. If data show that a potential hazard exceeds a health, safety, or environmental standard and if the registrant chooses to defend the registration, the pesticide enters an administrative review, currently called Special Review. During this review, USEPA examines the risks and benefits of the pesticide's use and decides whether the registration should be retained, modified, or cancelled.

A Review of Changing Policy

From the early 1900s when chemical pest control was in its infancy until the 1960s when pesticide use was growing rapidly, U.S. pesticide legislation encouraged adoption of the new technology by regulating product effectiveness, requiring labeling of contents, and issuing warnings to users about acutely toxic materials (Conner et al., 1987; National Academy of Sciences, 1980). The first U.S. pesticide law, the Insecticide Act of 1910, prohibited the manufacture, sale, or transport of adulterated or misbranded pesticides. FIFRA of 1947 required all toxic chemicals for sale in interstate commerce to be registered against manufacturers' claims of effectiveness by USDA.

FIFRA further required that the product label specify content and whether the substance was poisonous.

Concerns about the presence and safety of chemical residues in food emerged in the 1950s. The 1954 Miller Amendment to FDCA required that tolerances for pesticide residues be established for food and feed. The Delaney clause was passed as part of the Food Additive Amendment to FDCA in 1958.

The 1962 publication of Rachel Carson's *Silent Spring* focused public attention on the potential hazards of chemical use to the environment, when use was growing rapidly. Since then, public demands for protection from health and environmental hazards have forced many changes in FIFRA and regulatory institutions.

Authority for administering FIFRA and the pesticide regulatory functions of FDCA transferred to EPA when it was created in December 1970. FIFRA was amended in 1972 by the Federal Environmental Pest Control Act (FEPCA), which mandated reregistration of all previously registered pesticide products within 4 years using new health and environmental protection criteria. Materials with risks that exceeded those criteria were subject to cancellation of registration, but only after a comparison determined that risks outweighed benefits. The 1978 FIFRA amendments eliminated the deadline for reregistration but required an expeditious process.

Many legislators and their constituents, including pesticide registrants and environmental groups, became impatient with USEPA's slow completion of individual pesticide reviews and its progress in the reregistration process. FIFRA amendments were passed in 1988 to speed the reregistration process and provide EPA with additional financial resources. The amendments required that all pesticides containing active ingredients registered before November 1, 1984, be reregistered by 1995. The funds come from a system of reregistration and annual maintenance fees levied on pesticide registrations. Registrants have dropped many pesticide registrations rather than pay the fees or incur the costs of providing data required by the reregistration process.

The regulatory process has modified some pesticide registrations and removed some materials from the market. Important issues addressed under the revised regulatory process included farmworker safety, cancer risks, birth defects, and wildlife mortality. In recent years, groundwater quality, endangered species, and food safety have become particularly important pesticide issues. While the regulatory process has changed the mix of pesticides, the extent of pesticide use has primarily responded to such economic factors as input and output markets and commodity programs.

Recent Concerns

A major issue is the so-called Delaney paradox, where a no-carcinogenic-risk rule applies to granting residue tolerances for pesticides that concentrate in processed food and a benefit-risk rule applies to those that do not concentrate (National Academy of Sciences, 1987). Also, EPA has refused to grant tolerances to new pesticides that apparently have less carcinogenic risk than currently registered materials. As an example, the National Academy of Sciences (1987) discussed a tolerance that was denied for fosetyl Al (a fungicide with the brand name Aliette) on hops, even though fosetyl Al was estimated to have much less oncogenic risk than that of ethylenebisdithiocarbamate (EBDC) fungicides used on hops.

Most proposals to resolve the paradox involve a negligible-risk rule for carcinogens, often defined as a one-in-a-million or one-in-a-hundred-thousand chance of contracting a cancer in a 70-year lifetime. (This risk is in addition to the background cancer risk, from any cause, of 25%.) Some pesticides could be registered under a negligible-risk rule, but could not be under a zero-risk interpretation of the Delaney clause. The purpose of the new rule would be to reduce risk by registering safer pesticides than some currently being used and to simplify tolerance-setting by subjecting all foods and pesticides to the same rule. USEPA has written an administrative rule to interpret the Delaney clause with a negligible-risk rule instead of the zero-risk rule, but USEPA has been sued by environmental groups to keep the zero-risk rule.

Some legislative proposals (S. 1074 introduced by Sen. Kennedy and Sen. Dodd and H.R. 2342 introduced by Rep. Waxman) would resolve the paradox by applying a risk-only rule for all carcinogens on all food crops. Eliminating the benefit/risk comparison would make the new rule more restrictive than the current one where the Delaney clause does not apply. (The Kennedy and Waxman proposals would require total dietary carcinogenic risk for a particular pesticide from all sources or, in some cases, total dietary carcinogenic risk from all commonly used pesticides on each food to be negligible.) Other proposals (the President's Food Safety Initiative, H.R. 3216, introduced by Reps. Bruce and others) would apply a risk/benefit rule to all foods, resulting in a less restrictive rule where the Delaney clause now applies.

Currently (October 1991), the Delaney paradox remains unresolved. And there is increasing public awareness of potential hazards of pesticides in food, water, and the environment, political pressure for greater protection from hazards, and less public concern for production benefits. Some would apply risk-only rules to registration decisions. Because many people are uncertain about the severity of pesticide hazards in food and water, they cannot make informed choices to avoid or reduce such hazards. In the daminozide (trade name Alar) controversy, most people did not know which apples were treated with daminozide or how severe the risk really was. In some cases, people simply refused to eat apples or allow their children to eat them. In such cases, many people came to fear the unknown, perhaps overestimating the severity of pesticide hazards relative to other hazards, and to mistrust government's ability to control the hazards. These concerns should be major issues in debates over the reauthorization of FIFRA and amendments to FDCA.

Also emerging is an interest by some groups in the United States in restricting or reducing the total amount of pesticides used (Pimentel and others, 1991). Their goal is to reduce the adverse environmental and health effects of pesticide use. Denmark and Sweden have already instituted programs to reduce pesticide use by 50%. Pettersson (1991) said that quantity of active ingredient used in Sweden was reduced by 50% between 1985 and 1990 with little effect on acreage treated, which was attributed to reducing application rates, using more efficient application technology, and changing to new, lower-application-rate pesticides.

One potential reason for the interest in reducing pesticide use in the United States is that the regulatory process focuses on removing "unsafe" pesticides from the market or restricting how they can be used, not on limiting the extent of pesticide use. Many of the proponents of restricting pesticide quantity argue that some pesticides are overused; more efficient application technology, nonchemical practices, pest monitoring and economic thresholds, or crop rotations can reduce pesticide use with relatively small economic losses; and adverse environmental and health effects would be reduced significantly. Pimentel and others (1991) estimated that the economic loss of a 35–50% reduction in pesticide use would be minor—about $1 billion. Clearly, many in the agricultural community believe that substantial reductions in pesticide use would be more difficult and costly. Knutson and others (1990) discussed the potential for much more severe losses and estimated an $18 billion loss if all pesticides were banned.

A Weakness in Special Review

I think that the greatest weakness in EPA's approach to balancing risks and benefits is the pesticide-by-pesticide approach of Special Review, where individual pesticides are addressed when questions are raised about hazards. The problem is that regulatory decisions affecting alternative pesticides are interdependent in both risks and benefits. My view presumes, of course, that benefits should remain an important consideration in pesticide policy. Regulatory decisions are interdependent, because previous registration and reregistration decisions determine the availability of chemical alternatives to the pesticide under current review. Interactions in chemical use for such purposes as managing resistance can also create interdependence.

Estimating Benefits

A brief discussion of estimating benefits shows the role of alternatives, which leads to interdependence in regulatory decisions. The assessment of benefits is essentially the same as estimating the social welfare loss, excluding health and safety effects, of removing the pesticide from the market and switching to the best chemical or nonchemical alternative controls, if any.

The first step is to estimate the cost-effectiveness of an active ingredient compared to its alternatives, which requires estimates of the pesticide's use and of yield and cost changes caused by switching to alternatives. Data for benefit assessments are often scarce. In many cases, crop and pest control experts provide subjective estimates.

If the alternatives are less cost-effective, the cost per unit of output will increase, and there can be a variety of economic effects. When supply changes have no effect on price, the economic efficiency loss of an action is simply the value of production loss plus change in production cost caused by switching to alternatives.

When supply changes are large enough to change prices, a more extensive analysis is desirable, providing the necessary economic methods and information can be assembled. The social welfare loss is generally viewed as the net of consumer-plus-producer effects, but the magnitude of differential or distributional effects on different groups also becomes important. Because of higher prices and lower quantities, consumers pay a portion of the cost of the ban. The combination of price, yield, and cost changes can change planting decisions and the acreages of alternative crops grown. Since yields and costs would change only on acreage where the pesticide is used, farmers who do not use the banned pesticide might gain if prices rise, while users might gain or lose depending on how much prices and costs increase.

Differential yield and cost impacts among regions, owing to differences in pest infestations or soil or climatic factors that influence the effectiveness of alternatives, mean that some regions might gain and others lose. Higher prices reduce commodity program payments and offset gains in market revenues.

Some parties might feel that these differential or distributional effects are fair, while others might not. It is important to examine such differential effects, because fairness can become a major issue in regulatory decisions. The methods of estimating efficiency and distributional effects of pesticide regulatory decisions is an often-debated subject.

Implications of Alternative Pesticides

Regulatory decisions are interdependent for benefits, because previous regulatory decisions define the availability of chemical alternatives and influence the economic effects of later decisions. If chemical or nonchemical alternatives are nonexistent, ineffective, or too expensive, a pesticide can have substantial benefits. As the cost-effectiveness of alternatives increases, the benefits of a pesticide decrease. Thus, the availability of cost-effective chemical or nonchemical alternatives may be sufficient reason to remove a pesticide with health or environmental concerns from the market. When there are several cost-effective alternatives, the economic benefits of controlling a pest are much greater than the benefits of any single pesticide.

The interdependence exists for risks as well, because chemical alternatives to the pesticide in question also have risks. In the recent past, USEPA has assessed the hazard and exposure of chemicals in question, but not of their chemical alternatives. So, absolute risks were weighed against comparative benefits. It is imperative to compare the risks of alternatives, because some pesticide bans could ultimately increase health or environmental hazards while reducing the cost efficiency of crop production, an undesirable outcome. In those cases where a decision would reduce risks, would risks be reduced enough to justify the loss of production benefits or the distributional effects? It is often difficult to compare risks, because risk data are scarce for many pesticides. The necessity to compare risks has been recognized by USEPA in recent years and is an important concept in Administration proposals to modify FIFRA and FDCA.

The interdependence of regulatory decisions creates a potentially serious dilemma for the Special Review process. USEPA might find that one of the remaining chemical alternatives has greater risks than previously banned pesticides, but that it also has substantial benefits because the remaining chemical or nonchemical alternatives are less cost-effective. A better choice may have been to leave a previously banned material on the market and ban the one under consideration. Accelerating reregistration and Special Reviews to meet deadlines of the 1988 FIFRA amendments could force simultaneous but independent assessments of alternatives and aggravate the dilemma.

The sequence of pesticides assessed could substantially influence economic efficiency, income distribution, and risks borne by society over time. USEPA needs to review those pesticides that data indicate are the most hazardous and to examine risks of alternatives to reduce the possibility that a pesticide ban increases risk. USEPA might also simultaneously examine risks and benefits of most chemical and nonchemical alternatives used for a pest problem and determine an optimal strategy before deciding the fate of any single material. However, delaying a decision until risk data from all alternatives are available might still result in a ban of a pesticide under review while society suffers the adverse effects in the interim.

Summary

Pesticide use grew dramatically from the late 1940s to the early 1980s and then stabilized. Increased use of pesticides on major field crops has been a major factor. Two components are a dramatic rise in herbicide use on all crops and in pesticide use on corn and soybeans. Increased pesticide use is part of a larger technological change in agriculture that increased productivity by 2.2 times between 1947 and 1988, while weak evidence suggests that pest losses, as a proportion of production, have remained stable. Pesticide use appears to have grown to market saturation by 1980, so pesticide use is closely correlated to crop acreage. Growth in pesticide use has responded primarily to economic forces such as relative returns and farm programs, but pesticide regulation has changed the mix of materials used.

It is often argued that the use of pesticides grew because they have reduced costs of pest control while contributing to less-variable crop yields. Several studies cited previously indicate that, from the farmer's viewpoint, financial returns have justified pesticide use. Additionally, pesticide prices have fallen relative to crop and other input prices, which encouraged pesticide use. However, there is also an argument, supported by economic theory, that farm programs encourage more pesticide use per acre than is economically efficient. But in recent years, acreage restrictions have helped to stabilize pesticide use. Also, some analysts argue that the administrative freeze of program yields and increased planting flexibility allowed under recent farm legislation have reduced incentives for pesticide use.

However, increased pesticide use hasn't solved all pest control problems. One concern is that pesticides are overused, resulting in overly rapid development of pest resistance and mortality of beneficial species including natural enemies of pests. The result may be that farmers spend too much on pesticides and have greater pest losses than would otherwise occur. A response has been IPM and the use of economic thresholds to eliminate unnecessary, counterproductive pesticide applications and encourage nonchemical practices where economically feasible. IPM has been adopted most widely for high-value crops with high per-acre pesticide use, such as cotton, fruit, and vegetables.

Also important is the view that, from society's viewpoint, the health and environmental effects of some pesticides, including food safety, water quality, worker safety, and wildlife mortality, outweigh their production benefits. Adverse environmental and health effects often do not directly affect the farmer's decision to apply pesticides. Changing societal values toward pesticide risks and benefits have had a profound effect on pesticide policy. Pesticide regulatory policy was at first a response to the availability of the

new technology that encouraged adoption by attempting to assure product quality. However, public concerns, emerging in the 1960s, have changed policy to emphasize protection from various hazards, so most regulatory decisions involve a risk/benefit comparison. But there is continued public concern about pesticide hazards and EPA's ability to resolve pesticide controversies. One emerging view is that limits on or reductions in total pesticide use are needed to reduce environmental and health hazards. The result could be even stricter standards for health and environmental hazards or the institution of risk-only rules for registration decisions.

A major weakness in USEPA's approach to balancing risks and benefits is the pesticide-by-pesticide approach to Special Review. If there are several cost-effective chemical alternatives to control a pest, the first chemical reviewed will have few benefits because effective alternatives are available. The last available chemical alternative will have high benefits if no effective nonchemical alternatives are available. Also, chemical alternatives have risks, so a pesticide ban could increase risks while increasing costs per unit of output. Ultimately, the sequence of decisions could substantially influence economic efficiency, income distribution, and risks over time. To avoid such problems, USEPA needs to consider the *changes* in risks as well as the benefits when making regulatory decisions. USEPA might also simultaneously examine the risks and benefits of major alternatives to determine an optimal strategy.

References

Andrilenas, Paul A. *Farmers' Use of Pesticides in 1971—Quantities*. AER-252. U.S. Dept. Agr., Econ. Res. Serv., 1974.

Campbell, H.F. "Estimating the Marginal Productivity of Agricultural Pesticides: The Case of Tree-Fruit Farmers in the Okanogan Valley," *Canadian Journal of Agricultural Economics*, Vol. 24, No. 2 (1976), pp. 23–30.

Capalbo, S.M., and T.T. Vo. "A Review of the Evidence on Agricultural Productivity and Aggregate Technology," in *Agricultural Productivity Measurement and Explanation*. Eds. S. M. Capalbo and J. Antle. Washington, DC: Resources for the Future, 1988.

Carlson, G.A. "Long-run Productivity of Pesticides," *American Journal of Agricultural Economics*, Vol. 59, No. 3 (1977), pp. 543–48.

______. "Farm Programs and Pesticide Demand," unpublished manuscript, presented at Agricultural Economics Workshop, North Carolina State University, March 1990.

Carson, Rachel, *Silent Spring*. Greenwich, CT: Fawcett Publications, 1962.

Conner, John D., Jr., Lawrence S. Ebner, Charles A. O'Conner III, Christian Volz, Kenneth W. Weinstein, John C. Chambers, Alison A. Kerester, Stanley

W. Landfair, Risa H. Rahinsky, and Elizabeth M. Weaver. *Pesticide Regulation Handbook*. New York: Executive Enterprises Publication Co. Inc., 1987.

Crutchfield, S.R. *Cotton Agricultural Chemical Use and Farming Practices in 1989: An Overview of Results*. AGES 9076, U.S. Dept. Agr., Econ. Res. Serv., 1990.

Daberkow, Stan, and Katherine H. Reichelderfer. "Low-Input Agriculture: Trends, Goals, and Prospects for Input Use," *American Journal of Agricultural Economics*, Vol. 70, No. 4 (1988), pp. 1159–66.

Daberkow, Stan, LeRoy Hansen, and Harry Vroomen. "Resources: Low-Input Practices," *Agricultural Outlook*. AO-148. U.S. Dept. Agr., Econ. Res. Serv., Dec. 1988, pp. 22–25.

Duffy, M., and M. Hanthorn. *Returns to Corn and Soybean Tillage Practices*. AER-508. U.S. Dept. Agr., Econ. Res. Serv., Jan. 1984.

Eichers, Theodore R., Paul A. Andrilenas, and Thelma W. Anderson. *Farmers' Use of Pesticides in 1976*. AER-418. U.S. Dept. Agr., Econ., Stat., Coop. Serv., 1978.

———, Paul Andrilenas, Robert Jenkins, and Austin Fox. *Quantities of Pesticides Used by Farmers in 1964*. AER-131. U.S. Dept. Agr., Econ. Res. Serv., 1968.

———, Paul Andrilenas, Robert Jenkins, Helen Blake, and Austin Fox. *Quantities of Pesticides Used by Farmers in 1966*. AER-179. U.S. Dept. Agr., Econ. Res. Serv., 1970.

Eriksen, M.H., and K. Collins, "Effectiveness of Acreage Reduction Programs," in *Agricultural-Food Policy Review: Commodity Program Perspectives*, AER-530, U.S. Dept. Agr., Econ. Res. Serv., 1985, pp. 166–184.

Ferguson, W.L., "Cotton Pest Management Practices," in *Agricultural Resources: Inputs Situation and Outlook Report.* U.S. Dept. Agr., Econ. Res. Serv., AR-20, Oct. 1990.

Gianessi, Leonard, P., and Catherine R. Greene. "The Use of Pesticides in the Production of Vegetables: Benefits, Risks, Alternatives and Regulatory Policies," *Vegetables and Specialties Situation and Outlook Report*. TVS-245. U.S. Dept. Agr., Econ. Res. Serv., Sept. 1985, pp. 27–42.

Gill, Mohinder, and Stan Daberkow, "Crop Sequences Among 1990 Field Crops and Associated Farm Program Participation," in *Agricultural Resources: Input Situation and Outlook Report*. U.S. Dept. Agr., Econ. Res. Serv., AR-24, Oct. 1991.

Good, J.M., R.F. Hepp, P.O. Mohn, and D.L. Vogelsang. *Establishing and Operating Grower-Owned Organizations for Integrated Pest Management*. PA-1180. U.S. Dept. Agr., Ext. Serv., 1977.

Graebner, L., D.S. Moreno, and J.L. Baritelle. "The Fillmore Crop Protection District: A Success Story in Integrated Pest Management," *Bulletin of the Entomological Society of America*, Vol. 30, No. 1 (1984), pp. 27–33.

Greene, C.R., R.A. Kramer, G.W. Norton, E.G. Rajotte, and R.M. MacPherson. "An Economic Analysis of Soybean Integrated Pest Management," *American Journal of Agricultural Economics*, Vol. 67, No. 3 (1985), pp. 567–72.

Hawkins, D.E., W.F. Slife, and E.R. Swanson. "Economic Analysis of Herbicide Use in Various Crop Sequences," *Illinois Agricultural Economics*, Vol. 17, No. 1 (1977), pp. 8–13.

Headley, J.C. "Estimating the Productivity of Agricultural Pesticides," *Journal of Farm Economics*, Vol. 50, No. 1 (1968), pp. 13–23.

———. "Defining the Economic Threshold," *American Journal of Agricultural Economics*, Vol. 50, No. 1 (1972), pp. 13–23.

———. "Productivity of Agricultural Pesticides," *Economic Research on Pesticides for Policy Decisionmaking*. U.S. Dept. Agr., Econ. Res. Serv., Apr. 1971.

Hillebrandt, Patricia M. "The Economic Theory of the Use of Pesticides," *Journal of Agricultural Economics*, Vol. 13, No. 4 (1960), pp. 464–72.

Hyslop, J.A. *Losses Occasioned by Insects, Mites, and Ticks in the U.S.* E-444 U.S. Dept. Agr., Bur. of Ent. and Plant Quar., 1938, 57 pp.

Klassen, W., and P.H. Schwartz. "ARS Research Program in Chemical Insect Control," *Agricultural Chemicals of the Future*. Ed. J.L. Hilton. Totowa, NJ: Rowman and Allenheld Press, 1985.

Knutson, R.D., C.R. Taylor, J.B. Penson, and E.G. Smith. *Economic Impact of Reduced Chemical Use*. Knutson and Associates, College Station, TX, 1990, 72 pp.

Lazarus, W.F., and E.R. Swanson. "Insecticide Use and Crop Rotation Under Risk: Rootworm Control in Corn," *American Journal of Agricultural Economics*, Vol. 65, No. 4 (1983), pp. 738–47.

Lee, J.Y., and M. Langham, "A Simultaneous Equation Model of the Economic-Ecologic System in Citrus Groves," *Southern Journal of Agricultural Economics*, Vol. 5, No. 1 (July 1973), pp. 175–80.

Liapis, P.S., and L.J. Moffitt. "Economic Analysis of Cotton Integrated Pest Management Strategies," *Southern Journal of Agricultrual Economics*, Vol. 15, No. 1 (1983), pp. 97–102.

Lichtenberg, E., and D. Zilberman. "The Econometrics of Damage Control: Why Specification Matters," *American Journal of Agricultural Economics*, Vol. 68, No. 2 (1986), pp. 261–73.

Marlatt, C.L. "The Annual Loss Occasioned by Destructive Insects in the U.S.," *Yearbook of Department of Agriculture*, 1904, pp. 461–474.

Miranowski, J.A. *The Demand for Agricultural Crop Chemicals Under Alternative Farm Programs and Pollution Control Solutions*. Ph.D. dissertation. Harvard Univ., 1975.

National Academy of Sciences. *Regulating Pesticides*. 1980.

———. *Regulating Pesticides in Food: The Delaney Paradox*. 1987.

Pettersson, O. "Swedish Pesticide Policy in a Changing Environment," The Swedish University of Agricultural Sciences, Uppsala, draft, 1991, 24 pp.

Pimentel, D., "Agroecology and Economics," *Ecological Theory and Integrated Pest Management Practice*, Ed. M. Kogan, New York: John Wiley and Sons, 1986.

———, "World Food Crisis: Energy and Pests," *Bull. Ent. Soc. Am.* Vol. 20 (1976), pp. 20.

Pimentel, D., and C. Shoemaker, "An Economic and Land-Use Model for Reducing Insecticides on Cotton and Corn," *Environmental Entomology*, Vol. 3, No. 1 (1974) pp. 10–20.

Pimentel, D., and others. "Environmental and Economic Impacts of Reducing U.S. Agricultural Pesticide Use." *CRC Handbook of Pest Management in Agriculture*, Vol. 1, 2nd ed., Ed. David Pimentel, Boca Raton, FL: CRC Press, 1981, pp. 15–77. 1991, pp. 679–718.

Richardson, James W. "Farm Programs, Pesticide Use, and Social Costs," *Southern Journal of Agricultural Economics*, Vol. 5, No. 2 (1973), pp. 155–63.

Schwartz, P.H., and W. Klassen. "Estimate of Losses Caused by Insects and Mites to Agricultural Crops," *CRC Handbook of Pest Management in Agriculture*, Vol. 1, Ed. David Pimentel. Boca Raton, FL: CRC Press, 1981, pp. 15–77.

———. "Losses to Raw Agricultural Crops Attributed to Insect and Mite Pests," unpublished manuscript, 1990, 56 pp.

Smith, R.F., J.L. Apple, and D.G. Bottrell. "The Origins of Integrated Pest Management," *Integrated Pest Management*. Eds. J.L. Apple and R.F. Smith. New York: Plenum Press, 1976.

Stern, V.M., R.F. Smith, R. Van den Bosch, and K. Hagen. "The Integrated Control Concept," *Hilgardia*, Vol. 29, No. 2 (1959), pp. 81–101.

Stults, H., E.H. Glade, Jr., S. Sanford, and L.A. Meyer. "Cotton: Background for 1990 Farm Legislation." U.S. Dept. Agr., Econ. Res. Serv., Staff Report No. AGES 89-42, 1989, 82 pp.

Szmedra, P.I., R.W. McClendon, and M.E. Wetzstein. "Risk Efficiency of Pest Management Strategies: A Simulation Case Study," *Transactions of the ASAE*, Vol. 29, No. 6 (1988), pp. 1642–48.

Turpin, F.T. "Insect Insurance: Potential Management Tool for Corn Insects," *Bulletin of the Entomological Society of America*, Vol. 23, No. 3 (1977), pp. 181–84.

U.S. Department of Agriculture. *Agricultural Statistics*, issues from 1944–1962.

———. Agricultural Research Service. *Losses in Agriculture*. ARS-20-1, 1954.

———. Agricultural Research Service. *Losses in Agriculture*. Ag. Handbook 291, 1965.

———. Economic Research Service. *Agricultural Resources: Inputs Situation and Outlook Report*. AR-1, Feb. 1986; AR-5, Jan. 1987; AR-9, Jan. 1988; AR-13, Feb. 1989; AR-15; Aug. 1989; AR-17, Feb. 1990; and AR-20, Oct. 1990.

———. Economic Research Service. *Economic Indicators of the Farm Sector: Production and Efficiency Statistics, 1988*. ECIFS 8-5. Apr. 1990.

———. Economic Research Service. *Inputs Outlook and Situation Report*. IOS-6, Nov. 1984 and IOS-2, Oct. 1983.

———. National Agricultural Statistics Service. *Agricultural Chemical Usage: 1990 Field Crops Summary*. Ag Ch 1 (91), March 1991.

U.S. Environmental Protection Agency. *Pesticide Industry Sales and Usage: 1987 Market Estimates*. Feb. 1990.

13

Alar: The EPA's Mismanagement of an Agricultural Chemical

Janet S. Hathaway

Consumers Union found Alar in almost three-fourths of the apple juice it sampled in 1988 and 1989.[1] About a third of the apples tested by CBS's "60 Minutes" in May of 1989 contained the chemical.[2] Alar, a growth regulator which enhances firmness and color in apples and other fruit, is no longer sold in the United States for use on food. Uniroyal, the sole manufacturer of the chemical, voluntarily stopped most domestic sales in June of 1989[3] and in October announced a parallel cessation of overseas sales.[4] The chemical continues to be used, but only on ornamental plants and flowers. In March 1990, the Environmental Protection Agency set a schedule to phase down the legal limit for Alar residues in food (the "tolerance" level) and to make any detectable amount of Alar illegal in 1991.[5] What led to this precipitous decline in the use of a popular agricultural chemical? What can we learn from the Alar controversy about the power of consumers and environmental activists to reduce or eliminate the most dangerous pesticides in our food?

---

[1]Consumers Union, Press Release, "Consumers Union Announces Finds on Alar in Apples and Apple Juices, March 30, 1989.

[2]CBS's "60 Minutes", transcript of "What About Apples?" May 14, 1989.

[3]Press Conference Statement by James A. Wylie, Vice President and General Manager, Crop Protection Division, Uniroyal Chemical Company, June 2, 1989.

[4]Allan R. Gold, "Company Ends Use of Apple Chemical: Alar Manufacturer is Halting Most of Overseas Sales," *The New York Times*, October 18, 1989, p. A-18. *See also*, letter from Kenneth W. Weinstein, attorney for Uniroyal Chemical Company to Patricia A. Roberts, Office of General Counsel, U.S. Environmental Protection Agency, re: Voluntary Cancellation of Daminozide Registrations by Uniroyal Chemical Company, October 11, 1989.

[5]Environmental Protection Agency, "Pesticide Tolerance for Daminozide," 54 *Fed. Reg.* 6392, February 10, 1989.

See also EPA Press Release, "EPA Lowers Tolerances and Sets Tolerance Expiration Dates for Alar on Certain Food Commodities and Revokes Tolerances for Remaining Commodities," March 2, 1990.

History of Alar

Alar, or daminozide, was first registered for use on food in 1968.[6] The first study indicating the tumor-inducing activity of its breakdown product, UDMH, was released in 1973.[7] Numerous further studies, including ones published in 1977, 1978, and 1984, indicated that Alar or UDMH caused tumors in laboratory animals.[8] EPA first began an intensive review of the pesticide's risks in 1980 but shelved its investigations after closed meetings with Uniroyal. After the Natural Resources Defense Council (NRDC) sued EPA concerning its practice of holding secret sessions with pesticide manufacturers, EPA in 1984 reinstated its Special Review of Alar. In September 1985, EPA concluded that both Alar and UDMH were "probable human carcinogens" based on the studies it reviewed.[9] EPA estimated that Alar posed a dietary risk perhaps as high as one thousand cancers for every million people exposed.[10] EPA projected risks as high as four cancers for every hundred workers exposed while applying Alar to peanuts.[11] The agency stated that "continuing the current registrations for food uses of daminozide [Alar] presents unreasonable risks" and proposed cancellation of all food uses.[12]

Unfortunately, Alar was not banned. In 1986 EPA heeded the recommendation of a panel of scientists to not ban Alar but to merely request more cancer studies from Uniroyal. The Scientific Advisory Panel which called for further study has often been touted by the pesticide industry as a panel of independent experts. In fact, seven out of eight of the scientists serving on the Alar panel were later found by a Senate oversight subcommittee to have been "paid consultants to the chemical industry or to or-

---

[6]EPA, Daminozide Special Review Position Document 2/3/4, Draft, September 12, 1985, p. I-4.

[7]James V. Aidala, Specialist, Environment and Natural Resources Policy Division, Congressional Research Service, "Apple Alarm: Public Concern About Pesticide Residues in Fruits and Vegetables," March 10, 1989, p. CRS-7.

See also Toth, B., 1,1 Dimethylhydrazine (unsymmetrical) carcinogenesis in mice. Light microscopic and ultrastructural studies on neoplastic blood vessels. *Journal of the National Cancer Institute*, 50 (1): 181–194, 1973.

[8]EPA, Daminozide Special Review Position Document 2/3/4, Draft, September 12, 1985, [hereinafter called EPA, Alar Special Review] pp. II-1, II-2, II-14, and II-25.

See also Pepelko, W. *Memorandum—Evidence for Carcinogenicity of 1,1 Dimethylhydrazine (DMZ)*, Carcinogenic Assessment Group, U.S. Environmental Protection Agency, June 18, 1986. *IARC Monographs*, International Agency for Research on Cancer, Suppl. 7, 1987.

[9]EPA, Alar Special Review, pp. II-29 and II-30.

[10]EPA, Alar special review, p. II-47.

[11]EPA, Alar Special Review, p. II-52.

[12]EPA, Alar Special Review, Executive Summary, pp. 2–3.

ganizations supported by the industry, at the same time that they served on the panel."[13] NRDC and other environmental and consumer groups protested EPA's reversal of its decision to ban Alar, but to no avail.

The Alar Scientific Advisory Panel (SAP) concluded that the available data on daminozide and UDMH were insufficient to allow the performance of a quantitative risk assessment.[14] Other EPA experts came to a different conclusion, even when presented with the same data. In 1984, the EPA Carcinogen Assessment Group (CAG) determined that UDMH was a carcinogen and established a carcinogenic potency factor ("q_1*") based on the 1973 Toth study, the very study that the SAP so harshly criticized.[15] In 1987, CAG repeated its contention that existing evidence was more than adequate to classify UDMH as a "probable human carcinogen."[16] In addition to the EPA, both the International Agency for Research on Cancer (IARC) and the National Toxicology Program (NTP) concluded, again largely on the basis of the Toth study, that there was sufficient evidence that UDMH was a carcinogen.[17]

NRDC's Study of Children's Pesticide Risks: "Intolerable Risk"

In 1987 NRDC began a study to ascertain whether actual levels of pesticide residue in our preschoolers' diet pose a significant risk to children's health. NRDC collected government data on the diets of preschool children, actual pesticide residue levels, and risks posed by 23 of the approximately 300 pesticides legal for use on food in the United States.[18] NRDC selected 23 pesticides for this study based on availability of both toxicity

---

[13]"Government Regulation of Pesticides in Food: The Need for Administrative and Regulatory Reform," Report by the Subcommittee on Toxic Substances, Environmental Oversight, Research and Development to the Committee on Environment and Public Works, United States Senate, October 1989, pp. 33–34.

[14]Federal Insecticide, Fungicide and Rodenticide Act (FIFRA) Scientific Advisory Panel, Review of a Set of Scientific Issues Being Considered by EPA in Connection with the Special Review of Daminozide, October 4, 1985.

[15]U.S. Environmental Protection Agency, Office of Health and Environmental Assessment, *Health and Environmental Effects Profile for 1,1 Dimethylhydrazine*, EPA/600X-84/134, January 1984.

[16]Pepelko, W., U.S. Environmental Protection Agency, Carcinogen Assessment Group, *Memorandum—Evidence for Carcinogenicity of 1,1 Dimethylhydraxine (DMZ)*, January 9, 1987.

[17]International Agency for Research on Cancer, *IARC Monographs*, Suppl. 7, 1987.
U.S. Department of Health and Human Services, *Fourth Annual Report on Carcinogens*, 1985.

[18]The National Academy of Science estimates that 289 pesticides were legal for use on food in the United States in 1987. The Board on Agriculture, National Research Council, *Regulating Pesticides in Food: The Delaney Paradox*, National Academy Press, 1987, p. 51.

and residue data for each pesticide and information leading NRDC to expect significant use in foods commonly eaten by children. Of these 23 pesticides, the Environmental Protection Agency (EPA) identified eight as potential cancer-causing chemicals and 15 as capable of causing neurotoxic effects or damage to the nervous system.

NRDC estimated the health risk to preschoolers during their first years of life (ages 1–6 years) by examining actual exposure rates together with the risks of the pesticides. Consumption rates were derived for 27 fruits and vegetables most frequently eaten by children. Preschoolers' dietary exposure to the 23 selected pesticides was determined by combining children's consumption rates for the 27 food types with concentration of the 23 pesticides actually found in these foods.[19] Pesticide exposure estimates were then combined with data on the cancer potency or neurotoxicity of the pesticides to determine preschoolers' risk of developing cancer or experiencing a disruption in central nervous system function.

To develop an adequate database of preschooler exposure to pesticides, NRDC used consumption data from a nationwide food consumption survey of children and adult women conducted in 1985 by the U.S. Department of Agriculture (USDA).[20] The data on the residues of the 23 pesticides were derived from analyses of over 12,000 samples of the 27 fruits and vegetables obtained by regulatory programs of the Food and Drug Administration (FDA) and the EPA.[21] In the cases of daminozide/UDMH and the ethylene bisdithiocarbamates (EBDCs), NRDC relied on residue data submitted to the EPA by the pesticide's manufacturer.[22]

On February 26, 1989, CBS "60 Minutes" aired a program examining pesticide use and children's risk of contracting cancer. The "60 Minutes" report was largely based on the NRDC "Intolerable Risk" report. NRDC concluded that preschoolers were being exposed to hazardous levels of pesticides in fruits and vegetables. We estimated that between 5,500 and 6,200 of the current population of American preschoolers may eventually

---

[19] *Intolerable Risk: Pesticides in Our Children's Food*, The Natural Resources Defense Council, Washington D.C., February 27, 1989, pp. 15–23.

[20] USDA, Human Nutrition Information Service, *CSF II—Nationwide Food Consumption Survey: Continuing Survey of Food Intakes by Individuals, Women 19–50 Years and Their Children 1–5 Years, 6 Waves*, 1985.

[21] FDA, *List of Pesticides, Industrial Chemicals and Metals, Data by Fiscal Year, Origin, Sample Flag and Industry Product Code*, 1985 and 1986.

[22] EPA, Office of Pesticides and Toxic Substances, *Memorandum—Daminozide Special Review*. Phase III 1986 Uniroyal Market Basket Survey. May 18, 1985.

EPA, *Ethylene Bisdithiocarbamate (EBDC) Pesticides: Proposed Regulatory Options for the EBDCs*, 1989. The EBDC residue estimates are based on field trial data for maneb, mancozeb, and metiram, adjusted by percentage of crops treated. Exposure estimates for zineb were based on maneb, which is a chemically similar compound. To better approximate residues in food eaten, washing and cooking factors were applied.

get cancer solely as a result of their preschool exposure to eight pesticides or their "metabolites"[23] through eating fruits and vegetables during their preschool years.[24] These estimates were based on scientifically conservative risk assessment procedures. Our study found that more than 50% of a person's lifetime cancer risk from exposure to carcinogenic pesticides used on fruit is typically incurred in the first 6 years of life.

NRDC also found that more than 90% of the cancer risk we examined was contributed by UDMH, the potent cancer-causing metabolite of Alar. The average preschooler's UDMH exposure during the first 6 years of life alone was estimated to result in a cancer risk of approximately one case for every 4,200 preschoolers exposed. This risk of cancer was *240 times* greater than the cancer risk considered "acceptable" by EPA following a full lifetime of exposure.[25] For children who were heavy consumers[26] of the foods that contained UDMH residues, NRDC predicted one additional case of cancer for approximately every 1,100 children—910 times EPA's "acceptable" risk level.

NRDC also found that at least 17% of the preschool population, or three million children, receive exposure above levels the federal government considers safe to neurotoxic organophosphate insecticides just from eating raw fruits and vegetables. High-level exposures to these insecticides can cause nausea, convulsions, coma, and even death. The lower exposures received by preschoolers through their diets may impair learning and memory and otherwise alter neurological function.

Reaction to NRDC's "Intolerable Risk" Study

Government reaction to the NRDC report was predictable—EPA, FDA, and USDA were ubiquitous on the conference and media circuits, assuring the public that the food supply was safe.[27] Though agency representatives

---

[23]Metabolites are the chemical breakdown products resulting from the basic pesticide when the pesticide or the food in which the pesticide is found is heated, stored, processed, or simply subjected to digestion in the human body.

[24]These estimates were based on the eating habits of preschool children who participated in the 1985 USDA survey and who responded to the survey three or more times over the course of a year. These data were used to approximate average daily exposure over the year. However, cancer risk estimates were also made based on daily intake for all preschoolers in the survey and result in an estimated 5,700–6,400 additional cancer cases (2.6×10^{-4} to 2.9×10^{-4}) in the preschool population.

[25]EPA considers one cancer for every one million people exposed a "negligible" risk of cancer.

[26]NRDC considered children to be "heavy consumers" of a pesticide if they were above the 95th percentile of exposure to the pesticide through consumption of the foods we studied.

[27]Carole Sugarman, "Agencies Say Apples Safe, Chemical Not Imminent Risk," the *Washington Post*, March 17, 1989.

publicly decried what they considered "scare tactics" on the part of NRDC, EPA more quietly concluded that new data received from Uniroyal about the carcinogenicity of UDMH confirmed that this was indeed a potent tumor-inducing agent. In a document sent to the apple industry's trade association with little fanfare several weeks before the "60 Minutes" program on pesticides was scheduled to air, EPA stated once again that UDMH "risk estimates, based on the best information available at this time, raise serious concern about the safety of continued, long term exposure."[28] The EPA letter also concluded that dietary cancer risk to the average adult was approximately 50 cancers for every million people exposed and that risk to children was higher still.[29] Though an EPA official testified before Congress that NRDC's report "seriously misleads the public,"[30] EPA's own cancer risk estimates were in fact very similar to those NRDC revealed. Based both on the earlier cancer studies and on the new cancer studies submitted by Uniroyal late in 1988, EPA proposed, once again, to cancel Alar's food uses[31] and to list UDMH as a hazardous waste.[32] Obviously, despite all the posturing for the media's microphones and cameras, EPA did not consider Alar or UDMH safe.

Why Was Alar Banned? Why Did the Public Care?

The key to raising the public's attention to the dangers of Alar and other pesticides in food was the widespread coverage of NRDC's report in the popular media. NRDC has frequently conducted careful and credible studies, but probably never before has an environmental organization's study been so extensively discussed on television, in popular magazines, and in local newspapers. Had our study appeared only in *The New York Times* and the *Washington Post*, it would not have received attention from the

---

[28]Letter from Dr. John A. Moore, Acting Administrator, EPA, to International Apple Institute, re: Alar Decision, February 1, 1989, p. 3.

[29]EPA letter to International Apple Institute, February 1, 1989, p. 2. EPA never publicly issued a calculation of the lifetime risk from Alar for childhood exposure. However, government data indicated that preschool children's exposure to apple juice (the major dietary contributor of UDMH) was 18 times that of average adults. Therefore, it is quite likely that if EPA had calculated the risk to average preschool children, its risk estimate would have been even higher than NRDC's estimate of 240 cancers in a million.

[30]Press release of Dr. John A. Moore, Acting Deputy Administrator, U.S. Environmental Protection Agency, "Preliminary Assessment of 'Intolerable Risk: Pesticides in Our Children's Food,' a Report by the Natural Resources Defense Council," March 7, 1989.

[31]EPA, Pesticide Fact Sheet for Daminozide (Alar), Preliminary Determination to Cancel Food Uses, May 15, 1989.

[32]Letter from Sylvia K. Lowrance, Director, EPA, Office of Solid Waste, re: proposal to list UDMH as hazardous waste, May 1, 1990.

millions of Americans who care about food safety and children's welfare but who get their news from television and magazines. NRDC scientists and lawyers knew that our study was a well-documented indictment of the current federal pesticide regulatory program. We felt that our study deserved the widest possible dissemination in the media.

The public reacted to the Alar story with considerable anger and fear. Sales of apples and apple products declined rapidly,[33] and demand for organic foods was spurred.[34] Many conferences have been devoted to debating what led to consumers' outrage. My conclusion is that five factors were the decisive ones. First, this study was about a risk to children. Even people who will tolerate high risks for themselves are distressed at the prospect of high risk for their children. Second, the government has a legal obligation to regulate pesticides, and the government had evidence that Alar was a problem for more than a decade. Government's failure to carry out the law frustrates and angers the public, particularly when the problem —like the use of dangerous pesticides in agriculture—is unlikely to be solvable solely through the actions of private individuals.

A third element provoking the public's outrage was the fact that, in response to the report, the government and the apple industry appeared to equivocate or even lie to the public. EPA said that the food supply was safe virtually in the same breath as it said that Alar needed to be banned. The apple industry said that a person would have to eat thousands of pounds of apples a day to be at risk, even though both NRDC and EPA agreed that such a statement was a complete fabrication. Understandably, people resented having their intelligence insulted by industry and government.

Fourth, the report emphasized cancer, a dread and deadly disease which eventually afflicts a quarter of all Americans. Everyone knows someone who has struggled with cancer, and anyone with a shred of compassion wants to prevent the sufferings and death which result from cancer. Finally, the food which contributed most to the high cancer risk was the apple, which is a symbol of wholesomeness ("An apple a day keeps the doctor away.") and even of patriotism ("as American as apple pie"). The thought that our government was allowing cancer-causing residues on this most healthy of foods struck some as practically subversive.

While Alar is no longer a food safety problem, many other dangerous pesticides remain. The public has learned that when it flexes its muscles about pesticides, even the chemical industry cowers. It is high time for the entire nation to insist on a much safer food supply.

---

[33]"Sales of apple products dropped 30 percent in February. . . ." Carole Sugarman, "Apple Processors Urge the EPA to Ban Alar," the *Washington Post*, May 12, 1989.

[34]"California Certified Organic Farmers, an independent organization, has . . . been 'besieged by the public since the Natural Resources Defense Council Report came out,' Mr. Scowcroft said." Marian Burros, "Organic Food: Now the Mainstream," *The New York Times*, March 29, 1989.

Part IV

History, Public Attitudes, and Ethics in Regard to Pesticide Use

14

Values, Ethics, and the Use of Synthetic Pesticides in Agriculture

Hugh Lehman

Introduction

This chapter is divided into four parts. In the first part I make some remarks aimed at reducing obscurities associated, by many people, with the concepts of value and obligation. In the second part I am concerned primarily to distinguish value judgments and ethical judgments. In addition I discuss the nature of ethics and the value of systematic investigation of ethical issues. In the third part I review several ethical theories in order to formulate some basic ethical principles which may be applied in considering our obligations in regard to the use of synthetic pesticides. Only in the fourth part do I directly address the question, "What are our moral obligations in regard to the use of synthetic pesticides in agriculture?" Some people, impatient with the preliminary discussions, may think that we should have gone directly to a consideration of that question. However, if others are to understand and critically evaluate the claims made in the fourth part, the preliminary material is necessary.

These remarks are addressed to specialists in scientific disciplines who may not be familiar with ethics and value theory. For such individuals, the opening remarks serve to block out my perspectives on these and related matters from ontology and epistemology. Further, these remarks serve to call attention to assumptions which I make in thinking about our moral obligations. Making these assumptions explicit should contribute to understanding of these complex ethical issues. If we can find a set of assumptions on these matters that we share, then we can make progress in determining our obligations in regard to pesticide use. If we find that we disagree on some of these assumptions, we can pursue our disagreement rationally as a result of having made the assumptions explicit. Awareness of differences in opinion regarding these assumptions may suggest alternative and fruitful approaches to the central issues.

Philosophers will find these remarks on value theory and ethics incomplete. Several of the positions that I take here are controversial and I have not undertaken to defend them through extensive consideration of alternative positions, arguments, rebuttals, etc. Were I to do so any application of these ideas to consideration of the use of synthetic pesticides would be deferred indefinitely.

In the past few decades we[1] have acted precipitously in regard to the use of pesticides.[2] This has led, I believe, to putting the health of many human beings in jeopardy in addition to the accidental poisonings of which we are aware.[3] People whose health has been placed in jeopardy, have not, in many cases, been aware that they were being harmed and so have not accepted the risk voluntarily. Indeed, some of the people who will, most likely, suffer harm from our use of pesticides have not been born yet.

While many agricultural scientists continue to believe that proper use of agricultural pesticides does not pose significant threats to human health, and may even reduce human health risks through controlling organisms which produce substances which are toxic to humans, there is increasing concern both among scientists and the general public with respect to changes we are making to our environment.[4] We have changed the environment in which we live. We have changed the quality of elements which are essential for all life—of our air, water, and earth. Many people would regard purity of air, earth, and water as of fundamental importance. Given the choice, they might not be willing to trade these for cheaper food or other conveniences. However, most of us have not had the choice. Further,

---

[1]Saying "we have acted precipitously" or "we have changed the environment" (see next paragraph of text) is a simplification. Not all of us have used pesticides or changed our environment. Generally, I am thinking of those communities of people who have used, or permitted the use of, large quantities of synthetic pesticides without careful consideration of long-term consequences of such use. In the industrialized world, for example, we (the citizens) allowed widespread use of pesticides such as DDT long before we began to think carefully about ecological, environmental, or other potentially serious consequences of such use.

[2]Rapid increases in pesticide use are documented by Craig Osteen in chapter 12.

[3]For discussion of human pesticide poisonings see chapter 3 by David Pimentel et al. (in this volume).

[4]In lectures to my class in Ethics and Agriculture, agriculture scientists frequently suggest that concerns of the general public in regard to the effects of pesticide use on health are exaggerated or unwarranted. They call attention to the low levels of residues and to the health risks which arise from substances other than synthetic pesticides or from activities other than eating food containing residues. In the first paragraph of chapter 15 (this volume) Carolyn Sachs refers to similar claims by representatives of chemical companies and others. She also documents the increasing public concern regarding health effects of residues in food and water. Olle Pettersson notes, in chapter 8 (this volume), that there is a general assessment (presumably among scientists) that negative health effects of minor.

there is a spreading perception that we are damaging our environment in ways which jeopardize at least the quality of human life in the future and perhaps also the very possibility for continued existence of our species. Through our political processes we have simply allowed such changes to occur. However, as a free people, we are entitled to consider whether we wish to change practices in which we have engaged in the past and, if we so decide, to modify those processes. These questions are presently being publicly debated. Decisions will be made. I believe we are more likely to decide and act rightly if we consider these matters carefully in light of our values and moral principles than if we act otherwise.

Part I. Values

Section A. Remarks on the Nature of Values, and Value Judgments

Let us start by listing examples of judgments which are or which logically imply value judgments. I shall choose (without citing explicit references) examples similar to those expressed by agricultural scientists in early drafts of papers prepared for this volume. Other examples could be given.

1. Maximum potential yield is the yield produced under the best agricultural practices. (Saying that some agricultural practices are the best presupposes an evaluation. It is not unusual for agricultural scientists to refer to some agricultural practices as good.)
2. Use of pesticides sometimes produces adverse effects on the quality of soil. (Adverse affects are bad effects, a value judgment.)
3. Effects of pesticide residues on human health are of minor importance. (Here there is a value judgment expressed by the claim that certain effects are relatively unimportant. Further, the concept of health presupposes evaluations.)[5]
4. Many explanations as to why farmers use pesticides have some validity. (Here, the evaluation is of explanations rather than of agricultural practices or soil or health effects.)
5. We can do a better job of educating farmers concerning alternatives to the use of chemical pesticides.
6. Pest control experts are competent to determine whether certain uses of pesticides work. (Here there is an evaluation underlying the judgment that some people are experts. A person could not be an expert if he or she was not good at his or her particular type of job.)

---

[5]For a discussion of the value-laden nature of the concept of health see "On the Nature of Illness" *Man and Medicine* **4** (1979), by Bernard Rollin.

Apparently the term "values" is meaningful, yet many of us feel uneasy when reference is made to values. The unease may arise from either or both of two sources. First, we may be unsure what objects, if any, fall within the scope of this term. This is an ontological question. Second, we may have doubts as to whether assumptions regarding values can be supported by reasons or evidence. This is an epistemological question. Let us briefly consider each of these worries.

We shall approach the ontological question first. We shall do so by proposing a suggestion as to what is implied when an object is said to have value. However, prior to doing that I shall call attention to an assumption which, I believe, many scientists make, an assumption which I believe is mistaken. The assumption is that values and facts are mutually exclusive and that the real world consists exclusively of facts. But, if the real world consists exclusively of facts, then the term "value" refers to nothing at all and perhaps then any reference to values is, at best, misleading. If such a view were correct we should say that social scientists who believe that they are studying values are not doing so. Perhaps they are studying something else or perhaps they are studying nothing at all. People who hold that the universe consists of facts and that values are not facts would maintain that it would be better if we spoke of value judgments rather than of values. Clearly there are expressions of value even if the term "value" designates nothing.

I wish to challenge the contention that the real world consists exclusively of facts as opposed to values. To do so I shall call into question the assumption that the concepts of fact and of value are mutually exclusive. Values, in my sense of the term, exist because people (or other beings) value things. People (or other beings) value things either because of characteristics in those things or because of relationships amongst the things and the valuers. For example, a person may value certain sorts of plants because those plants provide nourishment for the person. Again, a person may say that classical music is good, that is, valuable, because listening to classical music causes him to have feelings which he cherishes. We may say that the real world consists of facts but this does not imply that values are not real also.

I have explained the noun "values" by reference to the verb "to value." Someone might ask that I explain the meaning of this verb. To do so would take us into the realm of psychology. To be brief, let us say that to value an object is to be motivated to make it real or to experience it or to possess it. To value something might be said to consist in having a positive attitude toward it. However, sometimes we speak of positive as opposed to negative values. To positively value an object is to be motivated to experience it, etc. To negatively value an object is to be motivated to prevent its becoming real or to avoid experiencing it or to rid oneself of it.

Value thus may be said to arise either through positive or negative attitudes. An alternative, though closely related, view of the nature of values implies that an object has value if it satisfies desires or if it would satisfy such desires.[6] On either of these views value is essentially connected to attitudes. An object has value either because it is desired or because it satisfies a desire. When the concept of value is explained along lines such as we have indicated it is clear that the real world does not consist solely of facts and not of values.[7] People (and other creatures) value things. They value things either because of the properties those things possess or because of the relationships in which those things stand to other objects. For example, it is a fact that certain people valued pesticides based on organochlorine compounds. They attempted to develop or acquire such compounds. Such compounds then were valuable. That is a fact.

Looking back to some of the examples of value judgments, we can say that in #2 it is implied that uses of pesticides have some effects which are negatively valuable, that is, which fail to satisfy desires of agricultural producers. Number 5 says that there are ways of educating farmers regarding alternatives to synthetic pesticides which satisfy the desire that farmers be knowledgeable concerning the existence of such alternatives. With respect to example #3, health is distinguished from disease by reference to desired physiological characteristics. This judgment is more complex in that it presupposes a grading of (the importance of) health effects. Such grading, according to the view of values proposed here, is ultimately based either on strength of desires or on capacity to satisfy desires.

We have been discussing the first of two factors which give rise to unease when reference is made to values. Let us now turn to the second. We made brief reference above to value judgments. I rejected the view that while there are value judgments there are no values. However, in considering evidence or reasons we must be concerned with value judgments or value statements, i.e., statements that such and such is good (valuable) or is worthless.

Some scientists may be tempted to conclude that serious ethical discussion concerning the use of pesticides is pointless since the basic principles of values and ethics which would underlie sound ethical arguments re-

---

[6]There is considerable philosophical literature regarding values. For extensive references and much more thorough discussion than I provide see *Introduction to Value Theory* by Nicholas Rescher (Prentice Hall, Englewood Cliffs, 1969).

[7]The above analysis of value may be said to be naturalistic since, according to this analysis, value is essentially tied to psychological states (states of objects found in the natural world). In this century there has been extensive discussion concerning naturalistic analyses. G.E. Moore claimed that such analyses committed what he called the naturalistic fallacy. For some discussion of this see *Ethics*, second edition, by William Frankena (Englewood Cliffs, Prentice Hall, 1973), p. 97f.

garding pesticide use have not been established. In my judgment such a conclusion is unwarranted. While some thinkers have claimed to have established certain ethical and value principles as true, we must agree that there is no concensus among thoughtful people as to which value and ethical principles are true.[8] However, this absence of consensus does not imply that serious ethical consideration is impossible or pointless. To see this, one need only note that serious debate concerning scientific conclusions is possible even though there is no concensus about many issues pertinent to that debate. For example, many statistical methods are widely used by scientists in assessing the strength of evidence for scientific conclusions even though there is no consensus among scientists concerning fundamental principles of scientific methodology.[9]

We have said that values arise out of people (or other creatures) being motivated to possess something, or experience it, or make it real. What is of value is the object of such a motive, i.e., the object that one wants to possess, etc. Now, if one thinks of value in this way, that there can be evidence for the existence of values does not seem particularly strange. The evidence that something is valuable is of essentially the same scientific sort as the evidence for other statements or principles affirmed in the social sciences. The idea that the evidence for judgments of value is essentially empirical scientific evidence was expressed in the work of the philosophers C.I. Lewis, John Dewey, and others. (See footnote 11.)

There are difficulties which arise with respect to evidential support for value judgments. These difficulties are comparable to the difficulties associated with knowledge of human attitudes or motivations in general. Some of these difficulties arise due to obscurities surrounding concepts of motivation and attitude. Some of the difficulties arise because there are often a number of desires which interact in respect to some item being evaluated. Consider example #2 of the six value judgments listed above. If one wanted to get evidence to confirm a value judgment concerning the overall value of a pesticide one would have to offset values that accrue

---

[8]R.M. Hare had claimed that a form of utilitarianism has been proved. See his *Freedom and Reason*, (Oxford, Oxford University Press, 1963), especially chapter 7. Alan Gewirth claims to have proved that all moral agents, e.g., most human beings, have moral rights. See his *Reason and Morality* (Chicago, University of Chicago Press, 1978). A number of moral philosophers have claimed that a number of moral principles have been established by the method of wide reflective equilibrium which was first formulated by John Rawls. See his *A Theory of Justice* (Harvard University Press, Cambridge, 1971).

[9]There have been many discussions concerning scientific methodology. At one extreme there is skepticism such as has been defended by Karl Popper. See *The Logic of Scientific Discovery* (1959, Basic Books). A review of alternative views is found in Israel Scheffler, *The Anatomy of Inquiry* (1964, Routledge and Kegan Paul). The controversy has continued, as may be seen by consulting articles in such journals as *Philosophy of Science* and *The British Journal for the Philosophy of Science*.

because the pesticide solves a pest problem with the (negative) values that accrue because the pesticide degrades the soil.[10] Getting evidence to determine the overall satisfaction of desires which results from such a case is a matter of great complexity. Those skeptical about the very possibility of scientific knowledge of attitudes (human or other sentient beings) are entitled to be skeptical about the possibility of knowledge of values also. However, even if it is impossible for us to have scientific knowledge of attitudes, we may have opinions about human attitudes which are evidentially justified in everyday commonsensical ways. Surely, some beliefs about the attitudes of our acquaintances are evidentially justfied. Whether such evidence is or could be made to satisfy scientific standards is a question which we shall not pursue here.

Section B. Objections to This Theory of Values and Replies to the Objections

Many questions might be raised with respect to the analysis of value that I have here proposed. Some people might call for a far more complete analysis of valuing than I have given here. I shall grant to such people that this account is seriously incomplete. However I shall not attempt to rectify this defect in this paper since to give a complete account would take many pages. Readers wishing such an account should turn to the work referred to in the bibliography of Rescher's work. (See footnote 6.)

Others might wonder whether this account of values is correct. They may note certain common beliefs about values which appear to be inconsistent with this account. In this regard, one objection is the belief that there is a distinction between what is valued and what is truly valuable. On the account given here it appears that whatever is valued by a valuer is valuable. Thus, it appears, there is no distinction between what is valued and what is truly valuable. For example, some people may suggest that it was a mistake to hold that pesticides such as DDT were really valuable. They may say that DDT was valued but was not really valuable. Since, such people will argue, on a correct analysis of the concept of value, there is a distinction between what is valued and what is valuable, this account must be incorrect.

To respond to this objection we would undertake some analysis of the expression "truly valuable" which would not imply that there are values which exist independently of what people value. Such an account might follow the thought of John Dewey, R.B. Perry, or C.I. Lewis and suggest that when we say that something is truly valuable we are making a prediction about what people (or other valuers) will value, or will value for

[10]Difficulties in determining overall value arising from use of a pesticide are discussed in chapter 12.

a significant amount of time, or will value over certain other things.[11] Another suggestion is that the distinction between what is truly valuable and what is not truly valuable may be made by reference to desires which persist in the face of scrutiny of circumstances which might be expected to extinguish the desire or to give rise to contrary desires.[12]

Some readers may find this response unsatisfactory. They may be inclined to postulate the existence of values which exist independently of valuers.[13] Consider again the example involving DDT. Some people would want to say that this substance was valued but was not truly valuable. These people maintain that the belief that DDT was valuable was an error, and further, that the belief was in error because while the belief attributes independently existing value to DDT, DDT lacks such value.

In defense of the view we have favored we may ask whether we should agree that DDT was worthless, that is, lacking in value. To say this is misleading. Much good was derived from the use of DDT. Rather than say that DDT was worthless we should say that while many people valued DDT they no longer do so. They valued it, that is, desired to use it, for some qualities which it possesses. They now recognize that DDT possesses other qualities and, in virtue of these, they no longer desire to use it; indeed, they desire that it not be used. We may express this change of attitude toward DDT by saying that DDT was not truly valuable. The belief that DDT is valuable for agriculture did not persist after we became aware of circumstances involving persistent harmful effects of the use of DDT (circumstances which might be expected to extinguish the desire that DDT be used in agriculture). Indeed, we may make a distinction between an object's having some value and its being valuable overall. DDT has

---

[11]C.I. Lewis said that "there is that most important and most frequent type of evaluation which is the ascription of the objective property of being valuable to an existent or possible existent. . . . Like other judgments of objective fact . . . determination of their truth or falsity can never be completed, and they are theoretically, never more than probable." In other words, such value ascriptions involve predictions which may or may not be confirmed. See C.I. Lewis, *An Analysis of Knowledge and Valuation* (LaSalle, Open Court Publishing Company, 1946), pp. 375–376. Dewey said, "Moreover there is a genuine difference between a false good, a spurious satisfaction, and a true good, and there is an empirical test for discovering the difference." Dewey suggests that in trying to determine whether a course of action is good we reach a point at which we have conflicting motivations as to what to do. The judgment that the act is good implies that that act constitutes a harmonious resolution of the conflict. Such a judgment may be mistaken; i.e., something we judge to be valuable may not really resolve the motivational conflict. See *Human Nature and Conduct* by John Dewey (New York, The Modern Library, 1930), p. 210f.

[12]For a discussion of this distinction see "The Science of Man and Wide Reflective Equilibrium" by R.B. Brandt, *Ethics*, Vol. 100, No. 2, January, 1990, p. 259f.

[13]For a recent work in which a philosopher maintains that value exists in objects independently of valuers, see *Environmental Ethics: Duties to and Values in the Natural World*, by Holmes Rolston, III (Temple University Press, Philadelphia, 1988).

some value but its use in agriculture (as opposed to its use for controlling disease) may well be negatively valuable overall; that is, overall it produces more consequences which people desire to avoid than consequences which people desire to have.

Essentially the same objection arises from another idea. People may value some object which does not exist; for example, a person may wish to be free when he or she is not free. Does this not show that values are not real? It is surely true, for example, that one does not become free simply by wanting to be so. However, one may respond to this objection as follows: We have suggested that objects acquire value through being the objects of certain motivations. To value an object is to be motivated to possess it, or to experience it or to make it real. An object, *if it exists*, acquires value through being the object of such a motive. So value can be real, i.e., can be present in real objects much as other relationships or qualities. This is not to say that when a person values an object, there is an object or that the object comes to exist because the person is motivated to possess it, etc. We can also say of unreal objects that they are objects of value, meaning in this case that valuers want to make them real or that they would cherish them if they were real. People can organize their lives with respect to unrealized and even unrealizable ideals, i.e., they can strive to realize them or to come as close as possible to doing so.

When we become aware that there are people who do not share our values, we are led to question whether our values or their values are correct. This may lead us to become doubtful about the possibility of supporting value judgments through evidence or reasons. Suppose, for example, I find that I value fruit which is free of pesticide residue while someone else values fruit which appears unmarred. Should this lead me to be completely skeptical about the possibility of supporting value judgments by evidence or reasons? While caution about the correctness of value judgments may indeed be warranted, complete skepticism, I shall argue, is not. In arguing for this I shall consider several things that might be meant by a claim that some values or value judgments are correct or incorrect. I shall not attempt to do so exhaustively.

First, however, we should note that given that values consist in a relationship between valuers and the objects they value, it is possible that a person's values are correct for him(or her) and mine are correct for me. To speak of values as simply being "incorrect" is, on the view advanced here, an oversimplification. Indeed, to say that a person's values are incorrect is a confusing way of speaking. Do we mean that his values are incorrect or that his value judgments are incorrect? A person may judge that he finds some object good and be mistaken in that judgment, i.e., he may not find that object good. However, a person may value some object, i.e., find it good, and mistakenly expect that he will continue to value it for a long time. Perhaps this is what is meant be saying that a person's values rather than his value judgments are mistaken.

Second, when a person says that someone else's values are incorrect, he may, as noted above, be making a prediction or be expressing an expectation. It might be a prediction that the other person will soon change his mind with respect to what he values, that he will soon cease to cherish unmarred fruit. Now this prediction might be correct and there might well be good evidence or reasons for thinking it is correct. For example, he (the person predicting a change of mind about unmarred fruit) may be aware that the pesticide used to achieve unmarred fruit also is contributing significantly to the reduction in numbers and possible extinction of bald eagles. He may be aware that the person admires bald eagles and so have good evidence for thinking that that person will change his mind about unmarred fruit when he is appraised that the pesticide is putting the continued existence of such birds in jeopardy.

In the past some people have made serious errors in appraising value judgments expressed by others. They have judged others' values incorrect for them and tried to get those others to change their values, often at great harm to those others. This fact does provide a good reason to be doubtful about our appraisals of the value judgments of others—especially others whose values differ considerably from our own. Awareness of the probability of such errors should perhaps make us doubtful as to the accuracy of our initial judgments concerning the correctness of others' value judgments. However, the fact that we have been wrong in some of our appraisals regarding other people's values and that it is often difficult to get evidence which provides a warrant for being sure about such appraisals does not mean that we never can have strong evidence that some such appraisal is correct.

Another thing that I might mean by saying that my values are correct is that it would be good if everyone had the same values that I do, that it would be good if they had the same objectives and in the same order or hierarchy. Clearly, such a view is almost certainly incorrect. But, again, this does not imply that we cannot have strong or good evidence for certain value judgments. The theory of values which I have been sketching suggests that the nature of such evidence is empirical or scientific. So far there has been no conclusive refutation of such a view.

While some scientists have expressed great skepticism about there being good evidence for value judgments, many other scientists are not at all skeptical about this. I have often heard crop scientists expressing views which imply that many members of the general public accept unwarranted value judgments about pesticides. Such scientists suggest that these members of the public are irrational because they believe that use of pesticides is bad. These scientists are maintaining, in effect, that those who hold that use of pesticides is bad are either making a mistaken value judgment or a value judgment which is not supported by the evidence.

Section C. Remarks on Some Difficulties in Determining the Value of Agricultural Uses of Pesticides

The analysis of value that we have provided enables us to make sense of an idea that is current lately in regard to the use of pesticides in agriculture. This is the idea that people's values have changed. In other words, what people find good or desirable (have a positive attitude toward) has changed. Whether this is true is a question that can, perhaps, be settled by research by social scientists. Allegedly, in the past, people valued fruit and vegetables that were not marred by the effects of insect pests. Such fruit and vegetables were obtained by methods that left various residues of pesticides in the fruit. Now, people (some people anyway) may be coming to value fruit and vegetables that contain less or even no such residues. Perhaps they value unmarred fruit. Perhaps they value residue-free fruit more than they value unmarred fruit (if they cannot have fruit that is both free of residue and unmarred).

Resolution of these issues by social science research may not be easy. There are a number of reasons for this. For one thing, people's values may change rapidly. By the time one devises a suitable instrument for measuring such a change, the values may have changed again. Indeed, what people value may fluctuate back and forth among a number of objects or qualities. When people hear of residues they may feel that what they want is residue-free fruit. But then when they see "organic" fruit marred and blemished, they may feel that what they want is fruit that is unblemished.

Further, what people value is manifest both in what they say and in the way they behave. However, it is possible that what people say they (positively) value is inconsistent with what their behavior indicates they value. This would be the case, for example, if people said that they wanted residue-free fruit but at the same time always purchased fruit containing residues (perhaps because it was cheaper or better looking).

The difficulty in determining what people value either because people are inconsistent in what they value or because their values change poses significant problems for agribusiness and government. Even if agribusinesses were willing to alter the nature of what they produce to bring it into line with people's values, they may have great difficulty in determining what those values are. And even if they can determine (quickly) what people's values are, agribusiness people may be unable to change their production practices as quickly as values change. Similarly, governments trying to respond to people's wishes in regard to regulation of pesticides may have difficulty in determining what they should do.

Part II. Ethics and Values

Section A. Moral Judgments and Value Judgments

We have been discussing value judgments, that is, judgments that something is good or not good. I now want to turn briefly to considering ethical, or moral judgments. Many people casually subsume moral judgments under the general category of value judgments. However, I believe that this is misleading. By moral judgments I refer to judgments such as that we are morally obligated or required to do or to abstain from doing something or other, or that we have a duty to do such and such or to abstain from doing such and such. Such judgments can be expressed by reference to what is morally permissible or acceptable. Of course, the terminology with which such judgments are expressed is variable. Sometimes such judgments are expressed without explicit use of terms such as "ought," "duty," or "obligation." Moral judgments cannot be identified simply by the occurrence of special terms or phrases. There is a long tradition in philosophy according to which judgments that something is a duty or an obligation are distinguished from judgments that something is good (valuable) or not good. It seems to me that this distinction is worth preserving. Failure to do so may lead to failing to distinguish distinct moral theories.

In Part I of this paper I listed six examples of value judgments commonly made by agricultural scientists. I shall now give several examples of moral or ethical judgments derived from the same sources:

1. Estimates of crop losses due to insects should be viewed with a significant degree of uncertainty.
2. The decision to measure reduction of pesticide usage by reference to tons of active ingredient was morally justified.
3. Governments may regulate pesticide usage even in the absence of complete scientific information regarding the consequences of such use.
4. The primary obligation of the pest control expert is to insure that members of the public are not seriously harmed by use of pesticides.
5. Greater economic resources should be devoted to research aimed at reduction in use of pesticides.
6. It is morally acceptable to use synthetic pesticides, subject to certain restrictions, in the production of food and fiber.

It is true that value judgments usually play important roles in moral theory. However, this alone does not make moral judgments into value judgments. Value judgments play several important roles in science. Such judgments enter into the determination of what problems are worth investigating, into the determination of what hypotheses are worth taking seriously when one is trying to solve a problem, and into the determination of what evidence is sufficiently strong to warrant claims to have knowledge. Of course, the fact that value judgments play such roles in science does not imply that scientific judgments are value judgments.

In this century there has been a positivistic tradition in regard to knowledge. It has been manifest in many ways. One way in which this tradition has been manifest has been in the assumption that there are two types of judgment. There are knowledge claims, such as are found in empirical science or pure mathematics, and there are emotive judgments. All non-scientific judgments were lumped together into this latter category. So, value judgments and moral judgments, since both were deemed not to be scientific, were dismissed as emotive judgments. However, I am not disposed to agree either that value and moral judgments should be so sharply distinguished from scientific judgments or that they should simply be lumped together as mere expressions of emotion.[14]

Ontological and epistemological issues, similar to those which we have just been discussing with respect to values, can also be raised with respect to duties or obligations. The view which we have defended with respect to values may be described as both realist and empiricist. By this I mean that we have defended the reality of value and the view that the evidence for value judgments is essentially empirical evidence. Of course, as we have noted, in maintaining that values are real, we have not maintained that values exist independently of valuers. Similarly, a realist ontology and an empiricist epistemology can be defended with respect to obligation and moral judgment. I have tried to show that unease arising out of ontological or epistemological concerns can be overcome by the above discussions. I do not think that it is useful to pursue such issues again with respect to moral obligation and moral judgment.[15] However, it is useful to undertake a brief consideration of the possibility of rationally resolving ethical disagreements and this requires a brief return to epistemological matters.

Section B. On the Rational Resolution of Moral Disagreements

Concerns with ethics that have been arising in many scientific quarters are, I believe, a concern to have some guidance as to what one ought to

---

[14]For an introductory discussion and critique of the idea that ethical judgments are merely expressions of emotion see chapter 9 of *Ethical Theory*, by R.B. Brandt (Englewood Cliffs, Prentice Hall, 1959).

[15]For an excellent defense of the view that the reasoning which enters into the justification of value or moral judgments is closely integrated with the reasoning that enters into the justification of beliefs about matters of fact, see *What Is and What Ought To Be Done: An Essay on Ethics and Epistemology*, by Morton White (New York, Oxford University Press, 1981).

do. Ethics, as I conceive it, is a discipline concerned with the justification for such guidance. We want to know whether it is permissible to manufacture and apply pesticides and if so which ones and to what extent and under what conditions.

Question might be raised at this point as to whether it is possible to discover what we ought to do in the world. This is a question concerning the possibility of evidentially justifying our beliefs about what we ought to do. We have indicated above that some thinkers hold that our moral beliefs can be justified in much the same way as scientific beliefs. Other thinkers subscribe to alternative theories of evidence for moral beliefs.[16] Of course, some people are skeptical with respect to ethical knowledge, that is, they do not believe that anyone can have knowledge of his or her moral obligations.

Knowledge, whether ethical knowledge or knowledge of other sorts, requires high standards of evidence. On some theories of knowledge, one must possess such strong evidence for a proposition that one allegedly knows, that there is no rational basis for doubting the proposition. On other views, the evidence must show that the proposition could not possibly be false. Clearly, consideration of these views would require considerable explanation and clarification and would be inappropriate in this context. One point does seem worth making. If we assume that there can be some evidence for claims concerning what is morally obligatory (or permissible), it may well be that we sometimes have enough evidence to rationally warrant believing some ethical proposition, because the evidence and reasons which support it are stronger than the evidence or reasons which support its contradictory. This point has a bearing on the possibility of rational resolution of moral disagreements. To see this let us consider a moral disagreement about the acceptability of a particular use of pesticides—let us say, a use of a particular pesticide to produce a particular crop on a particular occasion. We suppose that a decision must be made; we have not the luxury of waiting until the facts of the situation can be thoroughly determined by scientific procedures. Given that there is such disagreement, we may wonder whether it is possible to resolve such disagreement on rational or evidential grounds. The point here is that at any particular time, given that some action is necessary, there may be evidence which warrants moral beliefs, that is, beliefs concerning what we ought to do or to allow to be done, although such evidence may not be strong enough to justify saying that such beliefs are known to be correct.

---

[16]For a nonempiricist view of moral knowledge see *Reason and Morality* by Alan Gewirth (Chicago, University of Chicago Press, 1981). An alternative nonempiricist view concerning the rational justification of moral principles is expressed in *Moral Thinking: Its Levels, Method and Point* by R.M. Hare (Oxford, Clarendon Press, 1981).

Difficulties in resolving ethical disagreements on the basis of evidence or reasons appear to have led many agricultural scientists and others to draw two conclusions. The first of these is that it is not worth the effort required to engage in systematic investigation of the issues involved in moral controversy. The second of these is that we ought to be tolerant of moral beliefs and of practices which diverge from and perhaps even conflict with our own. I want to discuss both of these conclusions briefly.

The claim that it is not worth the effort required to engage in systematic investigation of the issues involved in moral controversy is a value judgment. It is not a value judgment with which I agree. People apparently make this value judgment when they believe that it is impossible to find evidence that would resolve moral controversies which is strong enough to convince any rational person. However, even if it is impossible to find such evidence, it may be possible to find evidence (that would resolve issues in a moral disagreement) which is *strong enough to convince all rational beings who share some basic moral beliefs*. Often moral disagreements occur among people who share basic moral and value beliefs. Systematic investigation of the issues involved in moral controversy can bring to light such shared beliefs. Where this happens, resolution of a moral disagreement on the basis of evidence or reasons may occur, and this is preferable to resolving disagreements through the use of irrational persuasion or force. Further, since, as noted above, we are concerned with moral beliefs on which we often must act, we believe that people are less likely to make serious moral errors if they act on the basis of ethical beliefs that have been subjected to rational scrutiny. For both of these reasons, it may well be worth the trouble to make the effort to systematically investigate moral issues. Indeed, in the recent decade or so there has been an increase of institutes or other social organizations devoted to such systematic investigation.

The claim that we ought to be tolerant of moral beliefs and practices which diverge from our own is a moral claim. Indeed, it is a moral claim which is, in my judgment, worthy of serious investigation. First, one would like to know the scope or extent of the tolerance which is advocated in this claim. Does it mean that one must be tolerant of any moral opinion whatsoever? For example, must we be tolerant of the producer who claims that he has a moral right to use large quantities of pesticides even though his doing so will destroy the livelihood of other people (perhaps through contamination of a water supply)? For example, must one be tolerant of the opinion that it is perfectly acceptable for a person to cause another person to die of cancer merely because the first person wishes to be confident that his crop will not suffer losses to pests over a certain level? Again, must one be tolerant of intolerant people? Second, one would like to know just what behavior is implied by this principle as being morally required of us. That is, we want to know what we must do to be tolerant.

It might well be the case that there are good reasons for being tolerant of divergent moral views within certain limits. However, someone who maintains that rational inquiry concerning moral opinion is not worth the effort seems to imply that it is not worth the effort to ferret out such reasons. We wonder then what evidence or reason can be available to such a person to justify the moral judgment which he expresses. If he (or she) maintains that no evidence or reasons are available to support it, then we wonder why he (or she) thinks we should agree with him (or her).

Part III. Survey of Ethical Theories

Section A. Consequentialism

Ethical theories may be classified into several families and there are a number of ways of doing this.[17] One major family of ethical theories may be called *consequentialist*. Consequentialist ethical theories, more than others, tie the content of moral principles closely to principles regarding what is valuable. According to consequentialist theories, one's moral obligations consist in acting so as to produce the greatest amount of value. (Value is often referred to as utility.)[18] The greatest value may be conceived in a number of different ways but it is common to think of it as the maximum difference between positive and negative value.

Our assumption concerning the nature of value, namely, that value is essentially connected to having or satisfying desires, yields the implication that value (positive or negative) occurs only for beings that have such desires. (Desires are sometimes referred to as preferences. One may hear a version of consequentialism referred to as "preference utilitarianism.") Given our assumption, there is no intrinsic value in the lives of insentient creatures, e.g., plants. Given our assumption about value and the assumption that insects do not have desires, it follows that there is no intrinsic value in their lives. Such value as their lives may contain is entirely instrumental value, that is, value which arises because the consequences of

---

[17]For an excellent introductory account of types of ethical theory see chapter 2 of *Matters of Life and Death: New Introductory Essays in Moral Philosophy*, ed. Tom Regan (Random House, New York, 1980). An older but excellent book, which is in some ways more complete, is *Ethics*, second edition, by William K. Frankena (Englewood Cliffs, Prentice Hall, 1973).

[18]If a person has a moral obligation then that person has a reason for acting, for doing whatever it is that the statement of the obligation requires. Whether such a reason for action can be reduced to a desire to achieve certain objectives is an issue about which philosophers disagree with each other. For some discussion of this issues see chapters 10 and 11 of *The Nature of Morality: An Introduction to Ethics*, by Gilbert Harman (New York, Oxford University Press, 1977).

their lives affect the satisfaction or frustration of desires of creatures that have desires. For example, a consequentialist who accepted the value assumption that we have made, would agree that in using pesticides we are obligated to consider consequences for the lives of organisms such as insects or biocontrol agents. Such lives would have to be taken into account on such a view because what happens to such organisms can contribute to satisfaction or frustration of desires of humans or other animals.

Varieties of consequentialist ethical theories may be distinguished by reference to the value assumptions with which they are combined. (Any consequentialist theory must be combined with some value assumption since, if it were not, the injunction to maximize *good* consequences would be meaningless.) Some people have assumed that life itself has intrinsic value.[19] On such a view it could be morally wrong to destroy a plant or a tapeworm regardless of whether that organism would have contributed to or interfered with the satisfactions of any creature with desires.

We have said that an object has positive value if a valuer desires to make it real, or possess it, or experience it. But how can we determine how much value an object has? Apparently not all valuable objects are equal in positive value (and not all negatively valuable objects are equal in negative value). If they were we could simply count each positively valuable object as +1 and each negatively valuable object as −1 and sum these numbers to arrive at the total value. Alternative courses of action could be evaluated to determine which would yield the greatest sum of values. Our obligation would be to act so as to produce the greatest sum.

One suggestion for measuring the value of objects would be to say that an object varies in positive value directly in proportion to the strength of the desire to possess it or make it real, etc., and similarly for negative values. For example, a person's desire to stay alive is normally stronger then his desire to read a particular book and so his life has greater value to him than the book does.

An obvious problem for such a view concerning values is to determine a satisfactory way to measure the strength of desires. What is needed is a satisfactory way of determining amongst a group of desires which desires are stronger than which others and which are equivalent in strength. Economic measures have been suggested. For example, one might say that the desire for object O(1) is stronger than the desire for object O(2) for a person if that person is willing to pay more for O(1) than for O(2). There are obvious problems for such an account. One person's desire for, say,

---

[19]Such a view was expressed by Albert Schweitzer. There is a brief discussion of Schweitzer's views in "The Search for an Environmental Ethics" by William T. Blackstone in the first edition of *Matters of Life and Death: New Introductory Essays in Moral Philosophy*, cited in footnote 17, p. 301f.

uncontaminated drinking water might be as great as a second person's desire, even though the first person is not willing to pay as much as the second person is because the first person is much poorer than the second person. Further, animals have desires; clearly, the above suggestion cannot be applied to determining the strength of animal desires. So far as I know, at the present time, there is no generally recognized way to measure strength of desires.

In spite of the difficulties associated with determining the value of objects so as to make measurable comparisons of value, it appears that it is often the case that a group of valuers can agree, among sets of alternative actions, as to which will lead to the most valuable outcomes. For example, we may compare two crop production practices each of which yields the same amount of product under circumstances which are the same except that in one case pesticides are applied more precisely whereas in the other they are sprayed on in some currently conventional manner. Clearly, the former practice would be better, because costs, environmental impact, and residues would be reduced. Under these circumstances if we have to choose which of these two practices is acceptable, we will conclude that the former is acceptable while the latter is not. This suggests that some form of consequentialism is at least partially correct. Thus, let us agree that in determining our obligations (in regard to use of synthetic pesticides) the value of the consequences of our choices in this matter is a relevant factor. *In trying to determine what we ought to do, it is appropriate to try to determine which course of action, among the alternative courses open to us, will yield the greatest amount of value.*

We should note at this point that our decision to identify value with respect to attitudes of valuers could have significant consequences. Some advocates of so-called "environmental ethics" have opted for a view which implies that value exists independently of valuers (beings who have either positive or negative attitudes toward objects). For example, on some views value may exist in certain ecosystems as a result of qualities possessed by those ecosystems and regardless of whether any being has a positive or negative attitude toward the possession or realization of those qualities.[20] A consequentialist who also subscribed to such a value theory would claim that in determining which actions produced the greatest value we would have to take such ecosystemic values into account.

Section B. Kantian Considerations

Philosophers debate whether ultimately all moral decisions can be correctly determined on consequentialist grounds. We shall not enter into this debate in this chapter. We shall, however, assume that other factors, in

---

[20]See, for example, the work of Rolston, cited in footnote 13.

addition to the value of the consequences, are relevant in determining our moral obligations. Those philosophers who think that all relevant moral reasons, i.e., reasons which are relevant for determining one's obligations, are reducible to consequentialist reasons may well hold that these additional factors are so reducible.

Our assumption that factors other than consideration of the goodness or badness of the consequences of an action or policy are relevant for determining whether the act or policy is morally acceptable arises from consideration of what obligations we owe to individuals. Consider a hypothetical case. Suppose that use of synthetic pesticides produced the greatest sum of good consequences and the smallest sum of bad consequences. Suppose further that the use of these substances imposed significant risk of serious harm on some individuals. Is it morally acceptable to use the pesticides given that such use imposes such risk on those individuals? Is it acceptable to impose great harm on some individuals when doing so produces the best consequences overall? Even worse, suppose that a large fraction of valuers desire to inflict pain or suffering on sentient creatures. Must the satisfaction of that desire be given any weight at all in determining our moral obligations? Consideration of these and related problems has led philosophers to investigate (and in some cases to support) nonconsequentialist ethical principles. Let us briefly consider some alternatives of this sort.

A second family of moral theories is traceable to the moral philosophy of the German philosopher Immanuel Kant. Like the consequentialists, Kant maintained that there is a single basic moral principle which he called "the categorical imperative." However, unlike the consequentialists, Kant maintained that in determining our moral obligations, consideration of what consequences our actions actually produce is irrelevant.[21] (While most interpreters of Kant would agree with what I have just said, R.M. Hare has maintained that Kant can be interpreted as a consequentialist, or at least that Kant's view is consistent with consequentialism.)[22] It is important to consider Kantian perspectives for reasons which we have mentioned in the Introduction. People who have been affected involuntarily by other's use of pesticides claim that such treatment is immoral. The moral theory to which they appeal in support of the view that they should not have been subjected to pesticides often appears to be a Kantian theory. Many scientists, when they begin to consider moral issues, fail to recognize this.

[21]Kant's work is difficult to read. For an introductory explanation, see "The Moral Perplexities of Famine and World Hunger" by Onora O'Neill in *Matters of Life and Death . . .*, ed. Tom Regan, cited in footnote 17.

[22]Hare argued for this in a paper presented to the Department of Philosophy at the University of Guelph in 1989. I don't know whether Hare has published his views on this matter.

The scientists argue in straightforwardly consequentialist terms and so fail to address the moral objections of their critics.

In one formulation, the categorical imperative may be expressed as follows: We ought never to treat rational beings merely as means to our own ends. Rational beings must always be treated as ends in themselves. Let us briefly consider the implications of this formulation. It appears that you have used another rational being merely as a means if through your actions you lead him to act other than he would have chosen had he been as well informed as you were concerning the nature or consequences of the act. For example, if you induce a person to do something which he would not have done had he known what you know, then you are not treating him as an end in himself. Further, you treat a person merely as a means if you force him to do something, for example, by a threat to severely harm him.

This Kantian principle, interpreted as above, conflicts with some principles of action which have been accepted as valid ethical principles in our society. In particular the standard of what we might call "openness" required by this Kantian principle is high. Often, for example, in commercial transactions such a high standard of openness is not expected. In some quarters the old principle "Let the buyer beware!" is essentially the governing principle. In accordance with this principle the seller may sell a product even though he knows of flaws in the design of the product in light of which he knows that use of the product may be dangerous or at least that use of the product will not conform to the objectives of the buyer.

In the business practices of our society the principle "Let the buyer beware!" has been modified. Government intervenes to protect consumers either by requiring some degree of openness with respect to products or by restricting the use of certain products, for example, by limiting the authority to dispense certain medicines or by insisting that certain products be made so as to satisfy certain standards. Such restrictions already are applied in the case of various agricultural products—for example, restrictions designed to limit pesticide residues that reach consumers. Some people are suggesting further restriction in agriculture; in particular, they are suggesting that we restrict those who are permitted to engage in agriculture for profit to those who have been properly trained to understand and abide by restrictions.[23]

While the principle "Let the buyer beware!" has been modified in our society, we may ask what moral basis can be given for such modifications? Is the moral basis a simple application of the categorical imperative? I think not. A straightforward application of the categorical imperative regarding openness, as interpreted above, goes well beyond what many of us would agree is acceptable government intervention. For example, a

[23]See, for example, the discussion in chapter 9 of this volume.

business is permitted to maintain trade secrets. It does not have to make public everything it knows about its products, manufacturing processes, etc. How are we to decide what degree of openness is required? The answer may be suggested by considering a second form of the categorical imperative.

This second formulation of the categorical imperative implies that an act is morally acceptable only if there is a rule according to which doing the act is permitted and which could be accepted as a valid rule by (each member of) a community of rational beings. Kant maintained that this version of the categorical imperative is equivalent to other formulation. Whether it is may be debated. Consideration of this formulation suggests that in trying to decide what actions are morally acceptable we formulate a rule to the effect that anyone may perform such and such an action in such and such circumstances and then consider whether we would be willing to have that rule as a law of our society. Some secrecy is permissible on such a view providing it is acceptable according to the laws of the society *if the laws satisfy a further condition.* (Clearly, not everything which is legally permissible is morally permissible.)

Kant suggests that to determine what moral judgments are correct, reference must be made to what rules would be unanimously accepted in a community of rational beings. To apply this idea we would first have to determine what rules a rational being would be willing to accept as binding on himself. Investigation of this matter again raises highly complex, controversial questions of ethical theory. There is considerable debate among philosophers concerning the nature of "rationality" and of what it would be rational to accept as binding on oneself.[24] Rather than undertake such a consideration let us suggest a plausible modification of Kant's formulation.

Kant's thought suggests that in determining what actions or policies are morally acceptable we must take due to consideration of the individual's capacity to direct his own life in accord with his own desires and beliefs. To express this idea, let us say that we have a basic obligation to respect individual autonomy. Consideration of how to do this raises questions concerning political practices. In the extreme, to respect individual autonomy might mean that each person determines what is right or wrong for

---

[24]The concept of rationality has entered into many efforts to formulate basic ethical principles. It enters into Kant's efforts to justify the categorical imperative. More recently it enters into the justification of contractualist views on ethical principles such as that of John Rawls. See his *A Theory of Justice* (Cambridge, Harvard University Press, 1971). Rawls views have been both very influential and subjected to extensive criticism. An interesting book concerned with the notion of rationality is *Paradoxes of Rationality and Cooperation: Prisoner's Dilemma and Newcomb's Problem*, ed. Richmond Campbell and Lanning Sowden (Vancouver, University of Vancouver Press, 1985).

himself. However, we do not accept this extreme view. It does not take due account of the fact that we each of us is a part of a world in which there are other individuals who are capable of directing their own lives and who, I would argue, are entitled to play a significant role in doing so. Such individuals are entitled to a chance to influence the social decision-making processes in the communities in which they live. An alternative way of expressing this is to suppose that individuals have certain basic political rights—the sort of rights that are established, at least to some extent, among the democratic nations of the world. Thus, let us suggest, *whether an act accords with laws enacted in a society with a political system which is democratic and respects basic moral and civil rights—rights such as those which form the basic ideals of the democracies of the world—is a relevant factor in determining our moral obligations.* Let us call this principle the principle of respect for individual autonomy.

A full elaboration of this principle would take us into complex issues of political theory. Here let us say that the reference to the basic rights and ideals of democracy should be understood as implying that no human beings should be subject to arbitrary exercise of force (whether in the form of economic compulsion or physical violence). In such a society no individual would be forced to subject himself or herself to risk of serious injury or death. Those with economic power would be required to inform affected people regarding enough aspects of prospective production, commercial, and other practices so that decisions regarding the acceptability of the practice would be informed decisions. Social consent to a commercial practice could be informed consent, as could be individual consent to participate in the practice. In such a society all human beings would be in a position to make decisions regarding the basic laws governing social practices which are as fully informed as possible.

The above Kantian principle makes reference to the concept of moral rights. Philosophers have spent considerable time in trying to explain and justify claims with regard to moral and political rights and in debating their independence from the principle that we must produce the best consequences overall. In this paper we shall not attempt to work through all the problems that arise in this effort but refer the reader to other works.[25] We have taken the concept of moral obligation or duty as the fundamental moral concept. If we are to make reference to moral rights, such as is done in the Kantian principle, we must explain the connection between moral rights and duties. This, however, is a complex topic. Here, I shall assume that if a being has a moral right then all other beings who are capable of

---

[25]One such work is *Taking Rights Seriously* by Ronald Dworkin (Cambridge, Harvard University Press, 1977).

having moral obligations have the obligation to respect that individual's rights.[26]

Let us conclude this section by considering the following question: To whom or what do we owe moral obligations? According to Kantian considerations we have obligations only to beings who may be described as capable of acting autonomously—capable of directing their own lives. Kant understood this notion of autonomy as applying only to human beings. To act autonomously, according to Kant, requires the capacity to understand rules applicable to behavior and to be able to regulate one's behavior in accordance with such rules. Only such beings are capable of determining whether some action which they are contemplating is consistent with laws that would be accepted as binding on all rational beings. Such a view implies that we have no moral obligations to nonhumans or other creatures (unless it should turn out that some nonhumans are indeed sufficiently rational).[27]

According to consequentialist considerations we have obligations to maximize value overall. Since nonhuman beings can be valuers, i.e., can have positive or negative attitudes toward objects, the class of beings to whom we owe obligations is broader. It includes all beings to whom we owe obligations as a result of Kantian considerations plus a certain range of nonhumans. The exact extent of this additional class of beings is unclear as a result of obscurities in the notions on which the concept of value rests. Mammals, birds, and reptiles can be described, reasonably accurately, as desiring to experience certain objects or to make them real, etc. Thus, on consequentialist grounds we may have obligations to such creatures. Let us refer to this broader class, whatever its exact limits, as the class of valuers.

Since we have allowed that both consequentialist and Kantian factors are relevant in determining our moral obligations and since the class of beings to whom we owe obligations according to consequentialist principles includes the class of creatures to whom we owe obligations according to Kantian principles, we shall have to say that in trying to determine our moral obligations with respect to pesticide use, we shall have to consider

---

[26]The survey of moral theories implicit in this discussion of is not complete. As formulated here, ethical theories are concerned with what people ought to do, that is, with moral duties or obligations. A traditional type of ethical theory is concerned with formulating ideals of human character, i.e., with moral virtues. On some views, the most satisfactory way of determining what a person ought to do in particular circumstances is to consider what a virtuous person would do in such circumstances.

[27]A brief selection of Kant's work in regard to our obligations to animals is found in *Animal Rights and Human Obligations*, second edition, ed. Tom Regan and Peter Singer (Englewood Cliffs, Prentice Hall, 1989). For a critical discussion of Kant's view plus discussions of alternative views regarding obligations to animals see *The Case for Animal Rights*, by Tom Regan (Berkeley, University of California Press, 1983).

potential obligations with respect to the broader of these two classes. We shall have to consider the consequences of our actions with respect to all valuers.

The class of beings to whom we owe moral obligations is indeed broader than many people normally recognize. It is arguable further that the class of valuers should include not only valuers that exist at the present time but, in addition, valuers that will exist in the future and whose lives may be affected by our actions. After all, to discriminate against future valuers appears as arbitrary and hence unjustifiable as is discrimination against valuers in other places at the present time. Recognition of this point, in conjunction with our assumption of consequentialist and Kantian moral principles, implies that we may have obligations to members of future generations. In particular, we have obligations not to poison them through leaving them to deal with water systems contaminated with toxins. We have obligations not to so impoverish the life-support systems on the Earth that they will be unable to sustain a tolerable quality of life.[28]

Section C. Criticism of Holistic Environmental Theories

Finally, we should briefly return to consideration of a family of theories regarding environmental ethics which contrast with both Kantian and the forms of consequentialist theories that we have been considering. According to both Kantian and these consequentialist theories we have direct moral obligations only to valuers—to beings capable of valuing things. By contrast, some theories of environmental ethics postulate that we have moral obligations directly to more inclusive entities or "wholes" which have biological individuals as well as other natural entities such as species, soil, streams, etc., as parts. These theories usually postulate that these larger wholes manifest certain valuable characteristics and that the value inherent in these wholes is, at least in part, logically independent of the mental states (positive or aversive attitudes) of biological individuals.[29] I refer to these theories as *holist* and to Kantian and consequentialist theories as *individualist*.

---

[28]We have said that we have moral obligations to all and only present and future valuers. It does not follow that we have no moral obligations *with respect to* insects or other creatures that are not valuers. What we do to insects will often have profound implications with respect to our obligations to valuers. The point is that we have obligations to other valuers with respect to insects, etc. On the view we are suggesting here, we do not have obligations to the insects. Further, we might have moral obligations, to other valuers, not to cause the extinction of certain species. This would be an obligation in respect to species but not an obligation to the species itself.

[29]Such a view is expressed in "The Search for an Environmental Ethic," by J. Baird Callicott, in *Matters of Life and Death: New Introductory Essays in Moral Philosophy*, ed. Tom Regan (New York, Random House, 1980).

While these holistic ethical theories have been useful in a number of respects, particularly in focusing attention on the complex interrelatedness of the biological networks of which we are parts, I object to them for the following reason. Consider a hypothetical situation in which a population of human beings is in jeopardy. They will be subject to malnutrition, disease, and early death unless they exploit some area of the Earth in a way which will degrade the ecosystems in that area. Suppose further that while there are alternative options for the survival of those people, such as through a redistribution of the wealth of other people, taking those options would lead to war with even greater death, injury, and illness to human beings (but not greater destruction of any ecosystems).

What are the implications of a holistic environment ethic in such a case? As with any ethical theory it is difficult to tell. However, there appear to be two alternatives. Either its implications agree with those of one of the major individualist alternatives such as we have been considering or they conflict with them. (Such a theory might agree with the individualist alternatives if, as might be argued, sacrifice of the ecosystems would lead to sacrifice of individual lives in any case.) Now, if the holistic theory agrees with the individualist alternatives then we need not consider it; we may regard it as a hypothesis of which we have no need. If, however, it conflicts with its individualist alternatives then it implies that it is acceptable to sacrifice individual human beings to achieve the alternative values of wholes. I regard such a consequence as mistaken and ethically unacceptable. In my judgment then, a holistic environmental ethic is either unnecessary or mistaken and we need not make holistic assumptions for determining our obligations regarding the use of synthetic pesticides.

Part IV. Moral Obligations in Regard to the Use of Synthetic Pesticides

Section A. Ought We to Discontinue Use of All Agricultural Pesticides?

Let us start by asking whether everyone ought to abandon use of synthetic pesticides in agriculture immediately. To consider this question we will, of course, have to engage in speculation. It seems unlikely that anyone has gathered the scientific data necessary to answer it carefully since it is not a prospect that anyone is considering seriously. However, the answer to this question is almost certainly negative. To prohibit all use of synthetic pesticides as of, let us say, tomorrow, would, I assume, result in enormous increases in crops lost to pests and consequently in serious harm and even death to human beings. Even assuming that we tried to reduce crop loss to pests by other means, the harm to humans would be enormous. We would not be able to mobilize alternative means of pest reduction on an adequate scale in time to prevent huge losses. While there would be some benefits to valuers, human and animals, and benefits to other parts of our environment from the reduced pesticide load, it seems unlikely that such benefits could compensate for the harm done. I assume that a gradual reduction or even withdrawal from pesticides could have essentially as great environmental benefits with far less harm to human welfare overall.

Someone might object that continued use of pesticides is unacceptable on Kantian grounds. After all, some people are subjected to pesticides involuntarily. However, this objection is not conclusive. We have not subscribed to a strict Kantian moral theory. We might argue that our use of pesticides has been acceptable given that they have been used in accord with our laws, which have been sanctioned by a democratic process. This reply, we must note, is not conclusive either. While in the democratic nation-states rights of individuals are widely applied, our societies are not perfect in this regard. Many individuals are politically weak or powerless in democratic countries. Further, these are often the individuals who are adversely affected by pesticide use.

These considerations suggest that continued use of synthetic pesticides be subject to certain restrictions. Such use ought to be subject to the principle of respect for individual autonomy. Farm workers should not be forced to subject themselves to high risk of serious injury or death due to exposure to pesticides. Prospective farm workers should be informed of known risks and, if they voluntarily accept the risks, should have the opportunity to exercise as much caution as possible. Safety gear should be available to them. Only workers who are fully informed and capable of understanding adequate safety precautions should be using dangerous chemicals—agricultural or other.

However, given that pesticide use is currently widespread, complete and immediate abandonment of all agricultural uses of pesticides is open to criticism on Kantian grounds also. Because of the drastic reductions in food and fiber production that such elimination of pesticides would entail, many people would suffer illness and probably death. A course of action that leads to such results is not compatible with the Kantian ideal of never treating people merely as means to ends to which they don't subscribe.

We could also ask whether everyone ought immediately to abandon use of synthetic pesticides for other (nonagricultural) purposes. Consideration of this suggests that the answer will depend on the purpose. Where synthetic pesticides may be used for eliminating or reducing the incidence of serious illness, we suspect that the continued use of these substances is acceptable providing that there is no alternative way of achieving this goal which is less harmful than use of pesticides. However, for other purposes—for example, use of pesticides for decorative purposes or to improve the appearance of fruits and vegetables (cosmetic purposes)—we suspect that an immediate ban on their use would not cause serious harm in general. Of course, some individuals would be harmed due to loss of income from the manufacture or use of such pesticides for gardens or lawns, but such harm would be minor. Where appropriate, individuals who suffer loss of business or income because of such a change in social policy could be compensated for their loss. Consideration of use of pesticides for decorative purposes calls attention to the vagueness in the range of the term "agriculture." Some parts of agriculture may include production of decorative plants or other nonessential products. In maintaining that we should not immediately end all use of synthetic pesticides in agriculture we are thinking of those parts of agriculture concerned with production of substances which are essential for satisfaction of human needs, e.g., of food, medicine, clothing, etc.

Section B. Ought We to Gradually Reduce Our Use of Agricultural Pesticides?

We might then ask whether we (everyone) ought to begin gradually reducing our dependence on synthetic pesticides. This question, like the above, is easily answered. Here the answer is almost certainly affirmative. The list of harms to people and other creatures that has resulted from the use of synthetic pesticides is impressive.[30] There is considerable evidence in support of the belief that continued use of synthetic pesticides at current or increased levels will cause equally great harm. Further, there is good reason to believe that development of alternative methods of controlling weed and insect pests will achieve the same benefits we currently derive with considerably less harm. There are strong consequentialist reasons for maintaining that we are obligated to reduce our dependence on synthetic pesticides.

Someone might ask whether Kantian considerations would yield a different conclusion, that is, whether Kantian considerations would imply that we are not obligated to reduce our dependence on synthetic pesticides. If we consider one significant aspect of much farm work in the past, prior to the introduction of pesticide use, we can see that this is a good question to ask. In the absence of synthetic pesticides a great deal of farm work was destructive of both body and spirit. It was physically hard and indeed, we suspect, usually reduced the individual's capacity to develop his or her distinctively human talents or to enjoy the great achievements of human culture. To force people to engage in such labor so that we (the rest of us) might have sufficient food and other agricultural products looks very much like treating farm workers merely as means to our ends. However,

[30]See "Is Silent Spring Behind Us?" by David Pimentel, American Chemical Society, 1987. Further, see chapter 3 of this volume.

while this is a serious question to raise, there is every reason to believe that with human ingenuity we can devise alternative methods of pest control which reduce our dependence on synthetic pesticides without reinstating the extensive drudgery formerly associated with farm labor. Gradually reducing our dependence on synthetic pesticides does not automatically violate our obligation to respect individual autonomy.

Section C. How Ought We Proceed at Present?

Given that we ought gradually reduce our dependence on synthetic pesticides, we may ask how to proceed? To answer this let us formulate a number of objectives or goals. In formulating these objectives we assume that we ought to be concerned both to produce the best consequences overall, i.e., to minimize harm while maximixing benefits, and to show due respect for individual autonomy.

> *First, we must continue to produce sufficient quantities of pure, nutritious food and other essential agricultural products so that everyone living at the present time can obtain the food, medicine, clothing, etc, necessary and sufficient for good health. A correlative of this is that the food, etc., must not only be available, it must be affordable.*
>
> *Second, adequate supplies of sufficiently pure water and air must be available and affordable.*
>
> *Third, we must not simplify the natural ecosystems within which we live to such an extent that we jeopardize our capacity to continue to satisfy the first and second objectives.*

These first three objectives can readily be defended by reference to consequentialist considerations. If people do not have sufficient food and other necessities for health, their overall welfare will be considerably reduced. Similar remarks apply to the availability of sufficient fresh air and water.

Given the assumption that it is possible to maintain production systems for food, fiber, etc., which are adequate to meet the needs of human populations without causing excessive degradation to natural ecosystems, and given that a sufficiently complex web of biota is essential for providing sufficient affordable food, water, fresh air, etc., the third objective is a corollary of the first two. Conceivably, however, human populations at some time would get so large that production of agricultural products in sufficient quantity could not be achieved without degrading the productive capacities of soil, water, etc. Clearly, we (human beings alive at present) are morally obligated to try to prevent such an overexpansion of the human population.[31]

When we consider the other basic ethical principle that we have adopted, namely, that we ought to respect individual autonomy, two further objectives are readily suggested. These are:

> *Fourth, we must achieve the above three objectives without subjecting agricultural workers to serfdom or comparable living or working conditions.*
>
> *Fifth, members of the public must have sufficient knowledge and political influence to have a significant influence in determining their own exposure to pesticides in air, food, and water.*

These objectives as formulated here are deliberately vague in certain respects. They don't say that the food has to be perfect or that the air and water must be totally free of contaminants. There is good reason for the vagueness. To attempt to obtain perfectly fresh air, water, or perfectly pure food might be worse with respect to one or more of our objectives. In the effort to achieve perfection with regard to purity of food, etc., we might fail to produce sufficient quantities of food, or the food might be too expensive for many people. If we assume that contaminants in food, water, and air lead to shortened lives and reduced quality of life, we can argue that the food, water, and air should be purified to the degree that any further purification would reduce benefits, increase harms, or decrease individual autonomy. If these three factors conflict with each other, that is, if increasing benefits also increases harm or reduces autonomy, then the conflict should be resolved democratically, that is, by the choice of the affected individuals.

Section D. Objections and Replies

In order to defend our five objectives in addition to indicating how they might be supported by appeal to our ethical considerations, we must consider possible objections. Let us here consider two possible objections to these objectives. First, it might be objected that what we are proposing here is that the use of synthetic pesticides be restricted. But such restriction means that agribusiness will not operate in a free market. Further, the critic may maintain that a free-market economy yields the greatest overall

---

[31]Clearly, I am assuming that we have not already reached the point where it is impossible to satisfy essential needs of presently existing human populations without destroying the capacity of the Earth's biological systems to sustain future human populations comparable in size to the human population at present. Should this assumption be mistaken, we (human beings living at the present time) would be morally obligated to try to reduce human population size through (severe) reductions in reproductive rates.

benefits (and least overall harm). This first objection rests on consequentialist grounds. Second, it may be objected that by restricting agriculture in accord with these objectives we will be telling producers of agricultural products how to operate their businesses and that we will be doing so for the good of others, i.e., to ensure that everyone has adequate food, etc. To tell people how to operate their businesses for the good of others is to violate the principle of respect for individual autonomy. The two objections imply that far from being implied by reference to our ethical principles, our objectives are incompatible with these principles.

Let us reply to these objections in order. With regard to the first objection, let us point out that the premise on which it rests, namely, that maximal good consequences are achieved in a free-market economy, is highly controversial. Considerable reasoning and evidence would be required before we would feel compelled to accept this premise. On the contrary, there is reason to believe that in an economy which is totally free of government regulation or other intervention in redistributing wealth, there will be a large number of people who are too poor to provide for their basic needs. We do not agree that in such circumstances the greatest goodness has been achieved. Consequently, we do not agree that in a free-market economy the greatest overall goodness is achieved. No doubt the defenders of free-market economies will not be satisfied with this response. However, a full debate on this matter is not appropriate in this context.[32]

We should point out further that in considering the above objectives we have not ignored economic considerations. Our objectives include the stipulation that pure food, air, and water be affordable. Of course, the notion of affordability is vague. How expensive should pure food, water, and air be? To answer this we would reply as above. The prices for these items have to be determined by reference to our ethical principles. The prices should be set so that we can achieve maximal good consequences without sacrificing individual autonomy. If the prices are too low, that will tend to reduce either the quantity or the quality of food below the minimally satisfactory level or will tend to produce poor working or living conditions for agricultural laborers.

According to the second objection, regulating a person's business for the good of others is a violation of our obligation to respect that person's autonomy. However, in reply to this we would suggest that the expression "regulating a person's business for the good of others" is indeed very general. Suppose that we prohibit killing people for profit. We could be described as regulating a person's business for the good of others. However,

[32]One discussion of the issue is found in "Should Business Be Regulated?" by Tibor Machan, in *Just Business: New Introductory Essays in Business Ethics*, ed. Tom Regan (New York, Random House, 1984.)

it is clear that such interference with commercial activity is not a violation of the principle that we must respect individual autonomy. Indeed, to abide by this principle we must regulate commercial enterprises such as this one. Similarly, we would argue that regulating agricultural business so as to achieve our objectives does not constitute failure to respect individual autonomy. Of course, we are not saying that using pesticides in food production is morally equivalent to killing people for profit. It is not. Nonetheless, as has been documented by Pimentel, use of pesticides causes considerable harm along with whatever benefits it produces.[33] If we can achieve the same benefits with less harm to people's health or to the environment, we are obligated to do so. We would agree that certain sorts of regulations of free enterprise would indeed constitute a failure in this regard. The term "regulating a person's business for the good of others" is so general that we cannot agree that it is always incompatible with our ethical principles to do so.

Consideration of our objectives can provide some guidance with respect to decisions that must be made concerning policies for controlling pests. There is considerable discussion these days of "integrated pest management." This term refers to any of the wide range of activities open to us to control pests—including use of synthetic pesticides, developing crops which are resistant to pests, using biological enemies of pests, etc. However, integrated pest management does not necessarily accord with the ethical principles which we have formulated. Integrated pest management could be used to achieve other objectives, for example, to increase profitability of agricultural production. The use of synthetic pesticides in itself is expensive and of limited effectiveness in controlling pests. Integrated pest management might increase effectiveness in pest control or reduce costs or both. However, to increase effectiveness or reduce costs of pest control does not necessarily lead to maximal benefits or maximal respect for individual autonomy.

Further, as we proceed to develop policies for integrated pest management, we will often have choices to make. Should we proceed with genetically engineered plants that are capable of tolerating synthetic pesticides or should we reject the use of such plants? In order to answer questions such as this one we can appeal to the above objectives. Of the two options presented, which is more likely to better achieve the objectives? When we have enough knowledge to answer this question we will know how we ought to proceed. In general, particular combinations of pest management policies should be evaluated with respect to our five objectives. Very likely, a combination that is right in certain circumstances would not be right in other circumstances.

---

[33]Pimentel; see footnote 30.

There is one final issue which I wish to bring up before concluding this paper. Some readers will notice that while the ethical principles we elaborated in Part III of this paper allowed for consideration of our moral obligations to animals, in general we have not done so. We have, of course, referred to the necessity of preventing the drastic harm that might be expected from oversimplification of ecological networks. However, the ethical principles that we formulated, and on the basis of which we supported five principles that should guide us as we move to reduce our use of pesticides, imply that we have moral obligations to individual animals.

Surely, our use of pesticides often injures or kills many mammals and birds. Further, since the injuries to or deaths of these creatures are through poisoning, there may be considerable suffering or discomfort for these creatures. Must we not take such creatures into account? Surely we must; however, there is reason to believe that taking them into account will not require much additional complexity in our thought. Where the pesticides we have used have caused suffering or death to mammals or birds, reducing the use of such pesticides should reduce the extent of harm inflicted on such creatures. That is, reduction in use of synthetic pesticides should lead to less human injury and less animal suffering and death also.

Of course, concern with moral obligations to mammals or birds does raise some complex and theoretical problems. Such creatures are themselves sometimes pests. Yet since we have obligations to them, on consequentialist grounds, we have to take benefits and harms to them into account when we are deciding on a pest management policy. Conceivably, there are circumstances in which the positive value which accrues to human beings is not sufficient to outweigh the negative value suffered by animals as a result of our behavior. Conceivably, there are situations where we should sacrifice our interests in favor of the interests of the pests. Consideration of this question does suggest a sixth objective for pesticide policy. It is:

> *Methods of controlling pests must be humane, i.e., must not cause excessive suffering to the pests.* (Excessive suffering to animals, in this context, means suffering to the animals which is greater than any reduction in suffering or increase in satisfaction gained by humans.)

Finally, let us mention one theoretical problem. In discussing our moral obligations to animals regarding pesticide use, I was arguing from a consequentialist perspective. However, some philosophers have argued that individual animals are entitled to the same type of respect as individual human beings, that is, that we must respect the individual autonomy of animals.[34] They would maintain that our obligations to individual animals

---

[34]Tom Regan has taken this position. See *The Case for Animal Rights* cited in footnote 27.

are much stronger than I have indicated, indeed, that it is as morally wrong to kill animals of certain sorts (many mammals, for example) as it would be to kill human beings merely because those humans are pests. Use of pesticides to control rodents, for example, would be immoral on their view (if we could survive without the food that the rodents would eat). We have discussed these ethical theories elsewhere and tried both to raise objections to the assumptions on which they rest and to argue that their conclusions regarding killing of animals are incompatible with other conclusions that they draw.[35] I do not agree that use of pesticides to control rodents is generally unacceptable on moral grounds.

[35]"The Case for Animal Rights" by Hugh Lehman, *Dialogue* XXIII (1984), 669–676. "On the Moral Acceptability of Killing Animals" by Hugh Lehman, *Journal of Agricultural Ethics*, Vol. 1. No. 2, 1988, 155–162. See also "Rights, Justice and Duties to Provide Assistance: A Critique of Regan's Theory of Rights," by Dale Jamieson, *Ethics*, Vol. 100, No. 2, January 1990, pp. 349–362.

15

Growing Public Concern Over Pesticides in Food and Water

Carolyn E. Sachs

Introduction

Public concern about pesticides in food and water has increased dramatically in the last decade. Food-safety and water-quality issues have received increased attention by consumers and interest groups. The growth of the environmental movement and Green politics have heightened public awareness and influenced policies relating to food and water quality. At the same time that the general public and public-interest groups are more concerned with pesticides in food and water, confidence in science and government regulatory processes has eroded. Thus, a number of public-interest groups are pressing for stricter government regulation of pesticides and development of alternatives to pesticides. But chemical companies, food industries, and farmers are arguing that consumers are overreacting to the dangers of pesticides and suggesting that consumers need more education about how food is grown, why pesticides are applied, and the minimal danger pesticides pose to their health and safety. This paper documents the level of public concern about pesticide use, discusses the public's confidence in government regulatory activity, explores the relation between science and public policy, and finally raises ethical issues relating to consumer concerns and public policy.

Level of Public Concern

Pesticides have been an integral part of our agricultural production system since World War II, but public concern about toxicity of pesticides was fairly minimal until the publication of Rachel Carson's *Silent Spring* in 1962. Carson convincingly decried the harmful effects of pesticides, especially DDT, on wildlife and human health and raised public awareness of the prevalence and potential harm of pesticides. During the 1960s, the public was more concerned about the effects of pesticides on wildlife than on farmers' or consumers' health. A 1965 study of Pennsylvania consumers' concern about pesticide use found that half of consumers were concerned about the danger of pesticides to wildlife (Bealer and Willits, 1968) (See Table 15.1).

Despite consumer concern, pesticides continue to be used in unprecedented levels on our food supply. In 1964, 225 million pounds of pesticides were used on major field crops compared to 558 million pounds in 1982. Herbicides account for the major increase in pesticide use between 1964 and 1982. The increase in pesticide use is tied to changing agricultural production systems that have become more concentrated, specialized, and capitalized (Sachs and Higdon, 1989) and therefore rely more extensively on pesticides.

As pesticide use has increased, the number of incidents of pesticide contamination has multiplied and public awareness and concern about pesticide use have become widespread. During the past 30 years, evidence about the environmental and health effects of pesticides has surfaced and resurfaced around particular incidents. Public awareness of pesticides and pesticide toxicity is often correlated with media exposure of pesticide problems. Problems of DDT were brought to light by Rachel Carson and more recently aldicarb or Temik in watermelons came to the public's attention, and Alar in apples was publicized by the National Resources Defense Council's report.

Several studies in the 1980s revealed that consumers are quite concerned about pesticide use. A 1984 survey of Pennsylvania consumers found that 71% of consumers were concerned about eating fruits and vegetables sprayed with pesticides compared to 41% of consumers reporting concern in 1965 (Sachs et al., 1987) (See Table 15.1). The same study found that consumer concerns about the danger of pesticides to farmers increased from 15% in

Table 15.1. Percentage of consumers with great deal or some concern with pesticide use, 1965 and 1984.

	1965	1984
Personally concerned about farmers using pesticides	31.6	76.0
Danger of pesticides to farmer	15.0	78.7
Danger of chemicals to wildlife	51.8	80.8
Danger to person who eats fruits and vegetables sprayed with pesticides	41.5	71.1

Source: Sachs et al., 1987.

1964 to 78.9% in 1984 and that concerns about danger to wildlife increased from 51.8 to 80.8 during the same time period. Another study in four U.S. cities reported that 83% of consumers had a high or medium level of concern about pesticides and chemicals. The most recent survey by Zind (1990) reported that 86% of consumers were concerned about chemical residues on fresh produce.

Several recent studies of food safety found that consumers considered pesticide residues to be an important food safety concern. A national study by the Food Marketing Institute found that pesticide and herbicide residues on food were considered a serious hazard by more consumers than were a number of other food-safety potential dangers (Table 15.2). Seventy-six percent of consumers considered pesticide residues on food to be a serious hazard. Surveys conducted in several states support the findings of the national study. Kansas consumers ranked pesticide residues as the third-most-important food-safety concern (Penner et al., 1985) and in 1989, 45% of Georgia consumers ranked food grown using pesticides as one of the three top food-safety concerns (Huang et al., 1990).

Another issue of concern to the public is the contamination of groundwater by pesticides. Contamination of groundwater by agricultural pesticides has been documented in 23 states (Cohen et al., 1984), and high levels of nitrates have been measured in wells in 32 states (O'Hare et al., 1984). Detection of pesticide residues in groundwater is increasing. The 16 pesticides that have been detected in groundwater in 23 states are the result of normal agricultural use as opposed to improper disposal or accidents involving pesticides (Panasewich, 1985). A recent study conducted in rural Iowa documented substantial public and farmer concern with groundwater quality (Padgitt, 1987). In one area of Iowa, 80% of the rural

Table 15.2. Concern about food safety issues.

	Serious hazard
Residues, such as pesticides and herbicides	76
Antibiotics and hormones in poultry and livestock feed	61
Fats	55
Cholesterol	51
Salt in food	43
Irradiated foods	43
Nitrates in food	38
Additives and preservatives	36
Sugar in food	28
Artificial coloring	24

Source: Food Marketing Institute, 1987.

population was concerned with the purity of their drinking water and in another county 66% were concerned with drinking-water quality.

Government Agencies and Pesticide Regulations

Consumers' concern about the impact of pesticides on their health and the environment may well be related to their lack of confidence in the government regulation of pesticides. Public confidence in government agencies' regulation of pesticides has eroded over time. In 1965, 97.7% of consumers in Pennsylvania agreed that the government adequately regulates chemical use in or on food in contrast to only 45.8% of consumers in 1984 (Sachs et al., 1987). thus, two decades ago, virtually all consumers had confidence in the government's ability to adequately regulate pesticides, but presently confidence in government's efficacy is lacking. A recent study asked shoppers who they relied on most to ensure that the products they buy are safe. Forty-five percent reported relying on themselves as individuals; 25% stated that they depended on the federal government; and 15% said they depended on consumer organizations (Food Marketing Institute, 1989). A 1989 survey of Georgia consumers found that 43% indicated that chemical pesticides should be banned or subjected to greater restrictions for use on fresh produce (Huang et al., 1990).

Lack of confidence in government agencies may be a result of (1) increased awareness of the potential hazards of pesticides on the part of consumers, (2) recognition that pesticide use has increased enormously, or (3) a result of the perception that the government is doing a poorer job of regulating chemicals in the food system.

As discussed in the previous section, consumers are definitely more concerned about pesticides in the 1980s than they were in earlier decades. As consumers become more concerned about pesticides, they expect more government regulation and other responses. Also, evidence that toxic pesticides are in the food supply raises consumers suspicions that government regulations have been inadequate to ensure the safety of their food supply.

Are government regulations adequate and are they enforced? Federal government regulation of pesticides is authorized by the Federal Insecticide, Fungicide, and Rodenticide Act (FIFRA), originally enacted by Congress in 1947, and by several sections of the Federal Food, Drug, and Cosmetic Act (FFDC). These acts authorize EPA to register pesticide products, specify the terms and conditions of their use prior to being marketed, and remove unreasonably hazardous pesticides from the marketplace (GAO, 1986a). FIFRA has been significantly amended to provide broader regulatory coverage that changes the emphasis from consumer protection and product performance to public health and environmental protection (GAO, 1986a). A balancing of risks and benefits is the major criterion for regulation. Thus in considering whether to ban or restrict a chemical, EPA must weigh the economic benefits of the use of a particular pesticide against the potential risks. The reliance on risk assessment is problematic due to limits of estimation techniques related to health effects, human exposure, and economic effects.

Amendments to FIFRA in the 1970s called for the registration of new chemicals and the reregistration of all pesticide products that were registered prior to 1975. The pace of reregistration of pesticides has been slow, and EPA has been criticized by GAO, Congress, industry representatives, and environmentalists for failing to reregister and remove pesticides from the market that pose unreasonable risk (GAO, 1986a). The task of reregistering approximately 50,000 pesticide products is difficult and time consuming and will undoubtedly extend into the 21st century. Thus, public concern about the inadequacy of government regulation may well be warranted.

Science and Pesticides

Consumers are often accused of being misinformed about the dangers of pesticides and of overreacting to claims of health or environmental damage from pesticides (Huang et al., 1990). This raises the question of whether there is a sound scientific base that consumers and the government can rely on to gain information about pesticides.

Government policies depend on scientific information to assess the risks and benefits of pesticides. Risk assessment is the primary analytic technique used by the federal government to inform regulatory decisions. Various critiques have noted that risk analysis is an evolving science that suffers from various sources of uncertainty itself. An extensive scientific and economic data base is necessary for risk assessment techniques to be useful. At present, there is insufficient scientific information available on a number of dimensions. First, there is inadequate data on the health effects of various pesticides. A 1983 study of 60 active ingredients in pesticides found that "48 percent of federally registered pesticides lacked data to assess their potential to cause tumors; roughly 38 percent lacked data on birth defects; 48 percent lacked data of reproductive impairment; and 90 percent lacked data on genetic mutations" (GAO, 1986b:23). Also, there is incomplete data on pesticide use and on residues on crops and in processed foods, and adequate information on dietary patterns (Archibald, 1989).

Consumers also depend on the government to use scientific tests to monitor the safety of our food supply in terms of pesticide residues on food. A recent GAO study found that FDA's pesticide monitoring program has major shortcomings (1986b). FDA does not regularly test for a large number of pesticides that are known as potential health hazards that may be present in food, does not prevent the marketing of food that contains illegal pesticide residues, and fails to penalize growers who market food with illegal pesticide residues. Scientific testing of food residues is not adequate due to limited samples, types of tests used, and irregular testing. For example, FDA's analytic methods are incapable of detecting two-thirds of the pesticides registered for use on foods (Mott, 1984).

Finally, scientific information can be interpreted in a variety of ways with quite distinct outcomes. A particularly compelling example of different interpretations of scientific data about pesticides is the case of government regulation of alachlor. Alachlor is a herbicide developed by Monsanto that is used primarily on corn and soybeans. Relying on data from the same laboratory studies that tested the carcinogenicity of alachlor, the Canadian government banned the pesticide while the U.S. government did not (Hoberg, 1990). Hoberg attempts to answer the question of why the two governments adopted divergent scientific interpretations. His conclusions are that rather than science driving policy, science is used to rationalize decisions made on other grounds. Antiregulation forces, including chemical companies and farmers, had greater stakes in the United States than in Canada. For example, alachlor is the largest-selling herbicide in the United States, but only the tenth-largest-selling herbicide in Canada; thus Monsanto put more effort into lobbying against regulation in the United States. The study concludes that politics and science are difficult to entangle in the case of pesticide regulations. Certainly science matters in pesticide regulation. When the scientific base is certain and shows a serious hazard, government regulators are constrained to act to regulate pesticides. However, at times of scientific uncertainty, governments are less constrained and may use scientific data to corroborate their political decisions.

Thus, there appears to be a woefully inadequate scientific base on which to make sound decisions about pesticide regulations and monitoring. The complex risk-assessment models that are used to determine regulation decisions are being implemented with inadequate data on a number of dimensions. Both proponents of more stringent regulations for pesticides and antiregulation constituencies have critiqued the adequacy of scientific information and risk-assessment methodologies. Presently, consumers are not able to depend on an adequate base of scientific data to guide decisions about the safety of a large number of pesticides.

Ethical Issues

Regarding pesticides, consumers doubt the government will protect their health, the food supply, and the environment. Three major ethical issues are raised. First, which interest groups are involved in influencing government policies—how do consumers fare in the trade-off between benefits and risks? Second, given the inadequacy of scientific information on the health and environmental effects of pesticides and the inherent inability of science to answer the question, How safe is safe enough? (Caplan, 1986), how should pesticide use be regulated to protect consumers and what should be the messages in consumer education? Finally, given the fact that food is unique compared to other consumer goods—in the sense that everyone must consume food to live—how can government policies in the face of uncertainty insure the health of disadvantaged people as well as highly educated upper-middle-class consumers?

1. Interest groups concerned with pesticide regulations include chemical companies, farmers, and environmental and consumer organizations. Chemical companies are interested in expanding the markets for their products and making profits. Farmers use pesticides to minimize the risk that their crops will be damaged, to save labor, and to produce products that will attain the highest price in the market. Environmental groups are primarily concerned with the impact of pesticides on wildlife, human health, and the natural world. Consumers are concerned with the safety of their food supply. Government policies are ostensibly designed to balance these interests with the primary goal of insuring a sufficient and safe food supply. There have been a number of critiques to the effect that consumers and environmentalists have not been sufficiently involved in decisions related to pesticide policies and food safety. Rather, overall agricultural policies primarily benefit farmers and agricultural input companies and encourage the widespread use of pesticides. Others have argued that our current farming systems have benefitted chemical companies and other large agricultural corporations to a much greater extent than farmers or consumers have benefitted (Lewontin and Berlan, 1986).
2. The second ethical issue revolves around the question of to what extent government should appeal to science to answer questions related to pesticide regulation and policies. At the current time, scientists do not have sufficient data to determine the health effects of a large number of pesticides. In addition, the data that does exist must currently be assessed in the midst of a cloud of uncertainty. As Caplan notes, "uncertainty about safety, whether it involves food or some other substance or process, is ultimately, often not resolvable on scientific grounds" (1986:182). Science can provide tentative information about the risks of certain prod-

ucts, but decisions requiring how safe is really safe require one to place values on profit, human health, and the natural world. Scientific effort in examining the health effects of pesticides is necessary but can never realistically be appealed to as the only basis for government regulations concerning pesticides.

3. The final ethical issue involves what Caplan refers to as "the special moral status of food" (1986:183). he argues that because food is among the most basic of human needs, government must take an active role in assuring that people can adequately have their food needs met. Appeals to educating consumers to enable them to make suitable choices are inadequate without stringent government regulation of pesticides. In fact, many of those most vulnerable to the effects of particular pesticides, such as infants and children, are not free to choose what types of food they consume. Also, food that is produced without pesticides is frequently higher priced and out of the reach of people with low incomes or in poverty. Thus, consumer education about pesticides is necessary but not a sufficient condition to protect consumers. The fact that consumers may choose to purchase organic food for a higher price does not protect infants, children, or the poor from pesticides.

Conclusion

Public concern about pesticide use has increased over the past several decades. Public awareness, incidence of pesticide poisonings, increased use of pesticides, and evidence of inadequate government regulation are factors that have contributed to the public's uneasiness about pesticides. In the 1960s, the public was primarily concerned about the impact of pesticides on wildlife. But in the 1980s over three-quarters of the public was concerned about the impact of pesticides on human health, farmers, and wildlife. Consumer advocacy groups have raised public awareness and are gaining some visibility in policy circles that have traditionally only catered to agricultural constituencies.

Government regulation of pesticides has proceeded slowly and is considered inadequate by the majority of the public. Despite amendments to laws that insist on stricter regulatory practices for pesticides, many pesticides that are currently being used have not been adequately tested for their safety relative to humans or the environment. The policy of testing pesticides one at a time rather than developing more comprehensive strategies for reducing use of toxic pesticides has contributed to the perception that the government is not doing enough in pesticide regulation.

Finally, science can provide information, but it cannot be the sole determinant of government policies relating to pesticides. Currently, inadequate information exists on the health and environmental effects of a number of pesticides. More funding should be provided to test the health effects of pesticides, but the findings of these studies will not move our food system in the direction of less dependence of pesticides in agriculture. In addition to providing scientific funding for the study of health and environmental effects of pesticides, more funding is needed for scientists and farmers investigating alternatives to pesticide-intensive agriculture. The federal government's funding of the Low-Input Sustainable Agriculture program is one step in this direction, but additional funding is needed. In addition, government policies that encourage and reward farmers for limiting pesticide use are necessary. Greater participation of consumers and consumer groups in developing food and agricultural policies is needed.

References

Archibald, Sandra O. (1989). "Regulating pesticides in food: The delaney paradox." *Land Economics* **65**(1):79–86.

Bealer, Robert and Fern Willits. (1968). "Worriers and non-worriers among consumers about farmers use of Pesticides" *Journal of Consumer Affairs* **2**:189–204.

Caplan, Arthur L. (1986). "The ethics of uncertainty: the regulation of food safety in the United States." *Agriculture, Change, and Human Values* **3**(1/2):180–190.

Carson, Rachel, (1962). *Silent Spring*. Greenwich, Connecticut: Fawcett Publications.

Cohen, S.X., C. Eiden, and M.N. Lorber. (1986). "Monitoring ground water for pesticides." In W. Garner, R.C. Honeycutt, and H.N. Nigg, eds. *Evaluation of Pesticides in Ground Water*. ACS Symposium Series 315, Washington, D.C.: American Chemical Society.

Food Marketing Institute. (1987). *Trends: 1987 consumer attitudes and the supermarket*. Washington, D.C.: Food Marketing Institute.

Food Marketing Institute. (1989). 1989 Trends: Consumer attitudes in supermarkets. Washington, D.C.: Food Marketing Institute.

General Accounting Office. (1986a). *Pesticides: Need to enhance FDA's Ability to Protect the Public From Illegal Residues*. Washington, D.C.: United States General Accounting Office RCED 87-7.

General Accounting Office. (1986b). *Pesticides: EPA's Formidable Task to Assess and Regulate Their Risks*. Washington, D.C.: United States General Accounting Office RCED-86-125.

Hoberg, George. (1990). "Risk, science and politics: alachlor regulation in Canada and the United States." *Canadian Journal of Political Science* **23**:257–277.

Huang, Chung L., Sukant Misra and Stephen L. Ott. (1990). "Modeling consumer risk perception and choice behavior: the case of chemical residues in fresh produce." Unpublished paper.

Lewontin, R.C. and Jean-Pierre Berlan. (1986). "Technology, research, and the penetration of capital: the case of U.S. Agriculture." *Monthly Review* **58**:21–34.

Mott, Lawrie. (1984). *Pesticides in Food: What the Public Needs to Know*. San Francisco: Natural Resources Defense Council.

O'Hare, M., D. Curry, S. Atkinson, S. Lee, and L. Cantor. (1984). *Contamination of Ground Water in the Contiguous United States From Usage of Agricultural Chemicals*. Environmental and Ground Water Institute. Norman, OK: University of Oklahoma.

Padgitt, Steve. (1987). *Agriculture and Groundwater Issues in Big Spring Basin and Winneshiek County, Iowa*. Ames, Iowa: Iowa State University.

Panasewich, Carol. (1985). "Protecting ground water from pesticides." *EPA Journal* **11**:18–25.

Penner, Karen P., Carol S. Kramer, and Gary L. Frantz. (1985). *Consumer Food Safety Perceptions*. Manhattan, KS: Kansas State University Cooperative Extension Service.

Sachs, Carolyn, Dorothy Blair, and Carolyn Richter. (1987). "Consumer pesticide concerns: a 1965 and 1984 comparison." *Journal of Consumer Affairs* **21**(1):96–107.

Sachs, Carolyn and Francis Higdon. (1989). "Pesticide use and structural changes in agriculture: the production of corn and apples." Paper presented at Agriculture, Food and Human Values Society Conference, Little Rock, Arkansas, November.

Zind, Tom. (1990). "Fresh trends 1990: a profile of fresh produce consumers." *The Packer Focus 1989–90*. Overland Park, KS: Vance Publishing Co.

16

Pesticides: Historical Changes Demand Ethical Choices

John H. Perkins and Nordica C. Holochuck

Framing the Argument

Controversy over pesticides and pest control is no stranger to American life. Since the 19th century, public concerns have periodically ricocheted around the use of pesticides and other means of pest control. Frequently the criticisms have asserted in a moral tone that the action in question was wrong in some fundamental fashion. Understanding the moral dimensions of the use of pesticides, however, requires an understanding of the context of pest control actions. This context is complex and consists of (1) modernization and its impacts, (2) changes in American agriculture after 1945, (3) the obligations of pest control experts as professionals, and (4) how entomology developed in the United States. We discuss each of these contextual areas before turning to an examination of major ethical issues involving insect control.

Pest Control and Modernization

The "modern world," based on science, industrial technology, and the individualism of liberal capitalism, has traveled far beyond its birthplace in 17th-century Western Europe. Modernism now pervades virtually all of Europe and North America, plus many parts of Asia, Latin America, and Africa. It is not that hunting-and-gathering societies and subsistence agrarian economies have completely disappeared, but they are now found as remnants of an older way of life rather than its foundation.

Relationships between the "modern world" and the ethics of pest control are quite simple in concept: those events crucial to the formation of the industrial capitalist societies also shaped our perceptions of organisms we call "pest problems." Moreover, the tools, including pesticides, with which people solve pest problems are characteristic of the modern world.

An important expansion of trade, travel, exploration, and colonization occurred between the 14th and 16th centuries. Europeans began to seek commerce, new lands, and ultimately, new areas for conquest. Since the 15th century, Europeans have swarmed over the world, their power and influence have become overwhelmingly dominant.[1]

Global expansion of European culture was critical to the shaping of pest control problems in several ways, some direct, some indirect. First, and most directly, the expansion of trade simultaneously brought other organisms to new lands. Species native to Europe accompanied Europeans to North and South America, Africa, Asia, and Australasia. And vice versa, organisms indigenous to the invaded and colonized areas often traveled to Europe. Thus the world came to know the problem of the introduced pest. Hessian flies, gypsy moths, Klamath weed, and many other species are all well known as introduced pests in North America. In many cases, the impacts were orders of magnitude higher in their new homes because these pests were no longer restrained by natural enemies as they were in their native lands. Thus European expansion and imperialism directly laid the foundation of some modern pest control problems.

European expansion also had indirect and far more powerful effects on the shape of modern pest control problems in that it spawned an entirely new lifestyle—that which came in tow with the Industrial Revolution. Increase in trade in early Renaissance Europe was the basis for accumulation of wealth among mercantile capitalist entrepreneurs. This group of merchants, along with artisans, craftsmen, and rulers who were based primarily in the emerging towns, created a way of life that was not based directly on the agrarian economy of late medieval Europe. Feudal land holdings and the divine right of kings were the foundations of an order of power and wealth based on land ownership and the Church. Townspeople, particularly merchants, ultimately challenged this older order and formed a capitalist class that triumphed and instituted a way of life we now know as liberal, patriarchal, capitalist democracy.[2]

In addition to supplanting a political order, the rise of merchant capitalism was crucial to the emergence of modern science and industry. Merchants were important sponsors of the new mechanical philosophy that began to develop in Renaissance Italy, France, England, and the Low Countries. These philosophers sometimes found their problems directly in

---

[1]Alfred W. Crosby, *Ecological Imperialism* (Cambridge: Cambridge University Press, 1986), pp. 105–131.

[2]Fernand Braudel, *Capitalism and Material Life, 1400–1800* (New York: Harper Torch books, [1967], 1973 English translation by Miriam Fochan), pp. 373–440; B.A. Holderness, *Pre-Industrial England* (London: J.M. Dent & Sons Ltd., 1976), pp. 116–132.

the matters concerning trade, such as developing a science of navigation.[3] More often, however, the support of science was important because it constructed a new worldview that allowed people to see nature as a mechanical, dead universe, rather than an organic, female entity. Nature was thus "killed" and "desanctified," a transformation that encouraged people to think of pests merely as objects to be controlled.[4] Moreover, God and the divine right of kings were deposed in favor of a philosophy that saw individuals, with natural rights, operating in a nature governed by natural, rational laws.[5] This philosophy of modern science in turn facilitated the breakdown of the old feudal order and the rise of an entrepreneurial bourgeoisie.

Merchant capitalism and this new philosophy of nature were the basis for an increasingly urban society based on manufacturing and trade. Particularly important for the new way of life were the developments after 1700 of new ways of making and working with metals, fuels like coal, and machines like the steam engine. Machines, metals, and fuels that amplified human labor became the basis of an industrial culture, first in England but later in Europe, North America, and Japan.

With the rise of trade and industry came profound effects on agriculture. Before the rise of modern cities, people farmed, hunted, and gathered. For the most part, what they raised or gathered was consumed very close to where it grew. Few found it either possible or necessary to trade in agricultural goods, and little or no money was involved in agriculture. Agriculture was essentially not commercial.

In the 14th century, England underwent what historian David Levine called "the crisis of feudalism," a trauma from which it never recovered. Cause-and-effect relationships were extremely complex, but a combination of rapid population growth followed by famine and plague established tensions and forced migrations that undermined the feudal political economy established in 1066.[6] Landowners saw possibilities to use their lands to produce goods for trade, often in far-distant markets, rather than merely for food and fiber for local consumption. Enclosure of land previously held

---

[3]Alan G.R. Smith, *Science and Society* (London: Harcourt, Brace, Jovanovich, Inc., 1972), pp. 29–42; Stephen F. Mason, *A History of the Sciences* (New York: Collier Books, 1962), pp. 112–123; A. Rupert Hall, Early modern technology, to 1600, in Melvin Kranzberg, and Carroll W. Pursell, Jr., eds., *Technology in Western Civilization* (New York: Oxford University Press, 1967), Vol. 1, pp. 79–103.

[4]Carolyn Merchant, *The Death of Nature* (San Francisco, Harper and Row, Publishers, 1980), 348 pp.

[5]John Locke, *An Essay Concerning Human Understanding* (ed. J.W. Yolton, New York, 1961).

[6]David Levine, *Reproducing Families* (Cambridge: Cambridge University Press, 1987), pp. 31–37.

in common began in England after 1400. Feudal lords turned cropland into pastureland for sheep and intensified production on arable land. Although the pace of enclosure and commercialization was slow at first, it picked up speed in the 18th century in England. By the 19th century, English agriculture was highly commercial. Nearly all produce went to markets rather than for local use.[7]

Transforming agriculture from subsistence to a commercial enterprise was deeply dependent upon the accompanying expansion of trade, the rise of mechanical philosophy and science, and the replacement of feudalism with liberal capitalism. All of these activities were intertwined with each other as they occurred in England first, and it may well be that none of them could have occurred without the others.

Modern pest control problems were shaped within this complex crucible of increasing trade, science, and capitalism. Together these factors led to increasing commercialization of agriculture, and pest losses in agriculture took on new significance. Pests had always caused losses, but in subsistence economies those losses were in yields harvested, not quantities of money. By the 19th and 20th centuries, however, especially in places where agriculture was completely commercial, pest losses threatened a decrease in income, inability to repay loans, and therefore bankruptcy and exodus from the farming business.

Commercialization, in other words, created strong perceptions in the farmer's mind of the need to control pests in order to survive. Profit maximization, another aspect of commercialization, sent a strong reinforcing signal that reliable pest control technology was essential.

As the transformation proceeded, the human population also began to rise much more rapidly. England may well have led the population increase, soon to be followed by Europe in general and, later in the 20th century, the rest of the planet.[8] Decrease of the death rate due to better nutrition, better sanitation, and finally better health care, most likely keyed the global population boom. With an expanding population, virtually all parts of the Earth were colonized for agricultural production, an activity that now occupies about one-third of the planet's land area.[9]

---

[7]B.A. Holderness, *Pre-Industrial England* (London: J.M. Dent & Sons Ltd., 1976), pp. 51–75; David Levine, *Reproducing Families* (Cambridge: Cambridge University Press, 1982), pp. 47–68.

[8]Colin McEvedy and Richard Jones, *Atlas of World Population History* (Harmsondsworth: Penguin Books Ltd., 1978), pp. 9–44.

[9]Of a total land area of 13.1 billion hectares, about 1.47 billion hectares are in crops and 3.22 billion hectares are in permanent pasture. If these two latter figures are taken as a crude measure of agricultural land, about 4.69 billion hectares (36%) of the Earth's land surface is devoted to agriculture (*World Resources, 1990–91* [New York: Oxford University Press, 1990] Table 17.1, p. 268).

Increased population, mostly in urban areas, was another factor encouraging practical pest control practices. The larger number of people cemented a dynamic interaction between the agricultural and industrial sectors of modern economies. Industry depended upon agriculture to provide food and raw materials and part of the market for manufactured goods. Agriculture depended upon industry for its markets and for needed inputs of production technology, including materials like pesticides.

This mutual dependency might have existed even if population had not increased, but with population increase, mutual dependency became ever firmer. Virtually all agricultural land was occupied and, with only minor exceptions, it is now a commodity that is traded and used within a capitalist framework. Farmers all over the world increasingly do not have the option of choosing between commercial production and subsistence. Consumers (including farmers) now number over five billion and are obligately dependent upon a continuation of the levels of yield that are usually achievable only with the high-yielding practices that depend upon industrially produced inputs.

Pesticides were one of several agricultural inputs developed in the context of the transformation of agriculture from subsistence to commercial modes. In turn, this transformation of agriculture was an integral part of the larger transformations that variously are called the Industrial Revolution, the scientific revolution, or the capitalist revolution that replaced peasant and feudal economies. In addition, the "population revolution" was a product of changes that substantially lowered the human death rate and initiated a population growth phase that has not yet ended in many parts of the world. It is the contention of this chapter that judgments about the ethics of pesticide use must recognize this overall framework within which pesticides were invented and used.

Within the scientifically literate population of the world, a strong presumption exists: changes in lifestyle and worldview associated with the industrial and other revolutions represent "progress" and are "good." If this assumption is uncritically accepted, it can lead to facile judgments to the effect that all or the vast majority of pesticide uses are ethically justified. Emergence of the modern world, however, had embedded within it a nature appreciation movement which sought to soften the mechanical images of nature developed in the scientific and commercial revolutions of the 17th century.[10] Nature writers of the 19th century began to develop themes about the beauty, tranquility, and the sacredness of nature.[11] This form of

---

[10]Carolyn Merchant, *The Death of Nature*, pp. 253–276.

[11]For example, see Gilbert White, *The Natural History of Selborne*, excerpts of which are in H.J. Massingham, ed., *The Essential Gilbert White of Selborne* (London: Breslich and Foss, 1983), 361 pp.; and George Perkins Marsh, *Man and Nature* (Cambridge: Belknap Press of Harvard University Press, 1965), 472 pp.

literature grew more prominent in the 20th century and currently serves as a major philosophical base from which to judge the moral quality of modern, industrial society. In the case of pesticides, the most powerful critique was administered by the modern nature writer Rachel Carson in her book, *Silent Spring* (1962).[12] A variety of failures of pesticides and the burgeoning environmental movement ensured that after 1962 pesticide technology would never again simply be taken for granted. The moral debate over pest control actions was changed irreversibly by Carson's book.

Changes in American Agriculture, 1945–Present

Scientific and commercial revolutions of the 17th century created the general context for choices about pest control and had their impacts in Europe, North America, and elsewhere. Specific choices made in different countries were shaped by the more general context and by factors peculiar to individual nations. For the United States, those choices were made during the industrialization of agriculture, which began in the 18th century and continues today.[13] These industrialization processes increased the productivity of both land and labor. By the 20th century they forced farmers to change from mixed commercial-subsistence farms into specialized businesses with a limited range of products, virtually all destined for market.

Industrialization was clearly visible in the United States by 1940, but it was in the years after 1945 that the structure of the agricultural industries changed most dramatically. Technical, political-economic, and organizational changes occurred as farms became fewer in number, larger, more heavily capitalized, specialized, and rationalized. New pesticides were part of the technical changes, and the incentives for their invention and adoption were tightly linked to the other dimensions of change.

Mechanization was the most important change in American agriculture during the 19th century. New steel plows, sulky plows and cultivators, reapers, threshers, combines, and other devices began to increase the productivity of labor.[14] Machines permitted one person to farm more land than he or she could have without the devices. Mechanization was encouraged by the availability of large tracts of land in the North American Midwest, and it enabled the cultivation of this land to be done by fewer people compared to earlier modes of farming.

In the 19th century, animals were the main source of power, but steam engines enjoyed a brief period of use. After 1900, however, gasoline and

---

[12]Rachel Carson, *Silent Spring* (Boston: Houghton Mifflin Company, 1962), 368 pp.

[13]Carolyn Merchant, *Ecological Revolutions* (Cahpel Hill: University of North Carolina Press, 1989), pp. 149–156.

[14]John T. Schlebecker, *Whereby We Thrive* (Ames: Iowa State University Press, 1975), pp. 97–112.

diesel engines and electrical motors increasingly mechanized the farm and substituted for both animals and the few steam engines in use. Tractors, spraying machines, harvesters, irrigation pumps, milking machines, and other equipment became the norm. Their purchase depended upon loans and cash incomes from commercial production.

After 1945, the well-established trends toward mechanization increased even further. Tractors and other machines got bigger, faster, and more efficient. In order to make the machines cost-effective, they had to be used over ever-larger pieces of land. Mechanization of agriculture therefore went hand in hand with a tendency for farmers to seek larger areas for their operating units. As machinery and farm acreages increased, more capital was tied up in investments, and the farmer was even more dependent upon a steady flow of income in order to service the debt on land and machinery. Mechanization and increasing size were a feature common to all sorts of agricultural operations, including cash grain farms, fruit and vegetable operations, cotton production, and various types of animal husbandry.[15]

Fertilizer use, especially nitrogen, also became increasingly common after 1900. Although mineral deposits such as sodium nitrate and organic deposits such as guano were used commercially in the 19th century, few farmers in North America employed them extensively at that time.[16] They were relatively expensive, and land was comparatively cheap, so most farmers simply used the abundant land and did not attempt to get yields higher than those permitted by natural fertility levels.

The Haber-Bosch process from Germany in the early 1900s, however, soon lowered the price of nitrogen to levels that made it uncompetitive *not* to use it. Nitrogen was usually the limiting nutrient in soils, so an economic source of it both attracted farmers and, through the process of the "agricultural treadmill," ultimately required that farmers use fertilizer to maintain competitive production.[17] Nevertheless, it was not until after 1945 that cheap and easy-to-use sources of anhydrous ammonia mandated the complete transformation to fertilizer use as a regular farming practice for all crops. Use figures for the United States demonstrate this dramatically: 62,000 tons of nitrogen were used in 1900, 419,000 in 1940, and 1,847,000 in 1954.[18]

Irrigation also became more common after 1900 and even more so after 1945. The first irrigation works in the United States tended to be in the

---

[15]Ladd Haystead and Gilbert C. Fite, *Agricultural Regions of the United States* (Norman: University of Oklahoma Press, 1955), 288 pp.

[16]Willard W. Cochrane, *The Development of American Agriculture: A Historical Analysis* (Minneapolis: University of Minnesota Press, 1979), p. 109.

[17]For an explanation of the agricultural treadmill hypothesis, see Willard W. Cochrane, *The Development of American Agriculture* (Minneapolis: University of Minnesota Press, 1979), pp. 387–393.

[18]W.B. Andrews, Anhydrous ammonia as a nitrogenous fertilizer, *Advances in Agronomy*, **VIII** (1956):61–125.

dry areas of the Southwest, and facilities were on line in California, for example, even before the conquest of the state by the Americans in the Mexican War (1846–1848). With increasing capital investment in agriculture after 1900, the use of irrigation spread to the semiarid regions of the Great Plains and even to the more humid regions of the East.[19]

Simultaneously with the transition to mechanization, fertilization, and irrigation, American farmers began after 1900 to switch more systematically to genetically improved crop plants. Even though a few higher-yielding varieties had been identified in the 19th century, it was not until the advent of Mendelian genetics and the development of hybrid maize that plant breeders began to turn out a steady stream of new cultivators. After 1945, applied breeders released a wide range of new crop plants that supplanted the older varieties in all fields of farm production.[20] Typically, plant breeders took their cues for what sorts of varieties were needed from the practices of farmers then raising the crop. This meant that as farmers adopted machinery, fertilizer, and irrigation, plant breeders adjusted their screening procedures to find the new varieties that would do well in those contexts. The new varieties that were considered desirable, therefore, were by design matched to the evolving technical framework of agriculture.[21]

Pesticides and other new methods of controlling pests also entered into agricultural technology along with new machines, fertilizers, irrigation, and new plant varieties. Discovery of the insecticidal properties of Paris green and London purple, both arsenicals, came in the mid-19th century. In 1892 scientists at the U.S. Department of Agriculture developed lead arsenate for the gypsy moth outbreak in New England. Calcium arsenate for boll weevil appeared in 1918, and about the same time para-dichlorobenzene, the first of the synthetic insecticides, was developed. Chemical poisons were also developed at about the same time for fungal diseases and weeds, but use of all pesticides was small and confined to high-value crops until after 1945.[22]

---

[19]Murray R. Benedict, *Farm Policies of the United States, 1790–1950* (New York: The Twentieth Century Fund, 1953), pp. 124–128.

[20]Jack Kloppenburg, Jr., argued that plant breeding, particularly in maize, turned seeds into a commodity (*First the Seed*, Cambridge: Cambridge University Press, 1988, especially pp. 39–65); Deborah Fitzgerald portrays the complexities of private compared to public maize breeding in *The Business of Breeding* (Ithaca: Cornell University Press, 1990), 247 pp.; and Barbara Ann Kimmelman describes how Mendelian genetics was appropriated for applied breeding work after 1900 (*A Progressive Era discipline: genetics at American agricultural colleges and experiment stations, 1900–1920*, Ph.D. thesis, University of Pennsylvania, 1987, 446 pp.).

[21]John H. Perkins, *Four Blades of Grass*, in preparation.

[22]Brief reviews of pesticide history are in E.O. Essig, *A History of Entomology* (New York: The Macmillan Company, 1931), pp. 403–501; Adelynn Hiller Whitaker, *A History of the Federal Pesticide Regulation in the United States to 1947*, Ph.D. thesis, Emory University, 1974, pp. 1–30; Thomas R. Dunlap, The triumph of chemical pesticides in insect control, 1890–1920, *Environmental Review* No. 5, 1978, pp. 38–47; and John H. Perkins, *Insects, Experts, and the Insecticide Crisis* (New York: Plenum Publishing Co., 1982), pp. 3–28.

Pesticides were not the only new technique for handling unwanted organisms. Biological control for insects and quarantine for all sorts of pests were well established in theory and practice by the turn of the century. Probably of even more importance, research scientists and farmers had known and used a variety of cultural practices to reduce or eliminate pest damage for many years. Alteration of planting dates for Hessian fly control on wheat, variation of variety and of row width for boll weevil control on cotton, and avoidance of wheat production in hot-humid climates for disease control were merely a few of the examples of how cultural practices were recognized as pest control practices by 1900.

Despite the advances made in pesticides, biological control, quarantine procedures, and cultural methods, pest control technologies still had severe limitations in the years after 1900. Discovery and development of the insecticidal properties of DDT and the herbicidal properties of 2,4-D during World War II, however, completely altered the vision of how chemicals might play a powerful role in agricultural production as well as in other areas of pest control such as public health, homeowner protection, and forestry. Only modest advances in pesticide technology were made during the war, but in the years after 1945 a flood of new chemicals and of new ways of using older materials flowed from the chemical and agricultural research stations. Chemicals became the dominant mode of controlling unwanted organisms in the years after 1950, a situation that has changed very little since that time.[23] Production and use of these materials quickly mounted into the millions of kilograms per year in the United States.[24]

Not only did the use of pesticides become predominant in pest control, the very successes of the chemicals tended to replace other methods of coping with pests. Biological control was relegated to a fringe research area of entomology in the years immediately after 1945. Soil insecticides made it possible to avoid crop rotations and thereby increase income in maize production. Herbicides tended to replace manual and mechanized weed control.[25]

Although in retrospect it is easy to see that in some cases the enthusiasm for chemical control was overblown and imprudent, that was not so clear at the time. Chemicals were essential to the successes of production methods characterized by mechanization and other heavy-capital inputs. Researchers and farmers alike thought they were making the correct *tech-*

---

[23]John H. Perkins, Reshaping technology in wartime: the effect of military goals on entomological research and insect-control practices, *Technology and Culture* **19**, No. 2 (1978): 169–186.

[24]Craig D. Osteen and Philip I. Szmedra, *Agricultural Pesticide Use Trends and Policy Issues* (USDA Economic Research Service, 1989), p. 8.

[25]John H. Perkins, The quest for innovation in agricultural entomology, 1945–1978, in David Pimentel and John H. Perkins, eds., *Pest Control: Cultural and Environmental Aspects* (Boulder: Westview Press, 1980), pp. 23–80.

nological choice, and a proper *moral* choice, when they switched into a regimen of pesticide control of bothersome organisms.

Genetic resistance to pesticides, especially of insects to insecticides; destruction of natural enemies; and pollution of the environment all were well-identified problems of pesticides by the 1960s. Several reform movements developed among research scientists to develop new methods of pest control that were less reliant on chemicals. Because of the serious problems surrounding insecticides, entomologists were most involved in attempts to forge new strategies. The concepts of integrated pest management (IPM) and efforts to manage whole pest populations over wide areas (total population management—TPM) were prominent in entomological circles during the late 1960s and 1970s.[26]

After 1974, developments in plant tissue culture, in manipulation of DNA, and in methods of transferring DNA interspecifically opened the doors to avenues of pest control through biological engineering. It became possible, for example, to make tobacco and tomato plants capable of synthesizing in their own cells the toxin from the bacterium *Bacillus thurengiensis*. Similarly, better methods of culturing fungi created the potential to package them as biological agents against weeds and plant diseases.[27] Despite the excitement generated, however, these developments in biological engineering did not represent a conceptual advance. The bioengineered products are essentially "chemicals" and therefore pesticides. To be sure, they are highly sophisticated pesticides, but the conceptual ways in which they are proposed to be used indicates little change in the theory of how to use toxic chemicals.[28]

Not only has the technology of agriculture evolved especially rapidly since 1945, the political economy of agriculture has also changed in important ways. Farming is a critically important arena for pest control innovations, because the major market for pest control innovations is in agriculture. Therefore it is important to understand how the agricultural industry has evolved in conjunction with pest control and other technologies.

---

[26]John H. Perkins, *Insects, Experts, and the Insecticide Crisis* (New York: Plenum Publishing Co., 1982), pp. 57–126.

[27]John H. Perkins and Richard Garcia, Social and economic factors affecting research and implementation of biological control, in T.W. Fisher, et al., eds., *Principles and Applications of Biological Control* (Berkeley: University of California Press, in press).

[28]John H. Perkins, The future history of biotechnology: a commentary, in *Proceedings* of a Conference of Biotechnology, Biological Pesticides and Novel Plant-Pest Resistance for Insect Pest Management, D.W. Roberts and R.R. Grandaos, eds., Boyce Thompson Institute for Plant Research, Ithaca, New York, 1988, pp. 168–175. For a recent critique of pest control through genetic engineering, see Richard Hindmarsh, The flawed "substainable" promise of genetic engineering, *The Ecologist* **21**:5 (1991):196–205. (We thank Jim La Spina for bringing this article to our attention.)

Agriculture during the 20th century, especially in the United States, was characterized by a number of developments. Consolidation and specialization were prominent. Driven by the need to obtain economies of scale for using new machinery and by the potentials for achieving higher farm income, the most technically progressive farmers expanded the sizes of their operations and reduced the range of crops they produced. Price-support programs after 1933 were supportive of consolidation and specialization for many crops.[29] Virtually all produce left the farm gate headed for processing plants and commercial markets. Often the markets were abroad, and the United States achieved a preeminent, if not dominant, position in a number of global commodity markets.[30]

Changes in agriculture were linked to a complex of financial supports from private lenders and the government. In addition, farming came to rely on a network of public and private agricultural research stations—for example, the private research done by chemical companies to develop new pesticides. Farmers operated in a highly competitive marketplace, so the abilities of individual farmers to succeed and prosper depended on obtaining timely financial credit and new technology.

Economically, farmers were each trying to improve their own standing in competition with their peers. They used new technology to lower their own production costs below those of their competitors. Socially, farmers became atomistic entrepreneurs. They wanted technology that they could use independently of their neighbors. Politically, farmers welcomed assistance that came from public laboratories and price-support systems, but they wanted to keep farm operations a private matter. They did not welcome regulation, and they resisted public support programs if the freedom to act as individuals was curtailed. Philosophically, farmers saw their job as controlling nature, not somehow subordinate or even coequal with the natural world. These were the cultural characteristics of American agriculture and successful production technologies tended to be compatible with such traits.

Pesticides were chosen because they fit well economically, politically, socially, and philosophically with the evolving complex of American agriculture.[31] The conclusion is of extraordinary importance in the debate about the ethics of pesticide use: if a moral objection is made against pesticide use, then it is very difficult to distinguish the target of attack, the chemical, from the entire complex of agricultural technology, of which the

---

[29]Willard W. Cochrane, *The Development of American Agriculture* (Minneapolis: University of Minnesota Press, 1979), pp. 286–289, 404–406.

[30]Willard W. Cochrane, *The Development Of American Agriculture*, pp. 267–273.

[31]John H. Perkins, *Insects, Experts, and the Insecticide Crisis* (New York: Plenum Publishing Co., 1982), pp. 3–28.

pesticide is a part. A complaint about pesticides may really be an attack on some other aspect of the complex. Understanding the moral and ethical dimensions of pesticide use, in other words, requires understanding the larger historical context, in which technical choices are made.

Professionalism and Pest Control Expertise

Two sorts of occupational classes have especially difficult ethical decisions to make about the ethics of pesticide use: the scientists who develop pest control technologies and the farmers who use them. Other occupational groups also have a series of ethical questions facing them, such as regulators, manufacturers, salespeople, and environmental activists. We will focus on the primary actors, scientists and farmers, in an effort to explore the ethical dilemmas facing them.

Scientists are especially important actors because they are in a professional/client relationship to farmers. Therefore it is important to understand the social dimensions of professionalism.

Professionalism is described in various ways, but Professor Jerome Ravetz summarized its most important features for scientists in a way that is quite useful.[32] He argued that scientific experts work in a way that is quite distinct from other groups who provide a service:

- Recipients, or clients, of the expert are dependent upon the expert in order to accomplish some task. It is not possible for the client to do what the expert can do, even if the client were to make time for and try to accomplish what the professional does. In other words, the service provided by an expert is qualitatively something that a nonexpert simply can't do.
- Clients are not capable of judging the competency of the services rendered. Only another trained scientist can tell if an expert opinion meets the standards for currency with existing knowledge and congruency with state-of-the-art practices in the field. Put briefly, only another pest control scientist can judge whether advice was competent.
- In some fields of scientific expertise, society grants a legal monopoly to an expert group so that only certified members of that group are allowed to provide expert services. Medicine has perhaps the strongest legal mandate to restrict the practice of health care to licensed physicians. Such restrictions are justified by a need to eliminate quackery and incompetence, but it should not be forgotten that the monopoly of right-to-practice also confers a power

[32]Jerome R. Ravetz, Ethics in scientific activity, in Albert Flores, ed., *Professional Ideals* (Belmont, CA: Wadsworth Publishing Company, 1988), p. 152.

to charge high, sometimes exorbitant, fees. Pest control scientists have never achieved a monopoly power over practicing pest control, but nevertheless society tends to grant people with pest control training a priority of credibility when it comes to deciding what to do about pests. For their part, experts try to use such deference to enhance their own prestige and social status.

- With the privilege of being the expert comes the obligation for the group of experts to guarantee that the individuals in the group are rendering high-quality services and protecting the interests of clients. Pest control experts have approached this part of their professionalism through such devises as the code of ethics of the American Registry of Professional Entomologists.[33] Because pest control experts are not licensed in the way doctors are, however, no easy decertification process exists for practitioners who embarrass the group. Pest control experts have only the power of persuasion with clients if the group believes some of its members are incredible and incapable of rendering service that protects the client.

The above four characteristics of professionals are necessary to consider in outlining the general social expectations that fall upon professionals, but they are not sufficient to understand the total context in which professional behavior must be judged. Also relevant are the historical sequences through which the profession developed, grew, and established a base in the political economy of the society.

The Development of Modern Entomology

We have discussed the historical changes affecting agriculture and the professional obligations of experts. We now turn to how pest control professionals became integrated with the larger historical changes in agriculture and pest control technology. Because of the severity of problems surrounding insect control, we emphasize the profession of entomology in this discussion.

Four major considerations from the history of American entomology are most important to consider. First, by the early 20th century, entomologists had identified five major practical control strategies for insects: biological control by finding and releasing exotic natural enemies, sanitation of crop residues, cultural alterations of crop production protocols, quarantines to keep exotic insects out, and chemical poisons.

A good entomologist knew the strengths and weaknesses of each method, and undoubtedly some practitioners used all of them in different situations.

[33]Entomological Society of America, Code of Ethics for Board Certified Entomologists, personal communication, January, 1992.

Entomologists had some understanding that different suppression methods could either augment or diminish the effectiveness of other control techniques, but no overarching theory purported to guide the integration or combination of different control methods in a systematic way. For the most part, entomologists diagnosed a situation and then recommended one of the different classes of suppression techniques. In most cases, the recommendation was for the control practice that was cheapest and easiest for the client to perform.

The second consideration of importance stems from the controversy that developed around chemical poisons. Although all insect suppression methods at one time or another became embroiled in disputes, insecticide use generated most of the arguments about whether entomologists were acting in ethically proper ways.

Development of lead arsenate in 1892 and its eventual adoption in fruit and vegetable production was the key issue. Elements of the medical professions and the public came to believe that entomologists were cavalier and insensitive about the health effects of lead and arsenic residues.[34] Thus the door was opened to the concern that entomologists might be acting with suspect ethical motives. Were they too attentive to the financial interests of farmers and thus guilty of ignoring the safety of consumers?

Arguments about the safety of residues of insecticides on food expanded when a wave of new inventions, the synthetic organic insecticides, came after DDT's invention in 1939. In addition, concerns about human health grew to include concerns about the health of the environment in general. Wildlife and fish kills began to make it look, at least to some people, like entomologists were in a conspiracy with the chemical industry to poison the biosphere in the search for commercial profits. Rachel Carson's eloquent *Silent Spring*[35] in 1962 was the focus of this view, and her writings made most of the profession of entomology bitterly defensive.

Not only did chemicals raise questions in the mind of the general public about whether entomologists were ethical, technical problems with insecticides began to raise questions about the competency of the science in the minds of the direct clients, farmers. Resistance of target insects to the chemicals and destruction of natural enemies, with resultant flarebacks of targets or outbreaks of secondary insect pests, both suggested the profession did not have an adequate knowledge base to control pest problems. While the suggestion did not immediately imply unethical or immoral behavior, it raised the question of whether the profession knew how to protect

---

[34]James Whorton, *Before Silent Spring* (Seattle: University of Washington Press, 1974), 288 pp.

[35]Rachel Carson, *Silent Spring* (Boston: Houghton Mifflin, 1962).

clients' direct financial interests. Persistence in recommending chemicals that didn't work would surely raise questions about morals, however.

Entomology's relationship with the state is the third consideration stemming from the history of this science. In a very important way, entomology in the United States was a science and profession that owed its existence to the state. Jobs in entomology for many years were almost exclusively limited to the posts in USDA and the land-grant colleges. Training in entomology, especially applied entomology, was equally tied to land-grant colleges. Most entomological research was financed with public moneys in either the USDA or the agricultural experiment station laboratories. Advice to the "public" was the near monopoly of the cooperative extension services. Only the chemical industry supplied rival research or expert advice about insect control, and the industry's concerns were rather narrowly focused on insect toxicology. The fact that entomologists predominantly worked in the public, nonprofit sector was important in setting the stage on which judgments of moral behavior are made.

Perhaps equally important was the other state role in entomology, the regulation of insecticides. Early battles over regulatory power (1910 and 1947) were aimed at protecting the consumers of pesticides, i.e., mostly farmers, homeowners, and others. After 1950, however, the battle over pesticide regulation moved to focus as much or more on environmental and health protection (1954 and 1972).[36] Although the chemical industry, not entomology, was the focus of these regulatory struggles, entomologists frequently offered expert testimony to the Congress and thus placed themselves in the arena with those making moral judgment.

The fourth historical consideration of importance to ethics in entomology centers on the wave of innovative activity that began in earnest in the late 1940s and continued vigorously into the 1970s and beyond. Before the 1940s, as noted earlier, entomologists knew a number of different suppression techniques but had no strategy for integrating them. Primarily because of the technical problems associated with insecticides (resistance and the destruction of natural enemies), a few entomologists began concerted efforts to invent grand strategies for controlling insects by combining different suppression practices.[37]

Integrated pest management (IPM) was a product primarily of the land grant university system and focused at first on how to integrate the uses of biological control and chemicals so that the combined suppression effects

---

[36]Adelynne Hiller Whitaker, *A History of Pesticide Regulation in the United States to 1947*, Ph.D. thesis, Emory University, 1974, 464 pp.; and John E. Blodgett, Federal *ad hoc* committees on pesticides, 1955–1969, unpublished manuscript.

[37]John H. Perkins, *Insects, Experts, and the Insecticide Crisis* (New York: Plenum Publishing Co., 1982), pp. 29–126.

were synergistic. In this strategy, the goal was management of the insect population at levels below an economic threshold rather than eradication and prevention of all damage. For the most part IPM was envisioned to be a control strategy used by individual farmers on a field-by-field basis.

Total population management (TPM) was a product primarily of the USDA laboratories and focused at first on how to integrate the uses of sterile male releases with chemicals so that the combined suppression effects supplemented each other. In this strategy, the goal was management of the insect population over large areas of land, not just on individual fields. The strategy presumed a collaboration of the state with organized farmers. Because of the sterile-male technique's abilities to reduce pest populations to very low levels, TPM strategists were willing to entertain the notion of complete eradication of a pest rather than merely its suppression below an economic threshold.

IPM and TPM were quite different from one another, but they also shared some similarities. Most important was that the state had key roles ranging from research to actual conduct of the control operation. IPM and TPM were not the first pest control activities to involve government action, but, compared to the chemical control strategy which preceded them, they both involved substantially more public investment than was considered "normal" for agricultural pest control. Farmers were the most important direct clientele for both the strategies, but the general public had indirect interests in the uses of each of them.[38]

Ethical Issues in Entomology

We have devoted substantial space to presenting the historical origins and context of pest control, because we believe that it is senseless to probe the moral dimensions of pest control actions without an awareness of the larger picture. These ethical questions will invariably be informed by this background.

Three broad arenas of ethical issues will predominate: (1) What is the proper code of conduct between individual entomologists and other individuals who are not experts in the study of insects? (2) What is the proper code of conduct between individual entomologists and the state? (3) What is the proper code of conduct between individual entomologists and nature? In this section, we will examine the complexities of these questions. We will then suggest some ways to answer them.

---

[38]IPM and TPM are both still present as strategies for pest control. They are both also still subordinate to the chemical control strategy, which dominates American pest control practices.

Individual Entomologists and the Rights of Clients

Undoubtedly the most common ethical dilemmas faced by professionals stem from the work they do as experts with clients and with members of the public. In these interactions, the professional usually has little or no training about how to think about matters involving moral judgment. So long as no one questions an expert's advice, any moral doubts will probably remain silent and personal. If the expert's advice is challenged, however the ethical dilemmas may become public and possibly painful. Moreover, the challenges are likely to cause a great deal of confusion in the resolution of the problems.

Consider the example of the program run by the California Department of Food and Agriculture to eradicate the Mediterranean fruit fly in the area east of Los Angeles during 1989–1990.[39] The eradication protocol involved spraying urban areas with malathion bait, and considerable opposition to the program erupted, particularly from the City of Pasadena. Just what was an entomologist who worked on the operation supposed to do when told by members of the public that the control program was resented and that the people running the program were doing something wrong? Was our hapless entomologist supposed to remain silent but continue supporting the operation because he/she knew best? Was the expert supposed to launch a stiff defense against the ignorance of the city dweller who objected to being in the treatment zone? Should the entomologist have collapsed under pressure and ended the program?

Sorting through these options quickly uncovers the most important point: the audience interested in pest control expertise is heterogeneous and complex. First, the expert must distinguish between who is a client and who is not. Second, the entomologist must distinguish between his/her obligations to clients compared to others.

It is in unraveling who is or is not a client that pest control comes face to face with one of the peculiarities of its history. Most entomologists work in the public sector, i.e., for local, state, or federal government. So who are the clients? Are the people who pay the expert's salary the client? If so, then the agency director or more generally the taxpayer is the client. Or are the direct users or beneficiaries of the expert knowledge the clients? Farmers generally care intensely about the actions of pest control experts, because farm income may be critically dependent upon the nature of the advice. Does this dependency make farmers the client whose interests the scientist is bound to protect?

Differentiating between the interests and rights of the taxpayer compared to the direct beneficiary is often the core of ethical disputes in entomology,

---

[39]Marcia Bariniga, Entomologists in the Medfly malestrom, *Science* **247**, (9 March 1990): 1168–1169.

because so many entomologists work in the public sector. Rachel Carson was quite harsh in her judgment of entomologists of the land-grant universities and the USDA who, in her eye, sold their integrity to the chemical industry in return for research grants. Similarly, opposition of urban dwellers to Medfly operations is undoubtedly driven by a sense that taxpayers are paying for a program that dumps poison all over their houses without any commensurate return benefit. Entomologists who recommend such programs are likely to appear to make unethical decisions to such people.

Equally complex situations stem from the work of entomologists in the private sector, generally the insecticide industry or, more recently, the biotechnology firms. In these cases, a traditional expert/client relationship applies, as envisioned by Ravetz. Under the standard interpretations of professional obligations, the expert would be expected to uphold the client's interests, the industrial employer.

If the client's product left a residue on food or was responsible for decimating populations of wildlife, the entomologist's duty would be to try as best as he/she could to defend the product or make its use less damaging. What the client/employer would *not* consider the expert's duty to be in such a case, however, would be to find a way to avoid the use of the product altogether.

Unfortunately, it is precisely the absence of obligation to look for alternatives that is most likely to bring a charge of unethical behavior to the industrial entomologist. The public may be willing to admit that an employee has an obligation to an employer and that a professional has a duty to a client, but only provided that fulfillment of such obligations is done in a context of maximum avoidance of harm to the public. When the expert has the skill and training to look for alternatives and doesn't, the public is very likely to respond with accusations of immorality.

Perhaps the onus belongs on the employer, not the entomologist, but such distinctions are of little comfort to the attacked expert. Moreover, lifting obligations from the individual expert entirely is the sort of action that creates moral morons.

What we have then is a situation in which entomologists from both the public and private sectors are under pressure from many directions. Direct clients, in the form of farmers and industrial employers, may demand loyalty, but the indirect clients (taxpayers and general public) will argue just as strenuously for a different set of judgments about ethical behavior. Especially vexing are arguments focused on the use of insecticides and the residues left on food and in the environment.

Entomologists under siege about the use of insecticides have often taken shelter and comfort in economic defenses. Economists maintain that new agricultural technology, such as new insecticides, has invariably aided consumers, who benefit from cheaper food. This argument purports to change the general public or consumer into a client and then maintains that the works of pest control experts have indeed upheld the interests of this amorphous client by creating a farm production system that delivers cheaper food.

Rational though the argument may be, its weakness is that the purported "client" has not given informed consent to being a client. Certainly no direct consent has been given by individual members of the public. Only the imperfect and indirect consent given by Congress in the form of allocations for research and of regulations for pesticides justifies a defense of pesticides based on "the client public benefitted by getting cheaper food."

Entomologists and the State

Let us now move to ethical problems that develop around the relationships between experts as a group and the state. The argument here is that pest control expertise is inevitably politicized because it serves as part of the basis of power in human society. Consequently, the moral dimensions of an entomologist's work derive partly from the role played by his/her expertise in the workings of political culture.

Entomology grew and prospered as a discipline and profession because it offered solutions to problems that came from the transformation of agriculture into industrial capitalist forms of production. "Industrial capitalist" requires a brief definition. Industrial refers to a number of properties: a division of labor, the emergence of management planning, the development of new institutions, and the deliberate search for new technology to lower production costs.

Farms became industrialized in the sense that by the early 20th century they showed increasing divisions of labor through tendencies to specialize in one or only a few commodities. Farmers increasingly had to have skills in planning the uses of machinery, new technology, precise cost-accounting systems, and sophisticated trading patterns. Industrial firms and the USDA–land-grant college system appeared as a new institutions that provided the new machines and technology and trained experts and farmers in planning skills. New technology was the route to successful competition in a market that showed no mercy to producers who could not lower the unit costs of production.

Farms became capitalist in the sense that they were dependent upon a system of private ownership of land, machinery and other technology, and management. Labor tended to be provided by the family of the owner-operator, but wage-laborers participated as another "input." Most American farms had "manager-capitalist" and "wage-laborer" combined in the same person, but even family farms frequently relied upon the "hired hand." Hired help sold their labor power to the manager-capitalist for a wage that was generally only somewhat over subsistence levels. For his/her part, the manager-capitalist hoped his/her skills would lead to a technologically progressive farm firm with a higher capitalized value and greater wealth for him/her and his/her family.

Industrial-capitalist forms of farm production replaced earlier subsistence and relatively noncommercial forms that operated with far less skill in terms of management, technology, and cost accounting. Older forms of farming were less efficient, a trait that made them virtually immoral in the eyes of a capitalist farmer. This transformation to industrial capitalism in agriculture was not morally neutral, because it meant the passing of one way of life in favor of another. Some farmers gained because they were skillful enough to play the new managerial game. Others lost because they couldn't.[40] Making ethical and moral judgments about what happened is complex, and pest control was right in the middle of the problem.

Entomology was critical to the transformation. Industrial-capitalist farming seeks to achieve higher production, so a farmer borrows money to invest in new production practices. The farmer not only wants higher yields but is actually dependent upon achieving them. Only enhanced efficiency will return the funds needed to retire the loan. If insects are bad, the farmer must be able to achieve control efficiently or ability to retire debt may be lost. The industrial-capitalist farmer can't just hope for better luck next year. He/she may be out of business if insects are too destructive.[41]

Entomology came of age at precisely the time the transformation of agriculture to industrial-capitalist modes was underway in the late 19th century. Land-grant colleges hired professors, students obtained training, jobs opened up in the college–experiment station–USDA complex, and practitioners organized themselves into professional associations (for example, the American Association of Economic Entomologists in 1889) and started learned journals (for example, the *Journal of Economic Entomology* in 1908). Congress and state legislatures appropriated the funds needed to make the profession viable because farmers had political clout and because pest damage was intolerable under the new modes of agricultural production. In its very grounding and being, therefore, entomological expertise was a component of the industrial capitalist revolution in America.

Where does this complex of political-economic considerations leave the entomologist who is seeking a code of ethical and moral behavior? Capitalism as a system of production, property ownership, social prestige, po-

---

[40]See Willard W. Cochrane, *The Development of American Agriculture. A Historical Analysis* (Minneapolis: University of Minnesota Press, 1979), pp. 378–395.

[41]John H. Perkins, The quest for innovation in agricultural entomology, 1945–1978, in David Pimentel and John H. Perkins, eds., *Pest Control: Cultural and Environmental Aspects* (Boulder: Westview Press, 1980), pp. 23–80.

litical power, and moral values is deeply embedded in American culture and the psyche of American citizens. Capitalism is such a complicated maze of social relationships that simple defenses or condemnations of it are unpersuasive. Nevertheless, an entomologist seeking a moral code is sure to be misled if she/he is devoid of any understanding of the political power relationships created in a capitalist economy.

Entomologists who wish to know the way to ethical behavior must be able to see what sorts of power relationships follow from their work. As a first question, every entomologist must know for whose benefit the work is done. In what ways will a new technology for controlling an insect advantage some people and harm others? Will those harmed be permanently put at a disadvantage compared to those helped? Will the new practices be a part of consolidating one group of client's wealth and power compared to another group's? Does any group have unequal access to expertise? Is that inequality unethical or does it have a moral defense? Have all parties that will be affected by the outcome of expert decisions given a direct, informed consent to being affected? If not, do mitigating circumstances avoid a necessity for obtaining their informed consent?

This list of questions is clearly preliminary, but it should give a sense of what is involved. An entomologist is not just an individual who serves other individuals or clients. Entomology, like all knowledge and technology, is a political-economic artifact and is part of a history in which some people benefitted but others did not. Claims to moral neutrality may be comforting, but a broader vision opens up new and complex vistas within which ethical behavior must be sought.

Entomologists and Nature

Not only is entomology not morally neutral in a *political economic* sense, it is also not neutral in a *political ecological* context. This aspect of entomology is the third problem confronting an ethical code for pesticide use and pest control.

At issue in political ecology is the human representation of nature and the scheme with which people govern the biosphere. Carolyn Merchant has offered an insightful analysis that identifies three approaches to environmental ethics: egocentric, homocentric, and ecocentric.[42] Her typology of approaches to the environment is useful in understanding how entomologists have treated nature. A moral code of conduct for pest control must incorporate a consideration of the person/nature relationship as well as of relationships between individuals and between an individual and the state.

---

[42]Carolyn Merchant, Environmental ethics and political conflict: a view from California, *Environmental Ethics* **12** (1990):45–68.

Egocentric ethics are grounded in individual rights and a sense that society benefits when the individual benefits. Merchant argues that egocentric ethics are highly compatible with laissez-faire capitalism in which individuals compete with each other, each seeking strictly to benefit himself or herself, and thus create the happiest conditions for all society. Government intervenes only to prevent the competition from turning to mayhem, murder, or the destruction of the Earth. Egocentric ethics are not selfish, because they envision a scheme in which pursuit of individual good creates, via the invisible hand of the market, a virtuous society for everyone with only a light touch of government to keep things orderly.

Homocentric ethics derive from a sense that a happy society is one that has the greatest good for the greatest number of people. The best route to such a utilitarian society is through social action by the state. Conservationism as developed by Gifford Pinchot was, in Merchant's analysis, an extension of utilitarianism to the natural environment through the wise use of resources. Well-trained experts in the employ of the state were best suited to devise suitable resource development schemes.

Ecocentric ethics differ markedly from both egocentric and homocentric ethics, both of which are centered on the needs and rights of people. Ecocentric ethics, in contrast, accord intrinsic dignity and rights to all living creatures and even to nonliving entities, on Earth and elsewhere in the universe. Merchant argues that the biological control of insects is one area in which ecocentric ethics predominated.

Merchant's scheme of environmental ethics can be used to analyze the arguments around pest control and pesticides. For example, the chemical industry since the early 20th century has used some government regulations in the sale of pesticides, but they have consistently resisted more labeling laws. Such laws require products sold as pesticides to bear true labels disclosing certain information. The industry has consistently argued, however, that the labels should be brief and that the law put as few restrictions on the use of the product as possible.

The pesticide industry operates consistently with an egocentric environmental ethics. It asserts that the pursuit of self-interest in the marketplace will promote not only the happiness of the individual but also the betterment of society as a whole. Government activity is acceptable so long as it is at a minimum consistent with human safety. Little or no need exists for government experts to provide extensive guidance, either to the industry or to farmers and other users of pesticides. In the industry's view, people have paramount rights, and pests and other creatures have value only in terms of their utility to humans.

As an example of homocentric ethics in pest control, consider the various schemes that have been launched for area-wide control of certain insects, such as boll weevil, screwworm fly, and Medfly. These programs were constructed by experts who worked for government, and they were conceived as a social action to provide the greatest good to the greatest number of people. Some of them, especially Medfly operations, may have caused inconvenience (but not harm in the eyes of the proponents) to some, but in the long run the whole society was better served. The programs were highly utilitarian, and no intrinsic rights were accorded any species other than humans.

Finally, consider the role of ecocentric ethics in pest control. Merchant, as noted above, cited biological control of insects as being influenced by ecocentric ethics, a judgment with which we partially concur. We would cite a different example, however, to demonstrate ecocentric ethics. During debates over proposals to launch eradication schemes of the boll weevil in the Southeast, certain entomologists raised arguments of two sorts: (1) success against the boll weevil would encourage human arrogance against other pest species and thus eradication was not desirable, and (2) people had no right to destroy a species because they could not create one.[43] Both of these arguments were based on a sense that other species have intrinsic rights.

A Case Study in Cotton Pest Management

Debates about the ethics of pesticide use frequently are disconnected from an agricultural or other specific pest control situation. Thus it is often difficult to understand the issues in terms that are meaningful to the protagonists. A closer look at U.S. commercial cotton growers' pest control activities should further clarify Merchant's analysis of environmental ethics.

In common with other field crops, the acceptance of DDT and other synthetic organic insecticides after 1945 revolutionized cotton farming.[44] Pest management practices varied throughout the Cotton Belt according to the population densities of insect pests, crop planting and harvesting dates, climate, and other factors. The chosen methods of pest control reflected the different circumstances American farmers faced, but chemical control predominated in most of the cotton-growing areas. Widespread uses of insecticides eventually resulted in new problems. By 1960, insects resistant to them and outbreaks of secondary pests threatened U.S. cotton production and prompted close examination of chemicals once considered fail-safe. To cope with insecticide crises, cotton farmers implemented both IPM and TPM crop protection strategies in their fields. They also continued

---

[43]John H. Perkins, Bool weevil eradication, *Science* 207 (7 March 1980): 1044–1050; John H. Perkins, The boll weevil in North America: scientific conflicts over management of environmental resources, *Agriculture, Ecosystems and Environment* **10** (1983):217–245.

[44]R.L. Ridgway, et al., *Cotton Insect Management With Special Reference To The Boll Weevil*, Washington: USDA, 1983, p. 5.

to rely on improved chemical controls. These innovations in cotton pest management were born out of combined scientific and political-economic interests. The intense search for effective cotton insect pest controls began during the 1950s, when for the first time, U.S. cotton growers faced not only insects but also increased competition from foreign cotton markets and the newly established synthetic fiber industry.[45]

Nowhere was the counterproductive use of pesticides more dramatically expressed than in the Lower Rio Grande Valley of Texas during the late 1960s.[46] Growers' desperation was evident: records indicate that the entire cotton crop of that geographical area, spanning approximately 700,000 acres in Texas and Mexico combined, was severely threatened.[47] The cotton farmers' only hope for economic survival was to withdraw from continued reliance on heavy pesticide use. Under the direction of Texas A&M University entomologists, farmers adopted a new pest control strategy based on the principles of integrated pest management (IPM).

In contrast with conventional chemical pest control, IPM philosophy is based on utilitarian, homocentric ethics. In addition, IPM has strong elements of ecocentric ethics. Rather than demanding total control over all living species in an agricultural ecosystem, IPM acknowledges that human technology has frailties, which mandate that people coexist with other species. In addition, to the extent that IPM schemes rely on biological control, the natural enemies become active partners in a human enterprise. Some IPM researchers have also accorded pest insects an intrinsic right to exist.

IPM programs in many cotton-producing areas have relieved the environmental stresses caused by the exclusive reliance on synthetic pesticides. Nonchemical means of pest control reported by cotton growers in 1989 included cultivation, stalk destruction, planting pest-resistant varieties, and using pheromone traps. Results of a recent USDA survey reports that cotton acreage beltwide examined by professional scouts increased by 37% in the last 7 years.[48]

---

[45]John H. Perkins, The boll weevil in North America: scientific conflicts over management of environmental resources, in *Agriculture, Ecosystems and the Environment*, 10 (1983): 217–245.

[46]John H. Perkins, *Insects, Experts and the Insecticide Crisis* (New York: Plenum, 1982), p. 42. See also Pest management in cotton, D.G. Bottrell and P.L. Adkisson, *Annual Review of Entomology* 1977, **22**:451–481.

[47]James R. Cate, Cotton: status and current limitations on biological control in Texas, in *Biological Control in Agricultural Systems*, ed. by M. Herzog and D. Hoy (Orlando: Academic Press, 1985), p. 537.

[48]Walter L. Ferguson, Cotton farmers control chemical use, in *USDA Agricultural Outlook*, AO-169, November, 1990, p. 21.

Total population management (TPM) has also served as a strategy for cotton insect control that is less dependent upon pesticides. TPM is based on the idea that the best strategy for controlling crop pests is to attack an entire insect population. Enthusiasm for the development of TPM in cotton pest control was bolstered by the success of USDA's 1954 screwworm eradication experiment, which marked the first major victory for TPM proponents.[49]

In 1958, due to the efforts of the National Cotton Council, Congress appropriated $1.1 million to establish the Boll Weevil Research Laboratory (BWRL) in Mississippi. Creation of the BWRL marked the road to maturation for TPM. E.F. Knipling, then director of entomology at USDA, was heavily influenced by the success of the screwworm eradication experiment and emphasized that the research directive at the laboratory should be geared toward the eradication of boll weevils. Less than 10 years later, the BWRL began the pilot boll weevil eradication program in the upland cotton areas of Mississippi. No longer would field-by-field pest management suffice; control efforts must be expanded to hundreds, perhaps thousands of square kilometers. Innovations based on this approach to pest management, proponents believed, would indeed eliminate most negative impacts of agricultural chemical use. The driving force behind TPM was a faith that scientific innovation can and will triumph over all agricultural pest problems.[50]

Research in TPM has resulted in significant adoption by southeastern cotton growers of pest eradication strategies. The USDA estimates that over 200,000 (total) acres of cotton in the Carolinas, Florida, Georgia, and Alabama are currently managed under the auspices of the Southeastern Boll Weevil Eradication Program, a cooperative effort between growers, the USDA, and state and local agriculture offices.[51] Whether these programs will actually reduce boll weevil populations to zero may be debated, but these TPM schemes undoubtedly led to a vast reduction in boll weevils over very large areas.

TPM in cotton is highly utilitarian. Proponents believe that the implementation of the philosophy will result in the greatest good for the greatest number of people. It is an example of homocentric ethics because the eradication schemes rely on substantial inputs from the state. Laws regulate the growing of cotton, and research and management for the efforts require

---

[49]John H. Perkins, The boll weevil in north america: scientific conflicts over management of environmental resources, in *Agriculture, Ecosystems and the Environment*, **10** (1983):217–245. See also Perkins's *Insects, Experts, and the Insecticide Crisis* (New York: Plenum, 1982) p. 103, for elaboration on USDA's screwworm eradication experiment.

[50]John H. Perkins, Boll weevil in North America . . . ," p. 225.

[51]Willard A. Dickerson, Goodbye boll weevil . . . hello again cotton, in *Agricultural Research* (Beltsville: USDA-ARS, 1987), October, pp. 9–10.

expertise based in USDA and land-grant universities. TPM accords no intrinsic rights to the target species and envisions no active partnership with natural enemies. In TPM schemes, human technology is not seen as frail. TPM is not ecocentric.

Despite the changes in cotton insect management due to IPM and TPM, chemicals remain highly important for cotton producers. Even in chemical control, however, marked changes have occurred within the past 20 years. Insecticide use on cotton in America fell from an estimated 73 million pounds of active ingredient in 1971 to approximately 17 million pounds 1984.[52] Cotton is the only major U.S. agricultural commodity to see such substantial reductions in pesticide use. Many arguments have been made to explain this dramatic reduction. The USDA reports that the reduction is due to a change in the mix of chemicals applied to crops. In 1977, largely in response to the ban of DDT, and pest resistance problems, 11 states petitioned the EPA for the emergency registration of the synthetic pyrethroids.[53] This new class of chemicals proved to be highly biodegradable, exhibited low levels of mammalian toxicity, and acted effectively when used to control a wide range of insect pests. Because pyrethroids were effective at much lower rates than the previously favored organophosphate and organochlorine pesticides, it follows that they were applied in smaller amounts. In the United States pyrethroids are currently used throughout the cotton belt mainly to control the *Heliothis* species. Some argue that the development and adoption of pyrethroids on cotton has thwarted efforts to implement IPM programs.[54]

Continued reliance on synthetic chemical insecticides as the primary means of insect control demonstrates an egocentric approach to environmental ethics. The use of chemical controls without regard for the impacts that they may have on agroecosystem elements other than the crop plant demonstrates the emphasis that market farming places on the success of individual farmers. Competition between farmers may increase with the variety and number of chemical controls available.

Future innovations in cotton insect pest management will likely correspond to some point in Merchant's scheme of environmental ethics. Further development in IPM, TPM, and chemical controls may all prove useful to cotton farmers. The farmers' choices of pest control strategies will likely continue to be guided by what they can afford and by what proves most profitable.

---

[52]Osteen and Smzedra, *Agricultural Pesticide use Trends and Policy Issues*, p. 14.

[53]Ridgway, et al., *Cotton Insect Management With Special Reference To The Boll Weevil*, p. 12.

[54]N. Morton and M. Collins, Managing the pyrethroid revolution, in *Pest Management In Cotton*, ed., Greene, p. 159.

Processes Leading to Resolution

Pest control experts seeking a moral code must grapple with the different ethical systems that can be the foundation for action. Before 1962 and Rachel Carson's *Silent Spring*, it was hard to see the relevance of environmental ethics for pest control, but she provided a philosophy of nature that made it impossible for pest control experts to work without considering their own personal stances. The public increasingly supports an ecocentric ethics, which means that pest control experts must come to grips with it, if only to be politically credible. For reasons that are more complex than can be outlined here, we also think the most exciting and productive research in pest control is likely to come from an ecocentric view of nature.

We have argued that understanding ethics in pest control requires an understanding of the characteristics of professionalism in science and of the historical development of the field. We have also argued that the framework within which specific ethical issues arise consists of professional/client relationships among individuals, the relationship of the experts to the state, and the moral content of the stance toward nature.

How can a pest control expert and the professional organizations of these experts proceed to establish codes of ethical behavior? In posing this question, we are suggesting that making ethical choices is both an individual exercise and a group process. Absolute unanimity on the nature of ethical behavior is unlikely to emerge, but the professions should seek to establish certain minimums that are agreeable to members and can win support from the general public and clients. Beyond the minimums, experts will vary in what they consider to be moral action. Nevertheless, it is incumbent upon individuals to articulate what they believe and why.

It is beyond our ability to provide a detailed scheme for establishing codes of conduct. However, we believe that ethical behavior is an issue in research, teaching, advising, consulting, and extension work, for both public-sector and private-sector professionals. Professionals have little trouble seeing the relevance of ethical codes for these activities, except in research.

Ethical questions, however, begin with research design. Although experts may at the time be face to face with a pest, they still have a relationship with an industrial-capitalist political economy, and they still have a stance toward nature and other species. What happens in research matters. It can ultimately affect other people and other species, either for better or for worse.

Experts need to reflect on the following:

- Client clarification. For whom do experts work? How can they tell? What difference does it make? How do they recognize the "public good?" What does it mean to think about "informed consent" when they are developing or proposing an activity? What do they do when it becomes evident that "informed consent" cannot be obtained from everyone, such as future generations?

- Clarification of political economic values. Do experts have assumptions about the meaning of "human nature" and how it relates to capitalist systems? Are people fundamentally competitive or are they better described as cooperative? Do research programs and advice reflect anything about assumptions of human nature? Does industrial-capitalism have any attributes that are upsetting? What are the strengths of industrial-capitalism? In what ways does work as an expert play a part in the functioning of industrial-capitalism?
- Clarification of nature values. Do pest control professionals accord other species any rights or integrity outside of their value or relationship to people? What would happen to research programs, teaching, and advising if other species, even pests, had intrinsic rights? Is the world a machine made up of parts, with all dignity residing in the human mind? What would happen if experts saw nature as having dignity because it had diversity? What would it take to see the world through a different set of assumptions than the ones currently held?

These questions are not easy to answer. The answers will always be arguable. We submit the questions in the hope that they will help guide us to a more profitable way to think about ethics and pesticides.

Acknowledgments

We thank David Pimentel, Hugh Lehman, Mark Levensky and anonymous reviewers for their helpful comments, support, and encouragement. We also thank Andrea Winship and other members of the library staff at the Evergreen State College for their help. Environmental Studies program staff Jane Lorenzo, Bonita Evans, Peggy Davenport, and Kiki Walker provided much assistance throughout this work.

Part V

The Benefits and Risks of Pesticides: Two Views

What are the benefits of pesticides compared with their risks? How can good results in pest control be insured while the public and the environment are protected against any dangers? EPA Journal asked two prominent spokesmen in areas associated with pesticides to express their views on these concerns. Their articles follow.

The first piece is by Nicholas L. Reding, executive vice president of Monsanto, a chemical manufacturing company, and a former chairman of the National Agricultural Chemicals Association. The second piece is by Dr. Robert L. Metcalf, Professor of Entomology, Biology, and Environmental Studies and Professor at the Center for Advanced Study at the University of Illinois at Urbana-Champaign.

17

Seeking a Balanced Perspective

Nicholas L. Reding

Recently, I received a letter from a high school student who lives in Pennsylvania. She was writing a research paper and wanted our help. The title was, "Pesticide Abuse and Pesticide Danger."

The letter bothered me deeply. First because the title summarized everything she knew about modern pesticide technology. Second, because she isn't alone in her views. For many people, pesticides mean either abuse or danger.

I don't agree with that view, of course. I see the commitment the industry has to testing its products, the emphasis on minimizing risks, the efforts to train pesticide applicators around the world, and the constant reappraisal of the industry's methods to keep improving. As my industry colleague, Dale Wolf of DuPont, said at last year's annual meeting of the National Agricultural Chemicals Association (NACA): "The highest priority of your companies and mine was, is and will continue to be the safe manufacture, transportation, use and disposal of agrichemicals." Those aren't hollow words.

In fact, I see a responsible industry that makes products that provide great benefits by controlling pests that attack crops, homes and health. And I see a scientific community that is beginning to put the possible risks of pesticides in a clear, less frightening perspective.

But I also try to understand why many people are concerned about our products. To a great degree, it's because of the success of the environmental movement in changing the way everyone from activists to industrialists views the world around them. We're more aware, more sensitive and more responsive. It's a positive change.

It's also the result of technological change. We can now detect materials in the environment that we never knew existed there before. Parts per billion, trillion and quadrillion are extremely minute traces of any material, but these words are the language of modern contamination. Our ability to understand what those traces mean isn't always so advanced.

Concern is also the result of extremely effective actions by activist groups. From Earth Day on, the mistakes, misjudgments and stumblings of all industries have been chronicled, spotlighted and rehashed at every opportunity—often, long after the effective changes have been made. It's all made to order for a news media which delights in high drama and controversy.

And the industry has brought some of the concerns upon itself. As criticism mounted, we often became reactive and combative. Or worse, we ignored legitimate concerns, even when we had the answers. We should have heeded Winston Churchill when he said, "I do not resent criticism, even when, for the sake of emphasis, it parts for a time with reality."

I'm not always so generous. I do believe that, at times, environmental crises over pesticides are manufactured for maximum effect. Moreover, some critics relish the fight more than the solution. But the vast majority of concerned people are sincere and deserve a response based on facts, not on hurt feelings.

The facts do support pesticides. This is not to argue that they are always safe, everywhere. Pesticides are chemicals designed to control insects, weeds, fungi, nematodes and other pests. They are biologically active and, to a greater or lesser degree, toxic. They must be used carefully and according to label instructions. But they can be and are used safely and produce benefits for millions of people.

My industry accepts its responsibilities in the area of product safety. Pesticides undergo incredible testing—often more than 100 different kinds of health and environmental studies which require thousands of individual analyses. These products must be effective while not posing unacceptable risks to humans, livestock, the environment or food. To establish that, we do tests on efficacy, crop safety, short- and long-term toxicology, metabolism in crops and animals, residue and environmental fate.

The Industrial Bio-Test (IBT) Laboratory scandal in the mid-1970s tarnished the reputation of pesticide testing, and IBT has become the rallying cry for other irresponsible charges against the industry. But the legacy of IBT is becoming history as new tests are completed. At Monsanto, we've strengthened our supervision of outside laboratories and moved a sizable percentage of testing to our own facility. We're proud of our tough standards for testing and the quality of our science.

Our industry also backs a strong, well-funded EPA. It's in the best interest of the manufacturer, the customer and the public that the EPA have the resources it needs to do an intense and thorough evaluation of all pesticide applications for registration. And the agency does a good job under tough conditions. It is expected to provide scientific standards and methods to what are often emotional or political questions. We don't always agree with EPA, and we'll defend our point of view vigorously when scientific questions are debated. But we respect and support its purpose.

Industry's responsibilities don't end after registration or at the point of sale. For example, the National Agricultural Chemicals Association (NACA) is sponsoring an education program for migrant workers who handle pesticides. Spanish-language brochures and radio and television spots give reminders on proper handling and hygiene. Some 100,000 brochures have been distributed so far, and dozens of radio and TV stations carry the public-service announcements.

Monsanto, like many other companies, is involved in training programs on proper use of chemicals elsewhere in this country and around the world. And NACA and individual companies support the National Agricultural Aviation Association in providing programs for training its members in the most modern, effective methods of applying pesticides. The program, called Operation SAFE, has been successful coast-to-coast.

The facts also support the benefits of using pesticides. Some 2,000 species of weeds, 1,000 species of nematodes and 10,000 species of insects compete with humans for food and fiber. While estimates vary, most experts say that without the use of pesticides, food supplies would decrease by 30 percent or more. Romantic notions to the contrary, we cannot return to the pesticide-free days of yesteryear and still provide food at low cost to millions, even billions, of people. As the world population continues to grow, the need to use modern agricultural techniques will increase, not decrease. The United States can produce a good part of that huge requirement, and modern technology can help other nations produce more.

Outside of agriculture, pesticides also provide benefits by protecting our homes from termites and other destructive pests that do billions of dollars of damage yearly. Pesticides also are necessary to provide protection from disease-bearing insects and contaminated water. These products are essential tools of modern life. Like all tools, they must be used correctly and with care, but they provide benefits that raise the quality of living for a growing number of people, worldwide.

The facts also support the view that these benefits are not gained only at the cost of assuming immense risks. Scientists are beginning to reassess the risks of pesticides and other chemicals. That's particularly true in the intensely emotional area of carcinogenicity.

Sir Richard Doll and Richard Peto of Oxford University, who analyzed cancer mortality rates for the Congressional Office of Technology Assessment, reported that the major causes of cancer were tobacco and diet. And by diet they did not mean chemical contaminants, if any, in food. Exposure to materials in the workplace, environment, food additives and industrial products, combined, totaled 8 percent. Constant effort is needed to reduce that percentage, but the facts do temper the myth that we live in a sea of manmade poisons.

Lewis Thomas, M.D., Chancellor of Memorial Sloan-Kettering Cancer Center, says flatly that there is no cancer epidemic. He fears that Americans are becoming a "nation of healthy hypochondriacs, living gingerly and worrying ourselves half to death."

Dr. Bruce Ames of the University of California at Berkeley says that Americans consume 10,000 times more cancer-causing chemicals in their daily diet from natural products than from manmade pesticides. He said, "I think we got off on the wrong track. We're concentrating almost exclusively on little bits of pollution and manmade things and completely ignoring enormous amounts of natural mutagens and carcinogens. I'm starting to question our whole way of thinking."

Dr. Ames is one of the few scientists to take on these issues head-on. He points out that aflatoxin found naturally on peanuts is a far more potent carcinogen in rats compared with EDB, a pesticide sometimes found in trace amounts in grain or flour. Aflatoxin is allowed in peanut butter at 15 parts per billion. Dr. Ames said that the risks "from eating the average peanut butter sandwich come out as more than eating the rare, highly contaminated muffin." And yet all of us should continue to enjoy peanut butter.

Perhaps the most startling and controversial view of cancer is provided by Edith Efron in her new book, *The Apocalyptics: Politics, Science and the Big Cancer Lie*. She challenges the methods used to "protect" Americans from cancer-causing substances. She says that the nation has used a hypocritical double standard in assessing risk. The book is thoughtful and thought-provoking. It raises a number of issues that need to be confronted by scientists and lay people alike.

There is a way to go, however. The publication of *The Apocalyptics* itself provides a commentary on perceived risks from chemicals, the risk not to health, but to reputation by challenging established views. The publisher sent copies for review to 16 distinguished scientists. All thought highly of the book; all refused to allow the use of their names.

The risks from pesticides need to be studied and re-evaluated constantly. But voices like those of Dr. Ames and Edith Efron also need to be heard if we are to put those risks into perspective. Otherwise, we may lose the very real benefits from pesticides while addressing not-so-real risks.

While that is under way, the public will continue to be concerned. Too many charges and too many deadlines have ingrained the fear of pesticides into the public's perception. But the time is right to work to reduce those fears with facts. All of us—government, industry and environmental groups—have a responsibility to fulfill, one that can best be undertaken in a spirit of cooperation and mutual respect. It's time to stop shouting at each other and begin to listen—hard. We're ready at Monsanto. Other companies will join. We would welcome the opportunity.

The Pennsylvania school girl who wrote her research paper on "Pesticide Abuse and Pesticide Danger" reflected some of today's thinking. Perhaps for her college thesis she'll write another paper on "The Benefits of Pesticides: A Balanced Perspective."

18

An Increasing Public Concern[1]

Robert L. Metcalf

The judicious use of modern pesticides is an important adjunct to modern agriculture and public health. None of us is eager to return to the standards of the Middle Ages when life had its full share of wormy apples and weevily biscuits, virtually everyone was lousy, and fleas and bedbugs were constant bedtime companions. The discovery of DDT, BHC and 2,4-D during the Second World War gave promise for greatly enhanced agricultural productivity, of banishing such villains as the house fly, the cockroach, the bedbug, and the louse, and of eradicating the scourges of malaria, typhus, and yellow fever.

Yet somehow much of it seems to have gone awry and we are still waiting for the EPA to put it right. As we approach the fiftieth anniversary of the discovery of these miraculous pesticides, there is steadily increasing public concern and mistrust about the hazards. The Council for Environmental Quality in a public survey in 1980 found that the level of public concern about toxic chemical wastes surpasses that for any other environmental problem and that more than 80% of those responding believed the government should screen chemicals for safety before they were marketed and that chemicals known to cause cancer should be controlled.

There is no such thing as an indispensable pesticide. The claims for DDT probably came as close as any; it was registered for use on some 334 crops and agricultural commodities in 1961, yet it was banned by EPA through an administrative order in 1972. Since that time we continue to hear that we can't grow corn without aldrin and heptachlor, we can't grow peaches without DBCP, we can't ranch in the southwest without 2,4,5-T, we can't produce sheep without 1080 predator poison, and we can't grow citrus and papayas without EDB.

---

[1]Reprinted from EPA Journal 10(5):30–31, 1984.

These pesticides have all had severe federal regulation and restriction, yet agriculture continues to produce vast surpluses, land is held out of cultivation, and most of us are better fed than ever before. The following examples demonstrate the growing need for careful benefit/risk evaluation and for prompt and decisive regulatory action. They are chosen from the many cases that required action by EPA scientists and administrators and by the Pesticide Science Advisory Panel over the past seven years.

In 1969, the Secretary of Health, Education and Welfare's "Commission on Pesticides and their Relationship to Environmental Health" emphasized the problems of widespread contamination by the persistent organochlorine insecticides. Toxaphene was suggested as requiring close surveillance. With restrictions of the other organochlorines, toxaphene became the most heavily used insecticide in the U.S.

Toxaphene was shown to be a carcinogen in laboratory animals by the National Cancer Institute in 1979, and residues were found to cause crippling bone deformities in fish at part-per-billion levels in water. After toxaphene residues were found to be accumulating in fish of the Great Lakes, there was pressure for its restriction but EPA did not ban the general uses of toxaphene until 1982 and then only after a U.S. Congressman added a cancellation order to a House appropriations bill.

Endrin is another of the "uncontrollable organochlorines" singled out by HEW for regulation in 1969. It is the most toxic of the group, so much so that it was registered as a rodenticide to kill field mice in orchards. Its use as a cotton insecticide caused so many damaging fish kills that its use east of the Mississippi River was finally restricted by EPA in 1981. Intensive agricultural lobbying preserved its registrations to control grasshoppers and cutworms attacking wheat in the Great Basin.

About 260,000 acres of wheat were sprayed with endrin by air in 1981 and partridge, grouse, ducks, and geese became contaminated with endrin residues well above the "safe level" and endangered species such as the bald eagle, peregrine falcon, and whooping crane were threatened. The 20 million migratory waterfowl passing through this area annually have extended endrin contamination to the 17 states of the Western flyway. At present endrin residues are widely distributed in the wildlife of the entire Great Basin ecosystem.

Heptachlor is another insecticide most of whose uses were cancelled in 1978. Curiously, one registration not cancelled was its use on pineapples in the Hawaiian Islands to control ants that upset the biological control of pineapple mealybugs. The results of this regulatory omission were spectacular. Pineapple tops were fed to dairy cattle as "green chop" and their heptachlor residues were concentrated in milk as a more toxic and more persistent chemical, heptachlor epoxide, that is a carcinogen in laboratory animals. Thus heptachlor epoxide residues were transferred to virtually all the inhabitants of the Islands. Mother's milk was found to be contaminated with residues of heptachlor epoxide and infants were ingesting several times the "acceptable daily intake" as determined by the Food and Agriculture Organization and World Health Organization, agencies of the United Nations.

The resulting brouhaha began with finger pointing and accusations by concerned citizens, the milk and pineapple industries, the State Department of Public Health, and the University of Hawaii. The issue is now in the courts.

Mirex, another persistent organochlorine, destroyed colonies of the imported fire ant when applied as a bait at minuscule doses. The Secretary of Agriculture in 1971 hailed mirex as the perfect pesticide: "It has no harmful effect on people, domestic animals, fish, wildlife or even bees, and it leaves no residue in milk, meat or crops." Armed with mirex the U.S. Department of Agriculture planned a massive eradication campaign against the fire ant to cover more than 100 million acres.

Mirex, as predictable from its chemical structure, is very persistent and biomagnified through food chains. Despite the low dosage applied, residues in the parts per million range were found in birds, fish, shrimp, and crab and in the fat of humans throughout treated areas. Mirex was determined to be a carcinogen by the National Cancer Institute in 1976, and after numerous skirmishes in the courts, EPA terminated the production and application of mirex in 1978.

In 1976 a new rodenticide, pyriminyl was widely marketed in the U.S. for the household control of rats and mice. It was advertised as almost a specific killer for rodents with very low hazard to man and higher animals. However, the rodenticide was marketed as a 0.5% active ingredient in 15 gram packets of peanut-flavored confection. Predictably, some of these were eaten and at least 30 persons, many of them children, were afflicted with severe and irreversible diabetes and damage to their nervous systems.

Belatedly, EPA scientists learned that pyriminyl had been test-marketed in South Korea as a rodenticide in 1975 and 251 cases of human poisoning with some fatalities were reported. With this evidence EPA was able to persuade the manufacturer to withdraw pyriminyl from the market in 1980.

Dibromochloropropane or DBCP was introduced about 1955 to control the soil-inhabiting nematodes that attack the roots of citrus, peach, grape, pineapple and annual root crops. It was particularly effective because it was not unduly hazardous to growing crops and it was thought to decompose in edible produce to harmless inorganic bromide.

Toxicological studies published in 1961 showed conclusively that exposure to DBCP caused severe atrophy and degeneration of the testes of mice, rats, and rabbits. These results were not communicated to factory workers until a group of them became concerned about their inability to father children. A private consultant hired by their union established that their infertility was due to exposure to DBCP in the workplace. A study by the National Cancer Institute in 1973 showed that both DBCP and the related nematocide EDB were active carcinogens producing stomach cancers in rats and mice, and warned of possible health hazards to humans.

As a result DBCP was targeted in 1976 as a candidate for re-evaluation and regulation. EPA demonstrated in a massive study of factory and farm workers that DBCP exposure was quantitatively related to decreased sperm counts. After exhaustive studies of benefit/risk and four public hearings, EPA finally suspended all uses of DBCP in 1981.

The preceding examples characterize pesticides whose benefits cannot match the risks they pose to human health and to the quality of the environment. Their demise was predictable. The entire philosophy of how we use pesticides in modern agricultural production is open to serious question.

As long ago as 1969, the Secretary of Health, Education and Welfare's Commission on Pesticides and Their Relationship to Environmental Health emphasized the problems of the widespread contamination of air, water, soil, food, and human bodies by persistent insecticides and pointed out "the absurdity of a situation in which 200 million Americans are undergoing lifelong exposure, yet our knowledge of what is happening is at best fragmentary." This absurdity is compounded many times today as the U.S. applies about 45 percent of all pesticide production to only 7 percent of the world's cultivated land.

The major difficulty with pesticides is that they are nearly all highly reactive chemicals that kill living organisms by reacting with some vital component of living tissue. Almost by definition they lack selectivity and their impact upon nontarget organisms such as fish, birds, bees, beneficial parasites, endangered species and even man can be devastating.

Consider the organophosphate parathions introduced as insecticides in 1946. Parathion poisoning is the major cause of the estimated 500,000 human illnesses and 20,000 deaths that occur annually from the use of pesticides, according to estimates of the World Health Organization. Yet the parathions are still produced and used worldwide at the rate of several hundred million pounds per year in appalling disregard for human welfare. There are dozens of effective and much safer substitutes.

The lack of selectivity of pesticides and their widespread overuse are causing immense problems to agriculture itself. A major consequence is the "natural selection" of resistant races of insects, mites, fungi, and even rodents and weeds that are no longer susceptible. This process has gone so far today that most insect pests exhibit multiple resistance not only to a few of the older organochlorines but also to the newer organophosphorus and carbamate insecticides. Some very important insects such as the house fly, the cotton bollworm, the Colorado potato beetle and the diamondback cabbage worm are resistant to all available types of insecticides and are virtually uncontrollable.

The existence of these "monster" insect pests, many of them unimportant until their natural enemies were decimated by the widespread use of broad spectrum insecticides—together with the environmental contamination and human health effects previously mentioned—has brought about an acute need for a new philosophy and methodology of pest control. This is called Integrated Pest Management and it seeks to combine all available techniques of pest suppression, crop rotations, resistant crop varieties, encouragement of natural enemies and diseases, together with the selective and judicious use of pesticides into a sound ecological framework.

Integrated Pest Management (IPM) has been endorsed by the USDA, by EPA, by the Council on Environmental Quality and by such United Nations agencies as the Food and Agriculture Organization and the World Health Organization. A central premise of IPM is to generally relegate the use of pesticides to emergency use when all else fails and to spray only when necessary.

Repeated successes with IPM programs in pest control all over the world have demonstrated that this ecological approach to pest control can reduce pesticide applications by 50 to 95% or more. This achievement promises to be one that all of us—farmers, conservationists, scientists, and concerned citizens alike—can live with. Additionally, IPM practices can materially reduce crop-production costs and prolong the useful life of present-day pesticides by decreasing the rate of selection of resistant species.

Index